Iron Catalysis

Design and Applications

CATALYTIC SCIENCE SERIES

ISSN 1793-1398 (Print)
ISSN 2399-4495 (Online)

Series Editor: Chris Hardacre (*The University of Manchester, UK*)

Catalysis is at the forefront of the chemical industry and is essential to many fields in the chemical sciences. This series explores all aspects of catalysis in authored and edited volumes drawing on expertise from around the globe in a focussed manner. Volumes are accessible by postgraduate students and professionals in academia and industry.

Published

Vol. 19 *Iron Catalysis: Design and Applications*
         edited by Jose M. Palomo

Vol. 18 *Photoorganocatalysis in Organic Synthesis*
         edited by Maurizio Fagnoni, Stefano Protti and Davide Ravelli

Vol. 17 *Hydroprocessing Catalysts and Processes:*
         The Challenges for Biofuels Production
         edited by Bo Zhang and Duncan Seddon

Vol. 16 *Electro-Catalysis at Chemically Modified Solid Surfaces*
         by Jacques Simonet

Vol. 15 *Noble Metal Noble Value: Ru-, Rh-, Pd-catalyzed Heterocycle Synthesis*
         edited by Xiao-Feng Wu

Vol. 14 *Enantioselective Titanium-Catalysed Transformations*
         by Hélène Pellissier

Vol. 13 *Gold Catalysis: An Homogeneous Approach*
         edited by F. Dean Toste and Véronique Michelet

Vol. 12 *Catalysis by Ceria and Related Materials (Second Edition)*
         edited by A. Trovarelli and P. Fornasiero

Vol. 11 *Supported Metals in Catalysis (Second Edition)*
         by J. A. Anderson

Vol. 10 *Concepts in Syngas Manufacture*
         by J. Rostrup-Nielsen and L. J. Christiansen

More information on this series can be found at http://www.worldscientific.com/series/css

(Continued at end of book)

CATALYTIC SCIENCE SERIES — VOL. 19

Series Editor: Chris Hardacre

Iron Catalysis
Design and Applications

edited by

Jose M. Palomo

CSIC, Spain

World Scientific

NEW JERSEY · LONDON · SINGAPORE · BEIJING · SHANGHAI · HONG KONG · TAIPEI · CHENNAI · TOKYO

Published by

World Scientific Publishing Europe Ltd.

57 Shelton Street, Covent Garden, London WC2H 9HE

Head office: 5 Toh Tuck Link, Singapore 596224

USA office: 27 Warren Street, Suite 401-402, Hackensack, NJ 07601

Library of Congress Cataloging-in-Publication Data

Names: Palomo, Jose M., editor.

Title: Iron catalysis : design and applications / edited by Jose M. Palomo, CSIC, Spain.

Description: New Jersey : World Scientific, 2021. | Series: Catalytic science series, 1793-1398 ;
    Vol. 19 | Includes bibliographical references and index.

Identifiers: LCCN 2020052515 | ISBN 9781786349613 (hardcover) |
    ISBN 9781786349620 (ebook for institutions) | ISBN 9781786349637 (ebook for individuals)

Subjects: LCSH: Iron catalysts.

Classification: LCC TP159.C3 I76 2021 | DDC 669/.1--dc23

LC record available at https://lccn.loc.gov/2020052515

British Library Cataloguing-in-Publication Data

A catalogue record for this book is available from the British Library.

For any available supplementary material, please visit
https://www.worldscientific.com/worldscibooks/10.1142/Q0283#t=suppl

Desk Editors: Britta Ramaraj/Michael Beale

Typeset by Stallion Press
Email: enquiries@stallionpress.com

Contents

Chapter 1 Design of Iron Nanostructured Catalysts   **1**
         Jose M. Palomo

1.1 Introduction   1
1.2 Preparation of Iron Nanoparticles   3
1.3 Preparation of Iron 1D Nanostructures   19
1.4 Conclusions and New Perspectives   25
Acknowledgments   25
References   25

Chapter 2 Keys and New Trends of Iron-Based
         Catalysts in Selective Oxidation of
         Propylene in Gas Phase   **35**
         Javier Fernández-Catalá, Jaime García-Aguilar,
         Diego Cazorla-Amorós, and Ángel Berenguer-Murcia

2.1 Introduction   36
2.2 Propylene Epoxidation   37
2.3 Iron-Based Catalysts as Candidates for Catalysts
     in the Selective Propylene Oxidation   40
2.4 Conclusion and New Perspectives   52
Acknowledgments   52
References   53

Chapter 3   Iron Oxide Nanoenzymes for the Treatment of
Polluted Water   **57**

N. Pariona, F. Mondaca, and A. I. Mtz-Enriquez

3.1 Introduction   58
3.2 Surface Properties of Iron Oxides   59
3.3 Peroxidase-like Catalytic Activity of Iron Oxides:
     Degradation of Organic Pollutants in Water   64
3.4 Light-Assisted Peroxidase-Like Catalysis   67
3.5 Conclusions   70
References   71

Chapter 4   Design of Artificial Iron Metalloenzymes
by Combining Proteins and Organometallics   **77**

Jose M. Palomo

4.1 Introduction   77
4.2 Synthesis of Artificial Iron Metalloenzymes by
     Using Protein Engineering Tools   79
4.3 Artificial Iron Metalloenzymes by Organometallics-
     Protein Conjugation   84
4.4 Iron NP Biohybrids as Novel Artificial Metalloenzymes   86
4.5 Conclusions   89
Acknowledgments   90
References   90

Chapter 5   Iron-Containing Enzyme Catalysts   **97**

Cesar Mateo

5.1 Introduction   98
5.2 Heme Proteins   99
5.3 Non Heme Iron Proteins   111
5.4 Non Heme Non-Sulfur Iron-Containing Proteins   117
5.5 Conclusion and Perspective   121
Acknowledgments   122
References   122

Chapter 6 Fe-Catalyzed C–H Activation/Functionalization   127
Melania Gómez-Martínez and Olga García Mancheño

6.1 Introduction   127
6.2 Homogeneous Fe-Catalyzed C–H Activation/
    Functionalization   129
6.3 Heterogeneous Fe-Catalyzed C–H Functionalization   179
6.4 Conclusion and Perspectives   184
References   184

**Chapter 7 Iron-Catalyzed C–H Functionalization
            Reactions *via* Carbene and Nitrene
            Transfer Reactions   203**
Claire Empel, Sripati Jana, and Rene M. Koenigs

7.1 Iron and Its Privileged Role   204
7.2 Synthetic Iron Complexes in Carbene
    Transfer Reactions   205
7.3 Iron-Catalyzed C–H Functionalization with Carbenes   208
7.4 C–H Functionalization with Iron Nitrenes   229
7.5 Conclusion and Perspective   245
References   246

Chapter 8 Iron Catalysis in Metal-Ion Batteries   253
A. Gomez-Martin and J. Ramirez-Rico

8.1 Introduction   254
8.2 Lithium-Ion Batteries   254
8.3 Fe-Catalysis in Carbon Anodes for LIBs   259
8.4 Fe-Catalyzed Graphitic Carbons in Metal-Ion Batteries   279
8.5 Concluding Remarks   284
References   285

Chapter 9 Iron-Catalysis in Environmental Remediation   299
Noelia Losada-García and Jose M. Palomo

9.1 Introduction   299
9.2 Degradation of BPA   301

9.3 Degradation of Chlorinated Organic Compounds 307
9.4 Remediation of Heavy Metals 316
9.5 Conclusion and Perspective 321
Acknowledgments 322
References 322

Index 331

https://doi.org/10.1142/9781786349620_0001

Chapter 1

Design of Iron Nanostructured Catalysts

Jose M. Palomo

*Department of Biocatalysis, Institute of Catalysis (CSIC),
Marie Curie 2, Campus UAM, Cantoblanco, 28049 Madrid, Spain*

josempalomo@icp.csic.es

Iron nanostructures have gained a tremendous attention in the last years as excellent catalysts applied in different areas. The cost-effective and environment friendly advantages of using iron catalysts have made them the alternative to other transition metals in many cases. In this way, the development of new strategies to improve the efficiency of iron as catalyst is mandatory. This chapter describes some of the most actual strategies to synthesize these magnetic iron nanoparticles and other iron nanostructures.

1.1 Introduction

Iron is the most abundant metal in the planet, cost-effective, environmentally friendly, and the remediation process is easily manipulated. For that reason, in the last years the use of this nonprecious metal has gained extraordinary attention, exploiting its potential as a catalyst in organic synthesis.[1-3]

This phenomenon has caused the continuous necessity to develop new kind of catalytic systems in order to improve the efficiency of iron as a catalyst.

One of the most emerging areas is the design of nanostructured metal systems. The use of these nanostructured materials offers several advantages, such as the high surface-to-volume ratio of nanomaterials compared to bulk materials that generally makes them attractive candidates for their application as catalysts.[4]

The particular conditions for the synthesis of iron nanostructures are critical in term of the particle size but also particle morphology and surface area, which exerts tremendous impact on their catalytic properties.[5,6] The control of obtaining small-sized nanoparticles is a critical issue for the production of excellent catalysts.[7]

Another important point of the iron nanostructures is the magnetism.[8] This capacity of the iron nanoparticles offers them the advantage—in comparison to other metal nanoparticles used as catalysts—of the facile separation from the reaction mixture through an external magnet reducing energy consumption, catalyst loss, and saves time in achieving catalyst recovery.

Many different strategies have been described in the preparation of iron nanostructures, where the corresponding iron species and nanostructures are obtained depending on the metal source and experimental conditions.[9,10] In this way, the most typical iron species obtainable as nanoparticles are iron oxides of Fe^{3+} species such as hematite (α-Fe_2O_3), maghemite (γ-Fe_2O_3), Fe^{3+}/Fe^{2+} species as the known magnetite (Fe_3O_4), Fe^{2+} species such as iron hydroxides (FeOOH) and metal Fe (0). The range of size of these synthesized nanoparticles goes from 5 to 100 nm.[10] These last iron nanoparticles (α-Fe) exhibit a core–oxide shell structure inevitably because of the extreme sensitivity of iron to oxidation under air conditions.

However, depending on the synthetic methodology, novel iron 2D nanostructures such as nanorods (NRs), nanowires (NWs), or nanoflowers has been reported.[11–16]

Some of the most recent strategies developed to synthesize these different iron nanoparticles are described in the following chapter.

1.2 Preparation of Iron Nanoparticles

Iron nanoparticles have gained important relevance in different areas over the last years, with special focus in the biomedical applications,[17–19] in particular on magnetic resonance imaging (MRI) contrast agents and drug delivery and interactive functions at the surface. These two applications have been well discussed in comprehensive reviews.[20,21]

The recent focus of these iron nanostructures on catalysis are based on their easy preparation, excellent reusability, high catalytic efficiency, and the convenient separation from the reaction solutions by external magnetic fields.[22]

Efficient and sustainable preparation of iron nanoparticles that enables fine control of the size, crystal structure, and surface properties has constituted the main focus of recent strategies.

1.2.1 *Synthesis of hematite (α-Fe$_2$O$_3$) nanoparticles*

From the different iron (iii) oxide species, α-Fe$_2$O$_3$ is the most important polymorph existing in nature as hematite. In particular, α-Fe$_2$O$_3$ nanoparticles are highly stable in an aqueous solution and show interesting magnetic properties because they can display antiferromagnetic, weak-ferromagnetic, or superparamagnetic properties depending on the conditions and morphology.[23,24] Therefore, these nanostructures have been broadly used in different areas and many different synthetic methods have been described.[25–27] In particular, the development of particular synthetic strategies permits us to control the size but also the morphology of the hematite nanoparticles. These iron species have a great tendency to generate nanoparticles with different morphologies (hexagonal, cubes, spheres, ellipsoids, rods).[28] Indeed, this iron oxide is formed mainly in other iron nanoparticles as oxide core, responsible in many occasions of the generation of other nanostructures.[29]

In particular, besides the methodology, the simple change in the iron salt as a starting material causes the formation of nanoparticles with different sizes and morphology.[30] In this case, depending on

J. M. Palomo

Table 1.1. Shape and phase evaluation of different iron oxide species.[30]

Precursor	Shape	As-prepared samples	Heated 250 °C in argon	Heated 450 °C in argon	Heated 450 °C in air
Fe(II)sulpahte heptahydrate	NRs	α-FeOOH + $5Fe_2O_3.9H_2O$	Fe_2O_3 + γ-Fe_2O_3	α-Fe_2O_3 + γ-Fe_2O_3	α-Fe_2O_3 + γ-Fe_2O_3
Fe(II)oxalate dehydrate	Nanohusks	β-FeOOH	γ-Fe_2O_3	α-Fe_2O_3 + γ-Fe_2O_3	α-Fe_2O_3
Fe(III)chloride hexahydrate	Cubes	α-Fe_2O_3	α-Fe_2O_3	α-Fe_2O_3 + γ-Fe_2O_3	α-Fe_2O_3 + γ-Fe_2O_3
Fe(II)nitrate dehydrate	Nanocubes	α-Fe_2O_3 + γ-Fe_2O_3	α-Fe_2O_3 + γ-Fe_2O_3	α-Fe_2O_3	α-Fe_2O_3 + γ-Fe_2O_3
Fe(II)gluconate dihydrate	Porous spheres	Amorphous	Amorphous	γ-Fe_2O_3	α-Fe_2O_3
Fe(0) pentacarbonyl	Self-arranged flowers	γ-Fe_2O_3 + oxyhydroxides	α-Fe_2O_3 + γ-Fe_2O_3	α-Fe_2O_3 + γ-Fe_2O_3	α-Fe_2O_3 + γ-Fe_2O_3

the salt and the final method, sometimes mixtures of hematite and maghemite (oxidized species of magnetite, γ-Fe_2O_3) were obtained (Table 1.1).[30]

For example, the formation of cubic nanoparticles was possible by using ferric chloride as a precursor at 160 °C, with an average size of 120 nm of pure hematite as shown by scanning electron microscopy (SEM) (Fig. 1.1A). However, when ferric nitrate was used, the nanocubes were much smaller, around 20 nm, although mainly consisted of hematite with some part of maghemite (Fig. 1.1B). Also the magnetism of both species was influenced by the iron species and the final nanoparticles size.[30]

Another strategy used ferric sulfate in a mixture of ethanol/water (4/1) with oleic acid heated at 160 °C.[31] Hematite nanoparticles with an average size of 8 nm were synthesized by this hydrothermal method. The magnetization (Ms) was 3.98 Am^2/Kg at 300 K, a high value for hematite nanomaterials and much higher than the Ms of bulk hematite materials (Ms = 0.3 Am^2/Kg).[32]

However, due to the high stability of this type of iron species, the synthesis of hematite nanoparticles based on biosynthetic strategies

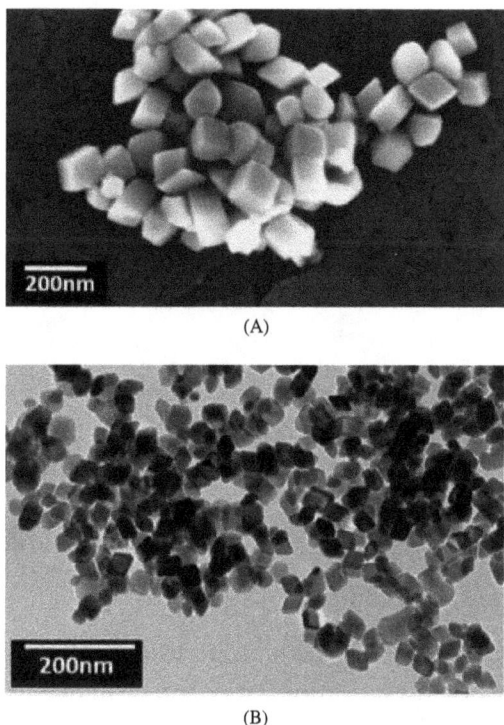

(A)

(B)

Fig. 1.1. SEM analysis of α-Fe$_2$O$_3$ nanocubes. (A) using FeCl$_3$. (B) using Fe(NO)$_3$.[30]

has grown in the last years. These represent efficient and more environmentally friendly protocols in comparison with more traditional strategies, where microorganism, bacteria, or plants have acted as reducing and capping agents.[33–35]

In this way, crystalline hematite nanoparticles with a size around 39 nm were synthesized by the use of aqueous extract of *Psoralea corylifolia* seeds with ferric chloride at 70 °C.[35] These were successfully used as catalysts in the degradation of methylene blue. Another strategy using the boiling solution of *Psidium guajava* leaf extract required the final calcination at 600 °C for obtaining the hematite nanoparticles. Transmission electron microscopy (TEM) analysis revealed hematite nanoparticles with a size around 34 nm (Fig. 1.2).[33]

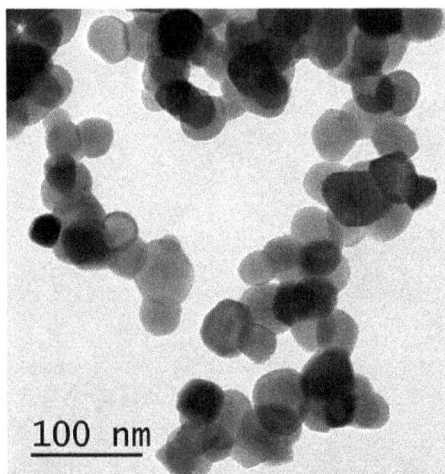

Fig. 1.2. TEM micrograph of α-Fe$_2$O$_3$ nanoparticles synthesized using *P. guajava* leaf.[33]

However, as it happens with the magnetite nanoparticles, the implementation of these nanomaterials in catalysis requires avoiding their strong agglomeration in solution. One actual interesting application is focused on the electrochemical capacity of these materials. For that purpose, the fabrication of nanocomposites formed by α-Fe$_2$O$_3$ supported on different materials (e.g., nanotubes, MOFs, graphene) has been performed.[36–38]

Compared with other conductive materials, graphene or chemically modified graphene are very interesting supports because of their highly flexible nature, large surface area, high electrical conductivity, and chemical stability. In this way, Lui and coworkers[37] synthesized an interesting hematite nanocomposite based on α-Fe$_2$O$_3$ nanorings on reduced graphene oxide (rGO). The addition of Na$_2$SO$_4$ was carried out in a sodium phosphate buffer solution containing ferric chloride was performed and after heating at 220 °C, the powder was mixed with aminopropyltriethoxysilane (APTES) and rGO acidic solution and supported hematite nanorings were synthesized hydrothermal treatment.[37]

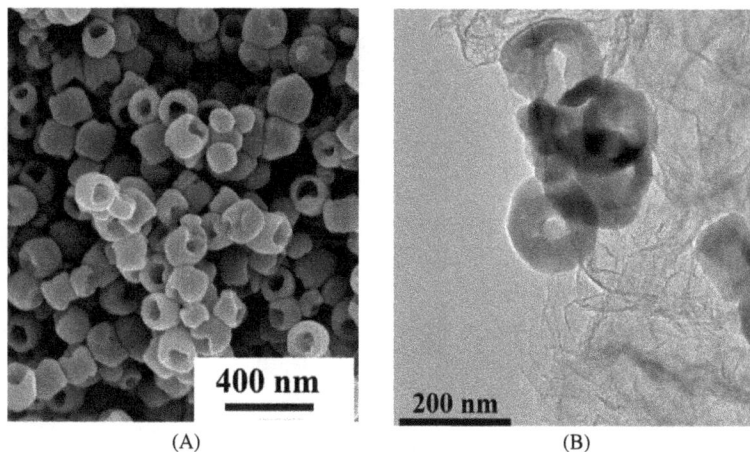

Fig. 1.3. Characterization of α-Fe$_2$O$_3$/rGO. (A) SEM image. (B) TEM image.

The presence of hematite was confirmed by X-ray powder diffraction (XRD) and the composite α-Fe$_2$O$_3$/rGO were obtained by SEM (Fig. 1.3A). These iron nanostructures presented uniform size and morphology (hollow structure with open ends), with a diameter around 150 nm (80 nm inner and 100 height) (Fig. 1.3B).[37]

These new α-Fe$_2$O$_3$/rGO composites showed excellent electrochemical performance with high specific capacitance, excellent rate capability, and good cycling life (about 10% decay in the first 1,000 cycles).[37]

Another interesting strategy has been recently described for the preparation of nanostructures of α-Fe$_2$O$_3$ doped with Pt.[38]

The synthesis was very simple by an efficient one-step solvothermal using iron and platinum water soluble salts. A range of amount of platinum was used in term of creating different nanostructures. Hematite nanoplates of around 100 nm were obtained by this strategy even without Pt in the medium (Fig. 1.4A), and in the hybrid materials Pt nanoparticles decorated Pt^{2+}-doped α-Fe$_2$O$_3$ nanoplates (Pt/Pt-Fe$_2$O$_3$ NPs) (Fig. 1.4B–D). These excellent hybrid hematite–platinum nanostructures were successfully used for photocatalytic water oxidation, a hybrid iron nanomaterial was produced by the

Fig. 1.4. The SEM images of the as-prepared (A) α-Fe$_2$O$_3$. (B) Pt/Pt-Fe$_2$O$_3$_0.05. (C) Pt/Pt-Fe$_2$O$_3$_0.2. (D) Pt/Fe$_2$O$_3$ NPs.

introduction of Pt in the iron nanoplates, improving the isolation effciency of the photo-induced carriers, photoactivity, and photostability for water oxidation.[38]

1.2.2 Synthesis of magnetite (Fe$_3$O$_4$) nanoparticles

Fe$_3$O$_4$ nanoparticles are the most commonly used magnetic nanomaterials. They have been synthesized by different approaches, such as thermal decomposition, coprecipitation, etc.,[39,40] where the control of size and morphology of the nanoparticles are key in the final catalytic application and also in the magnetic properties.[41]

However, the aggregation of Fe$_3$O$_4$ nanoparticles is inevitable in many cases, which leads to their wide-sized distributions, small surface areas, and hence low catalytic efficiencies.[42]

This phenomenon has been avoided by using different molecules (polymers, ionic liquids, surfactants, etc.) as additives in the

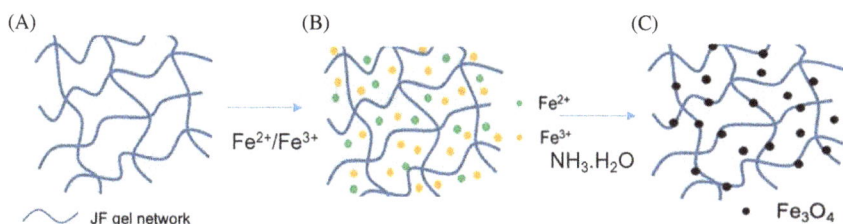

Scheme 1.1. Preparation of Fe_3O_4 nanoparticles–hydrogel nanocomposite. (A) JF gel. (B) the Fe^{2+}/Fe^{3+} loaded JF gel. (C) Fe_3O_4 nanoparticles hybrid.[43]

synthetic strategy, also making more sustainable the preparation of the nanoparticles.

One interesting example of that has been the use of a biological hydrogel.[43] Fe_3O_4 nanoparticles were successfully synthesized by coprecipitation method using jellyfish mesoglea (JF), a nanocomposite hydrogel, which permits one to obtain nanoparticles well dispersed on the gel structure (Scheme 1.1).

The gel network was formed by nanofibers, and the iron nanoparticles were generated from a mixture of Fe^{2+}/Fe^{3+} under green conditions.

The reaction mechanism in this synthetic method can be simplified as $Fe^{2+} + 2Fe^{3+} + 8OH^- \leftrightharpoons Fe(OH)_2 + 2Fe(OH)_3 \rightarrow Fe_3O_4\downarrow + 4H_2O$. The nucleation of the Fe_3O_4 nucleus is easier when the solution pH is lower than 11, while the growth of the Fe_3O_4 nucleus is easier when the solution pH is higher than 11.[44]

The content of iron in the nanocomposite was of 75%. XRD and X-ray photoelectron spectroscopy (XPS) analysis demonstrated the exclusive formation of magnetite nanoparticles. The sizes of the nanoparticles were quite uniform, with a diameter of around 20 nm. This gel–nanohybrid showed a saturation Ms of 32 Am^2/Kg, smaller than that of bulk magnetite (92 Am^2/Kg).[43]

This nanohybrid was successfully applied in catalyzing the oxidative degradation of xylenol orange with excellent reusability.[43]

The use of organic polymers such as polyethyleneglycol (PEG),[45] polyvinylpyrrolidone (PVP),[46] or propylene glycol has been recently described in the synthesis of magnetite nanoparticles.[47]

One interesting example is focused on the application of PEG in the synthesis of magnetite–Prussian blue (PB) nanoparticles.[45] PB nanoparticles have been utilized worldwide for fabrication of chemical sensors/biosensor.[48] Recently, a very interesting work has demonstrated the effectiveness of PB nanoparticles as versatile multienzyme-like nanomimetics.[48]

The synthesis of magnetic PB nanoparticles was attempted by using PEG as a protective reagent to keep nanoparticles from agglomeration in a one-pot hydrothermal reduction of ferric (III) chloride and potassium ferricyanide in ethylene glycol as solvent.[45] This methodology generated numerous spherical and cubic nanoparticles combined compactly together with different sizes (from 20 to 100 nm) (Fig. 1.5). Fourier-transform infrared spectroscopy (FTIR) analysis confirmed the formation of the nanohybrid Fe_3O_4–PB by the characteristic adsorption bands of 2,085 cm^{-1} (CN group), 496 (Fe–CN–Fe), and 596 (Fe–O bond).

α-Propylene glycol solution (PG) was applied to assist the magnetite nanoparticles synthesis in a facile one-step hydrothermal method.[47] This strategy resulted in the formation of nanoparticles with an octahedral morphology (Fig. 1.6). In comparison, the nanoparticles synthesized without the polymer presented much more disordered nanoparticles.

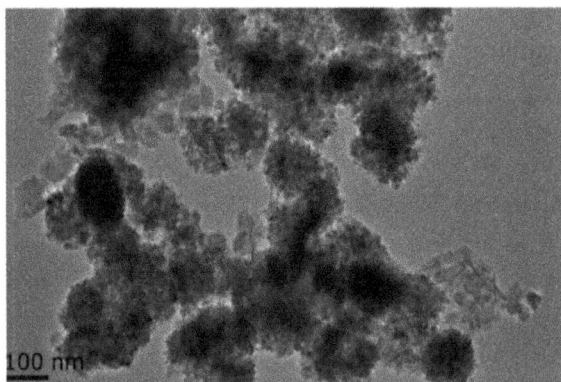

Fig. 1.5.   TEM image of magnetite–PB nanoparticles.

Fig. 1.6. SEM images of the synthesized Fe_3O_4 nanoparticles. (A) PG. (B) without PG.

The strategy used $FeSO_4$ as iron starting material dissolved in water. It was dropped to a solution of PG, and sodium hydroxide is added after that. The mixture temperature was increased 10 °C/min up to 200 °C, keeping that temperature for 20 h. The results showed the formation of Fe_3O_4 NPs with an average edge length of around 500 nm. The magnetic analysis revealed that the synthesized octahedral nanoparticles showed high ferromagnetic properties at room temperature with a Ms of 87.48 Am^2/Kg.[47]

Another simple strategy to synthesize small magnetite nanoparticles was performed by using tetramethylammonium hydroxide (TMAOH), a well-known phase transfer agent with surfactant properties.[49] The mixture of Fe^{3+} and Fe^{2+} chloride salts was used at mild conditions in the presence of ammonium hydroxide at pH 8. This methodology generated spherical magnetite nanoparticles of around 15–30 nm (Fig. 1.7).

Crystalline nanoparticle structure was confirmed by clear visibility of lattice fringes (width = 0.27 nm) (Fig. 1.7B). A surface area of hydroxyl (Fe–OH) groups could also be observed in the TEM analysis. These magnetite nanoparticles showed excellent removal capability for different contaminant metals.[49]

The use of surfactants was also applied by Safaie and coworkers in the development of green synthetic methodology to produce

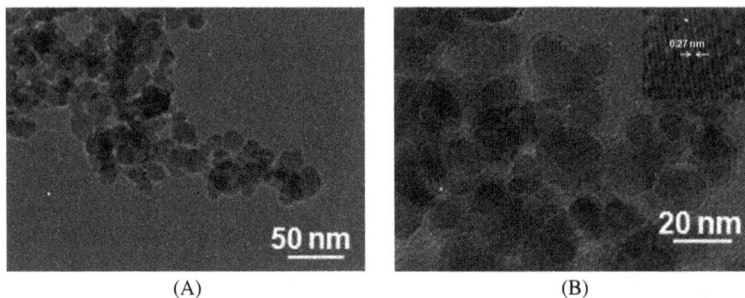

Fig. 1.7. TEM micrographs of magnetite nanoparticles. (A) 410 KX. (B) 1,020 KX, (inset 4,100 KX) magnifications.

magnetite nanoparticles.[50] The synthesis was carried out by simple precipitation at room temperature and aqueous solution combining iron $Fe(NO_3)_3$ and $FeCl_2$ in the presence of cetyl trimethylammonium bromide (CTAB) or sodium dodecyl sulfate (SDS) as additives. Nanoparticles with an average diameter size of 40 nm were obtained by using CTAB, whereas around 70 nm nanoparticles were synthesized using SDS. The magnetism of the nanoparticles was also affected by the synthetic protocol, exhibiting CTAB–Fe_3O_4 a higher ferromagnetic behavior over SDS–Fe_3O_4 nanocomposites (30.64 Am^2/Kg against 17.5 Am^2/Kg).[50]

However, even more efficient and simple methods to obtain smaller and superparamagnetic magnetite nanoparticles have been developed by the use of triethanolamine (TEA) as additive.[51]

First, TEA and Fe(III) salt were mixed together in water to create the Fe (III)–TEA solution. This solution was heated at 180 °C to obtain the Fe_3O_4 nanoparticles. In the process, iron hydroxide species were formed as intermediates to finally produce magnetite. Mossbauer spectra and XRD confirmed the existence of magnetite as the unique iron species. Small magnetite nanoparticles of around 10 nm were determined by TEM analysis (Fig. 1.8).

TEA has a critical role in the process acting as reducing, precipitating, and dispersing agents. Furthermore, hematite (α-Fe_2O_3) phase would be obtained without TEA addition.

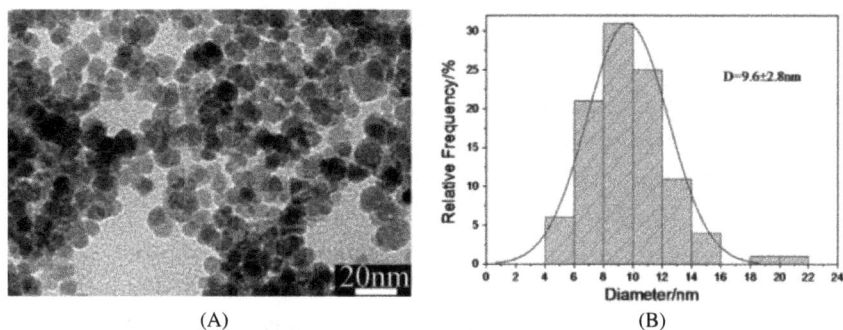

(A)   (B)

Fig. 1.8. Physicochemical characterization of TEA–Fe_3O_4 nanoparticles. (A) TEM. (B) particle size distribution.

This methodology gave also an important magnetism value for these new nanoparticles. This TEA–Fe_3O_4 nanoparticles exhibited superparamagnetic behavior with the saturation Ms in 70 Am^2/Kg, and 0 Oe, respectively. This value is near to the Ms value of bulk magnetite (92 Am^2/Kg).

Additional strategies based on sustainability have been developed using sugars, such as maltose, where superparamagnetic magnetite nano-particles of 12 nm were synthesized by a hydrothermal method.[52] At this point, glucose act as reducing agent and gluconic acid as capping.

On the other hand, the employment of different materials as templates for the creation of magnetite nanoparticles on the solid phase has gained great attention in recent years.[53,54] In particular, the application of graphene and derivatives (such as graphene oxide) has been strongly emphasized because it allowed the crea-tion of two-dimensional (2D) and three-dimensional (3D) magnet-ite hybrid nanostructures with great potential applications for their electrochemical and catalytic properties.[55–57]

Another strategy based on a high temperature coprecipitation system using ammonia solution and hydrazine permitted the synthe-sis of ultra small magnetite nanoparticles (3 nm) anchored on the surfaces of RGO nanosheets, which were tested as anodes in sodium–ion batteries.[58]

Combination with other metals is also a very powerful tool for catalytic performance. The combination of catalytic and magnetic properties of magnetite with Pd, Ni, Cu, or Au has been successfully applied in catalytic processes.[59–61]

In particular, the design of 3D porous graphene-based nanohybrids combining two metals in a 3D graphene nanostructure possesses higher catalytic activity[62] and some special properties.[63]

In this way, a good example has been developed by Chen and coworkers.[60] They synthesized *in situ* Fe_3O_4 and Pd nanoparticles on a self-assembled 3D graphene nanohybrids by a one-pot solvothermal method (Scheme 1.2). Firstly, the creation of a 3D graphene nanostructure was made by an ultrasonic process of an aqueous solution containing L-glutamic acid and graphene. Then, ferric chloride in ethyleneglycol was added to the 3D graphene solution, where iron was self-assembled to the solid and then palladium chloride was added, adjusting pH at 13. Both metals coexist in the graphene surface as hydroxide until nanoparticles are formed after thermal reduction.[60] The best performing catalyst contained a molar ratio of metals of 1:1.

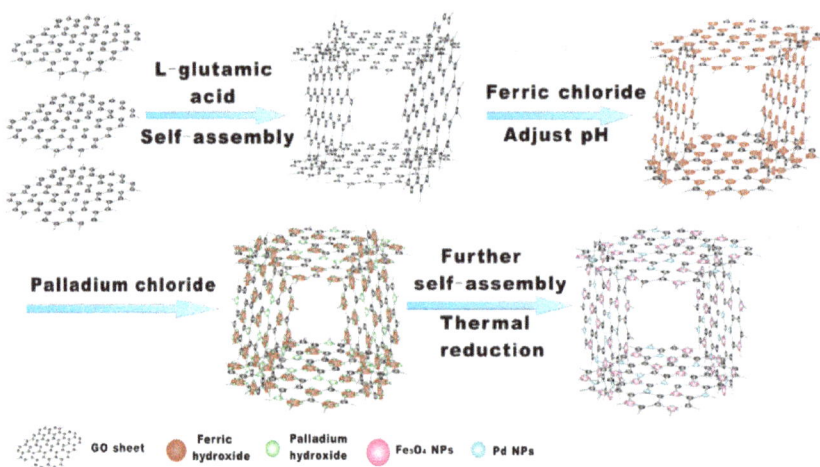

Scheme 1.2. Synthesis of GO–Fe_3O_4–Pd nanohybrid. Figure adapted from Ref. 60.

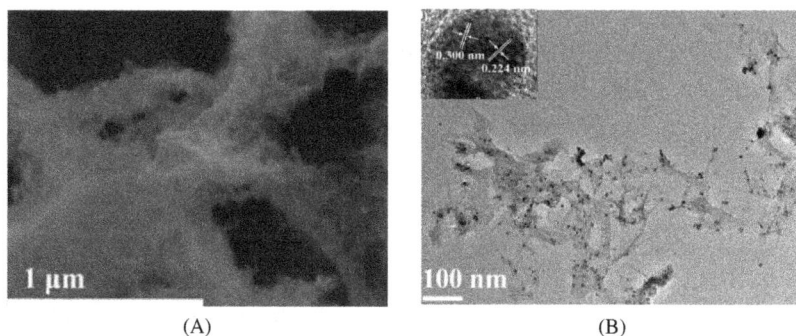

(A)   (B)

Fig. 1.9. (A) SEM images of the GO–Fe$_3$O$_4$–Pd. (B) TEM images of 3D-RGO–Fe$_3$O$_4$–Pd; inset shows HRTEM image.

The structure of the nanohybrid was evaluated using microscopy (Fig. 1.9).[60] SEM revealed that the interconnected 3D porous network of graphene sheets was well-defined and also some nanoparticles were visualized uniformly located on the graphene surface (Fig. 1.9A). The size of the nanoparticles anchored on the graphene sheets were around 8–10 nm, as shown in TEM (Fig. 1.9B). By the high-resolution TEM (HRTEM) it was clear that the nanoparticles grown on 3D–RGO sheets are Pd and Fe$_3$O$_4$ NPs, respectively.

These new graphene–magnetite nanohybrids showed interesting peroxidase-like and oxidase-like activity with a high reproducibility, stability, and reusability.

1.2.3 *Synthesis of Fe (0) nanoparticles*

Zero-valent iron (ZVI) nanoparticles are very interesting nanocatalysts in different applications, especially in environmental remediation.[64,65] This phenomenon is based on the highest capacity to reduce organic compounds as well as inorganic impurities because of the easy way to get oxidized to C2 and C3 oxidation states (standard redox potential, E^0_h = −0.44 V). Metallic iron easily acts as an electron donor: Fe^0 → Fe^{2+} + 2e$^-$.

However, this extreme sensitivity to the oxidation makes the synthetic ZVI nanoparticles inevitably possess a core–shell structure composed of a metal iron core encapsulated by an iron oxide shell. This oxide shell has an important influence in the final catalytic properties of the entire nanoparticles.[66]

For example, the oxide shell has been demonstrated to have an important effect on the oxidative reactivity of these iron nanoparticles, which arose from the combined effects of the incrassated iron oxide shell and more surface bound ferrous ions on the amorphous iron oxide shell formed during the water-aging process.[67]

The oxide shell might consist on a single phase, wüstite (FeO), magnetite (Fe_3O_4), maghemite (γ-Fe_2O_3), hematite (α-Fe_2O_3), and goethite (FeOOH) or (an unknown phase), or on several phases.[68] The thickness of this oxide shell goes from 2 to 7 nm depending on the method used for synthesizing the ZVI nanoparticles.

In this term, different methodologies have been described in the preparation of these types of iron nanoparticles, mainly based on two strategies: i) the so-called *bottom-up*, where ZVI are constructed via self-assembling and positional assembling[69] and ii) the so-called *top-down*, where nanoscale ZVI nZVI are synthesized through breaking down or restructuring a bulk material to the nanoscale with the aid of physical or chemical methods.[70] Conditions such as pH of preparation, aerobic or anaerobic conditions and amount of reducing agent have been demonstrated to have important influence in the synthesis but also in the catalytic applications.

However, for high catalytic performance, small-sized nanoparticles (≤ 15 nm) are desirable.

Also the consideration of reducing the hazardous conditions—usually involved in the nanoparticle synthesis—is an advantage on the development of biosynthetic approaches. Leafs or plants extracts have been used to synthesize ZVI NPs at mild conditions. This is a promising approach for a possible high-scale production, although in most of the cases the iron nanoparticles synthesized has a diameter size around 30–50 nm, some result in the production of amorphous nanoparticles.[71]

Therefore, mild conditions using sodium borohydride as a reducing agent constitutes the most applied strategy. Anyway it is difficult to develop methodologies where very small-sized crystalline

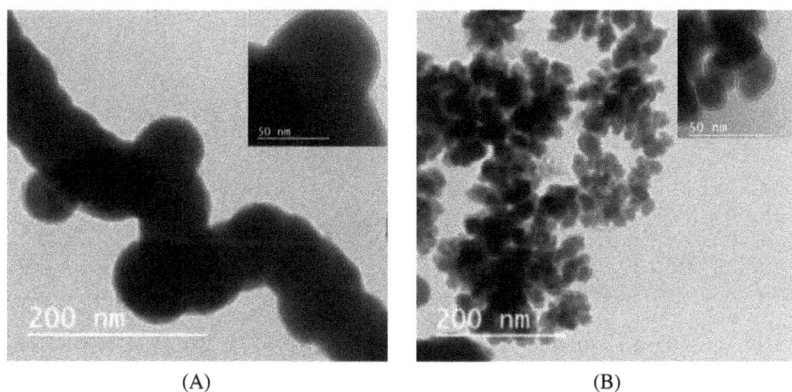

Fig. 1.10. TEM images showing the nZVI particles at different concentrations of NaBH$_4$. (A) without NaBH$_4$. (B) with NaBH$_4$.

nanoparticles can be obtained. For example, the addition of NaBH$_4$ (100 mM) into nZVI suspension showed the disintegration of nZVI (60–100 nm), resulting in the formation of much smaller particles (15–40 nm) due to the chemical etching of outermost surfaces (i.e., magnetite).[72] In this case, the oxide shell was conserved by around 2.5–3.5 nm (Fig. 1.10).

The interesting application of this approach is that the NaBH$_4$ significantly enhanced the reactivity of nZVI in oxygen, avoiding the rapid passivation of nZVI surface in these ambient environments (21% O$_2$ condition). This showed a very active catalyst in the reduction of aromatic compounds with excellent stability.[72]

This method generated smaller nanoparticles but the agglomeration process of the nanoparticles increased. In this way, the application of additives or, more fascinating, the synthesis of nanoparticles on the solid phase of different surfaces is actually a very interesting application for generating advanced materials.[73–75]

A very simple method consisted in the creation of ZVI nanoparticles on the graphene oxide surface by a simple solution of ferric chloride, graphite oxide, and with PVP in aqueous solution at room temperature, finally reducing by sodium borohydride.[75]

The formation mechanism is based on an ionic interaction between negatively charged surface and positive iron, which finally

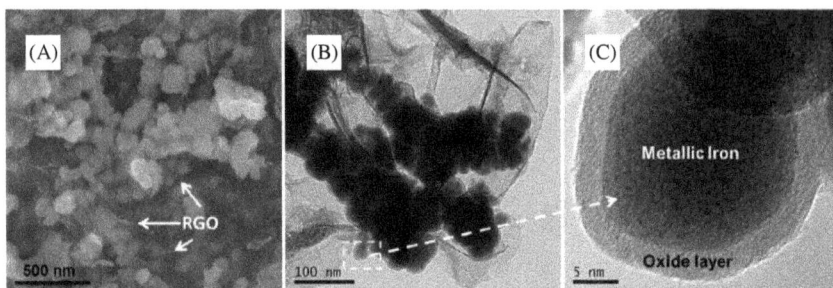

Fig. 1.11.   Characterization of GO–Fe nanohybrid. (A) SEM. (B) TEM. (C) HRTEM.

was reduced to form the magnetite nanoparticles. This methodology generated nanoparticles of ZVI core covered with an oxide layer (Fe_2O_3) responsible for the final formation of nanonecklaces (Fig. 1.11). TEM analysis showed that individual nanoparticles with a size of 15 nm were formed (Fig. 1.11C) although the final nanostructure showed a diameter of around 20–50 nm (Fig. 1.11B).[75] This nanostructure showed interesting enzyme-like activity quite useful in the biosensor industry.

Together with the catalytic capacity, the magnetic properties of the molecules are also important. A new method to prepare airstable superparamagnetic iron nanoparticles trapped between thermally RGO (TRGO) nanosheets has been reported.[73]

Very small ZVI nanoparticles of around 5 nm of diameter were synthesized according to the described method using TRGO impregnated with ferric nitrate as an iron precursor.[73] Different iron concentrations were used in the preparation of these nanohybrids with a clear influence over the final iron nanoparticle morphology (Fig. 1.12). TEM analysis revealed the formation of nanoparticles in the Fe (5)/TRGO nanohybrid with a core-shell structure, core of ZVI with a shell of Fe(III) oxide (Fig. 1.12A). However, by doubling the iron concentration, the formation of peculiar ring-like assemblies composed of ultra small iron nanoparticles (5 nm) was observed for Fe (10)/TRGO nanohybrid (Fig. 1.12B). The rings have diameters of 20–100 nm. X-ray experiments demonstrated that

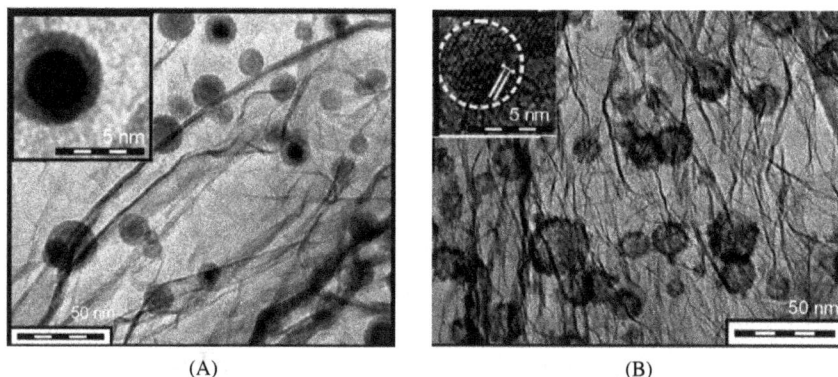

Fig. 1.12. Characterization of Fe/TRGO nanohybrids. (A) TEM image of the Fe (5)/TRGO hybrid system (inset HRTEM image of a single iron nanoparticle with a core–shell structure). (B) TEM image of the Fe (10)/TRGO hybrid system and (inset HRTEM image).

the nanohybrid was composed exclusively by α-Fe without any oxide shell.

This Fe (10)/TRGO nanohybrid exhibited remarkable magnetic properties, showing a saturation MS at room temperature of 185 Am^2/Kg,[73] the highest values reported for iron-based superparamagnets.

1.3 Preparation of Iron 1D Nanostructures

1D nanomaterials have been observed as iron nanostructures apart for apart from nanoparticles.[76] In particular, magnetic iron NRs or NWs recently attracted much attention due to their enhanced optical, magnetic, catalytic, and mechanical properties over spherically shaped nanoparticles.[77,78] Several synthetic protocols have been reported in order to produce Fe(0) NWs, Fe_2O_3, or a combination of Fe/Fe_2O_3 or Fe/Fe_3O_4 NWs, with an average diameter of 30–100 nm with length of 800 nm to micrometer scale as the best results in literature.[77–79]

However, akaganeite (β-FeOOH) is a very interesting one because of its applications in photocatalysis. In the same way as iron oxides, this hydroxide trends to generate nanostructures higher than the simple nanoparticles, especially NRs.[80,81]

In particular, interesting photocatalytic applications have been observed by the creation of these β-FeOOH nanostructures on different material surfaces.[82]

One very interesting strategy is based on the creation of β-FeOOH NRs by a mineralization process on a material surface previously coated with dopamine. The creation of the NRs was performed in aqueous media using ferric chloride as a precursor at 60 °C on a polydopamine (PDA) coated surface. XRD revealed the formation of this iron oxo-hydroxide as the unique species, while SEM analysis revealed the dependence on the material used of the size of the NRs. For example, using polypropylene nonwoven (PPNW) as starting material, β-FeOOH NRs of 180 ± 20 length and 60 ± 8 nm diameter were obtained (Fig. 1.13A).

Higher size NRs were formed using polypropylene microfiltration membrane (PPMM), which were mainly decorated on the outer surface because the membrane pores (~200 nm) are smaller than the mineralized NRs (370 nm length) (Fig. 1.13B).[80]

These iron NRs were quite stable and were used as excellent photo-Fenton catalysts in the degradation of different dyes.[80]

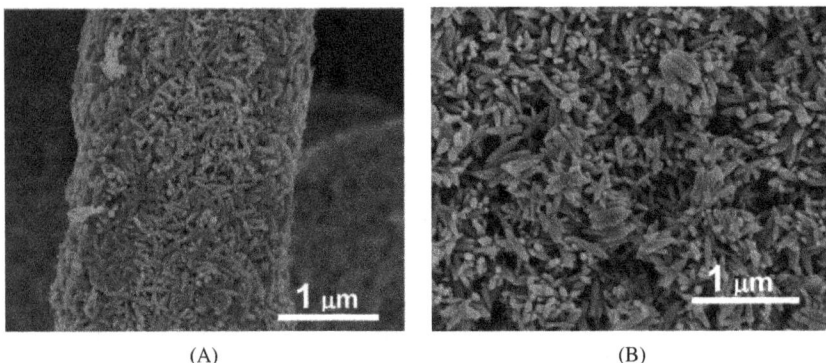

(A)   (B)

Fig. 1.13. FESEM images of the β-FeOOH NRs on different PDA-coated materials. (A) PPNW. (B) PPMM.

In an other case, β-FeOOH nanostructures were synthesized on TiO_2 nanofibers at 90 °C.[81] The nanofibers were previously synthesized from PVP and tetrabutyl titanate.[81]

β-FeOOH nanostructure is an ideal material to combine with TiO_2 nanofibers for photocatalytic activity in the visible range. Although both show similar performance levels and a simple synthetic process, β-FeOOH presented some advantages over hematite, obtaining a wide variety of morphologies without adding additives and not leading to the collapse of the hierarchical structures.[81]

Also in the fabrication of this nanostructure, the amount of iron salt resulted is an important factor for the final morphology, obtaining flasks, particles, or needles. In particular, microscopy analysis revealed that β-FeOOH nanoparticles of around 50 nm were synthesized in a nanofiber of around 200–300 nm of diameter (Fig. 1.14).

This β-FeOOH–TiO_2 heterostructure affords multiple functions of photocatalysis and sensing, which are highly promising for environment monitoring and clean up applications.

From the catalytic point of view, the creation of nanostructures with lower size is interesting. In this way, recently our group has designed for the first time the creation of novel nanobiohybrids containing very small siderite ($FeCO_3$) NRs ($FeCO_3$ NRs) (Fig. 1.15).[11]

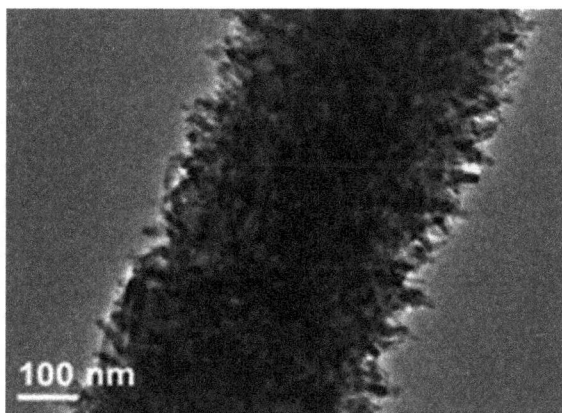

Fig. 1.14. TEM image of β-FeOOH–TiO_2 nanostructure.

J. M. Palomo

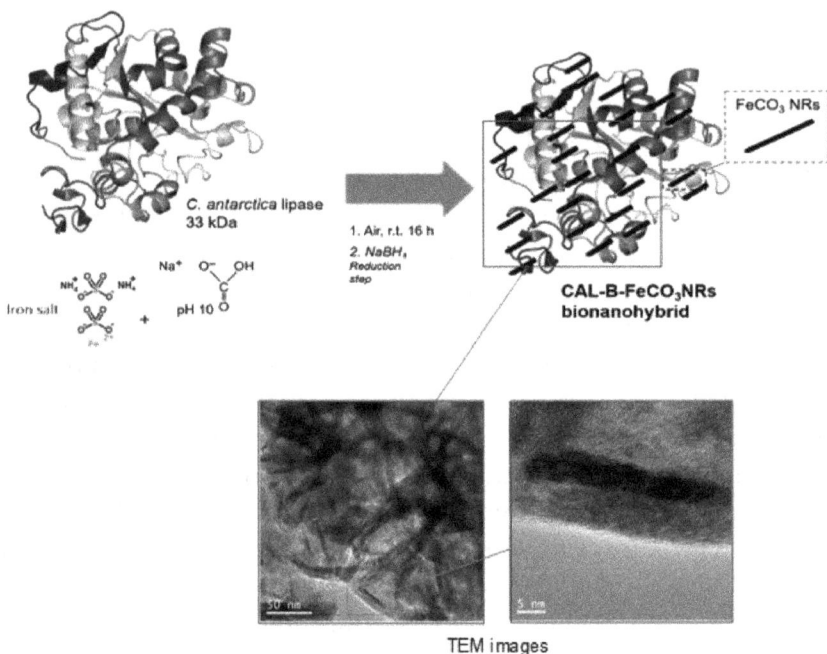

Fig. 1.15. Synthesis of CAL–B–FeCO$_3$ nanorods bionanohybrid.

This simple strategy is based on the use of an enzyme and iron salt to create a heterogeneous nanomaterial at room temperature in aqueous media where the size and diameter of the NRs are dependent on the experimental conditions used.[11]

By using a final reducing step, it was possible to obtain a biohybrid containing iron (ii) carbonate NRs of 5 × 40 nm with superparamagnetic properties. This methodology allows us to achieve this nanomaterial in multimilligram scale, which could be scale, up to grams. Interesting results in catalytic application, were successfully obtained in the reduction of arenes in water or C–C bond reactions at mild conditions.[11]

One interesting example described the creation of different types of iron NWs controlling the experimental conditions during the synthesis (Fig. 1.16).[83] α-FeOOH NWs with diameter of 30 nm

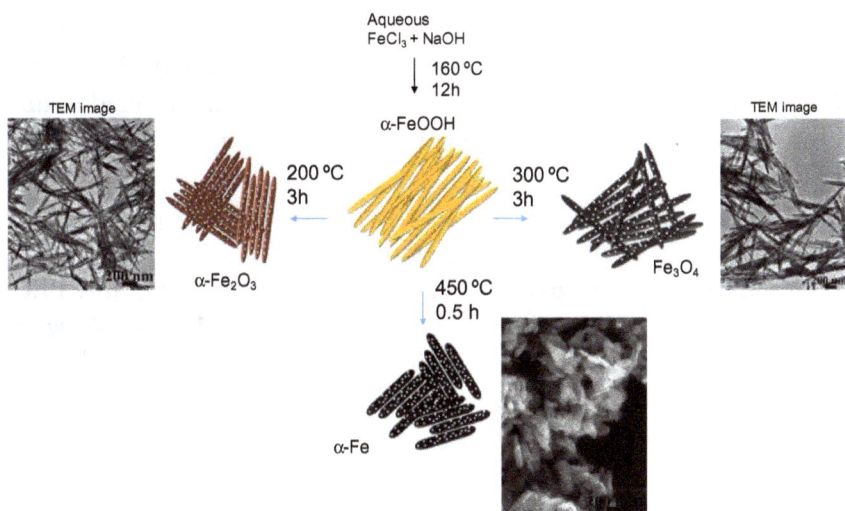

Fig. 1.16. Synthesis of different iron NWs.

and length of 550 nm were synthesized by hydrolysis of ferric chloride using NaOH and 160 °C as conditions. Thermal treatment at 200 °C of this iron NWs permitted the transformation of this species in hematite NWs with the same morphology and slightly smaller size, whereas increasing T at 300 °C slightly longer magnetite NWs or even smaller α-Fe NWs by performing the reaction at 450 °C (Fig. 1.16). In each case, XRD demonstrated the formation exclusively of one iron species in each case.[83]

Hematite NWs have also been synthesized by an electrospinning method using PVP from ferric nitrate by incubation at 550 °C, with diameter around 50–60 nm.[79] Reduction process of this hematite NWs generated magnetite NWs by heating at 250 °C for 1 h in H_2 atmosphere, whereas Fe(0) NWs were obtained when heating at 350 °C for 1 h in H_2 atmosphere. The magnetite NWs showed a diameter of 50–60 nm, whereas the iron oxide presented an average diameter of 30–40 nm. For these NWs, magnetite exhibited excellent microwave absorbing properties.[79]

Another type of nanostructure that iron is able to produce are flower like nanomaterials. Nanoflowers are a newly developed class of nanoparticles showing structure similar to flowers and gaining much attention due to their simple method of preparation, high stability, and enhanced efficiency. These materials are applied in different areas such as biosensor, enzyme purification, remediation, or bioimaging.[84]

In this case, recently has been described several protocols of controllable synthesis of iron nanoflowers.[15,16, 85–87]

A simple solvothermal strategy using ferric chloride and PEG has been described for the preparation of magnetite nanoflowers (Fe–NFs) with diameter range from 70 to 250 nm depending on the reaction conditions (Fig. 1.17), where the reducing agent or reaction time are key parameters in the synthetic control.[15]

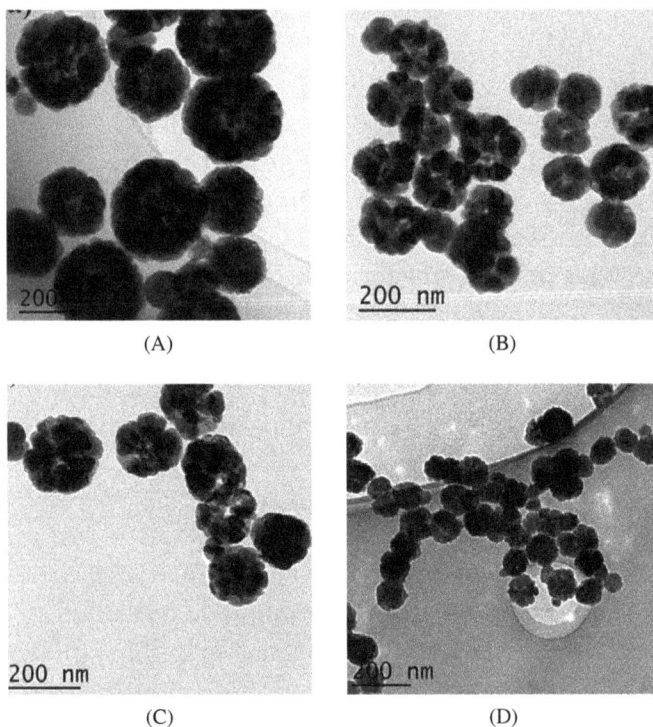

(A)   (B)

(C)   (D)

Fig. 1.17. TEM images of a series of Fe-NFs with a tunable diameter range. (A) 200–250 nm. (B) 100–150 nm. (C) 75–125 nm. (D) 70–80 nm.

Another interesting approach has been recently developed in order to create catalytically potential FePd nanoflowers for the oxygen reduction reaction.[16]

1.4 Conclusions and New Perspectives

This chapter describes several actual strategies to synthesize iron nanostructured materials. The extraordinary sensitivity of iron to oxidation makes possible the fabrication of different iron species and morphologies depending on the experimental conditions. The preparation of iron nanoparticles, other 1D and 2D structures has been well discussed. Therefore, we have a large portfolio of methodologies to create a tailor-made iron nanomaterial depending on the application to be used. In particular, most of the iron nanomaterials presented in this chapter has been successfully applied in a catalytic application in different conditions and different reactions.

However, considering the environmental point of view in the application of transition metals in catalysis, nanostructured iron materials represent one of the alternatives to the most used metals such as Pd, Au, or Ag in the near future.

Acknowledgments

This work was supported by the Spanish Government the Spanish National Research Council (CSIC) and SAMSUNG Electronics by GRO PROGRAM 2017.

References

1. A. Fustner, Iron catalysis in organic synthesis: A critical assessment of what it takes to make this base metal a multitasking champion, *ACS Cent Sci.* **23**, 778–789 (2016).
2. B. Plietker, *Iron catalysis in organic chemistry*, Wiley-VCH, Weinheim (2008).
3. I. Bauer and H. J. Knölker, Iron catalysis in organic synthesis, *Chem. Rev.* **115**, 3170 (2015).

4. S. Shylesh, V. Schnemann and W. R. Thiel, Magnetically separable nanocatalysts: Bridges between homogeneous and heterogeneous catalysis, *Angew. Chem. Int. Ed.* **49**, 342 (2010).

5. N. Narayan, A. Meiyazhagan and R. Vajtai, Metal nanoparticles as green catalysts, *Materials* **12**, 3602 (2019).

6. X. Xie, Y. Li, Z. Q. Liu, M. Haruta and W. Shen, Low-temperature oxidation of CO catalysed by Co_3O_4 nanorods, *Nature* **458**, 746 (2009).

7. L. Bai, S. Zhang, Q. Chen and C. Gao, Synthesis of ultrasmall platinum nanoparticles on polymer nanoshells for size-dependent catalytic oxidation reactions, *ACS Appl. Mater. Inter.* **9**, 9710–9717 (2017).

8. S. Singamaneni, V. N. Bliznyuk, C. Binekc and E. Y. Tsymbal, Magnetic nanoparticles: Recent advances in synthesis, self-assembly and applications, *J. Mater. Chem.* **21**, 16819 (2011).

9. J. Jeevanandam, A. Barhoum, Y. S. Chan, A. Dufresne and M. K. Danquah, Review on nanoparticles and nanostructured materials: History, sources, toxicity and regulations, *Beilstein J. Nanotechnol.* **9**, 1050–1074 (2018).

10. L. H. Reddy, J. L. Arias, J. Nicolas and P. Couvreur, Magnetic nanoparticles: Design and characterization, toxicity and biocompatibility, pharmaceutical and biomedical applications, *Chem. Rev.* **112**, 5818 (2012).

11. R. Benavente, D. Lopez-Tejedor and J. M. Palomo, Synthesis of a superparamagnetic ultrathin $FeCO_3$ nanorods–enzyme bionanohybrid as a novel heterogeneous catalyst, *Chem. Commun.* **54**, 6256–6259 (2018).

12. D. Zeng, T. Zhou, W.-J. Ong, M. Wu, X. Duan, W. Xu, Y. Chen, Y.-A. Zhu, D.-L Peng, Sub-5 nm ultra-fine FeP nanodots as efficient co-catalysts modified porous $g-C_3N_4$ for precious-metal-free photocatalytic hydrogen evolution under visible light, *ACS App. Mat. Inter.* **11**, 5651–5660 (2019).

13. Q. Hu, G. Li, Z. Han, Z. Wang, X. Huang, X. Chai, Q. Zhang, J. Liu, C. He, General synthesis of ultrathin metal borate nanomeshes enabled by 3D bark-like N-doped carbon for electrocatalysis, *Adv. Energ. Mat.* **9**, 1901130 (2019).

14. X. Yang, J. Hu, R. Wu, B. E. Koel, Balancing activity and stability in a ternary Au-Pd/Fe electrocatalyst for ORR with high surface coverages of Au, *Chem. Cat. Chem.* **11**, 693–697 (2019).

15. M. Saeed, M. Z. Iqbal, W. Ren, Y. Xia, C. Liu, W. S. Khanac and A. Wu, Controllable synthesis of Fe_3O_4 nanoflowers: Enhanced imaging guided cancer therapy and comparison of photothermal efficiency with black-TiO2, *J. Mater. Chem. B* **6**, 3800–3810 (2018).

16. C. Lian, Y. Cheng, L. Chen, X. Han, X. Lei, Y. Liu and Y. Wang, Synthesis and electrocatalytic properties for oxygen reduction of Pd_4Fe nanoflowers, *Chem. Commun.* **54**, 7058–7061 (2018).

17. S. J. Soenen, W. J. Parak, J. Rejman and B. Manshian, (Intra)cellular stability of inorganic nanoparticles: Effects on cytotoxicity, particle functionality, and biomedical applications, *Chem. Rev.* **115**, 2109–2135 (2015).

18. T. Skorjanc, F. Benyettou, J. C. Olsen and A. Trabolsi, Design of organic macrocycle-modifiediron oxide nanoparticles for drug delivery, *Chem. A Eur. J.* **23**, 8333–8347 (2017).

19. W. Wu, Z. Wu, T. Yu, C. Jiang and W. S. Kim, Recent progress on magnetic iron oxide nanoparticles: Synthesis, surface functional strategies and biomedical applications, *Sci. Technol. Adv. Mat.* **16**, 23501 (2015).

20. N. Lee, D. Yoo, D. Ling, M. H. Cho, T. Hyeon and J. Cheon, Iron oxide based nanoparticles for multimodal imaging and magnetoresponsive therapy, *Chem. Rev.* **115**, 10637–10689 (2015).

21. H. Arami, A. Khandhar, D. Liggitt and K. M. Krishnan, In vivo delivery, pharmacokinetics, biodistribution and toxicity of iron oxide nanoparticles, *Chem. Soc. Rev.* **44**, 8576–8607 (2015).

22. S. Behrens and I. Appel, Magnetic nanocomposites, *Curr. Opin. Biotechnol.* **39**, 89–96 (2016).

23. D. K. Zhong, J. Sun, H. Inumaru and D. R. Gamelin, Solar water oxidation by composite catalyst/α-Fe_2O_3 photoanodes, *J. Am. Chem. Soc.* **131**, 6086–6087 (2009).

24. M. Tadic, N. Citakovic, M. Panjan, Z. Stojanovic, D. Markovic and V. Spasojevic, Synthesis, morphology, microstructure and magnetic properties of hematite submicron particles, *J. Alloys Compd.* **509**, 7639–7644 (2011).

25. M. Khalil, J. Yu, N. Liu and R. L. Lee, Hydrothermal synthesis, characterization, and growth mechanism of hematite nanoparticles, *J. Nanopart. Res.* **16**, 2362 (2014).

26. D. Cardillo, M. Tehei, M. Lerch, S. Corde, A. Rosenfeld and K. Konstantinov, Highly porous hematite nanorods prepared via direct spray precipitation method, *Mater. Lett.* **117**, 279–282 (2014).

27. H. Liang, W. Chen, Y. Yao, Z. Wang and Y. Yang, Hydrothermal synthesis, self-assembly and electrochemical performance of α-Fe_2O_3 microspheres for lithium ion batteries, *Ceram. Int.* **40**, 10283–10290 (2014).

28. X. Hu and J. C. Yu, Continuous aspect-ratio tuning and fine shape control of monodisperse α-Fe_2O_3 nanocrystals by a programmed microwave–hydrothermal method, *Adv. Funct. Mater.* **18**, 880–887 (2008).

29. M. Krajewski, W. S. Lin, H. M. Lin, K. Brzozka, S. Lewinska, N. Nedelko, A. Slawska-Waniewska, J. Borysiuk and D. W. Beilstein, Structural and magnetic properties of iron nanowires and iron nanoparticles fabricated through a reduction reaction, *J. Nanotechnol.* **6**, 1652–1660 (2015).

30. F. N. Sayed and V. Polshettiwar, Facile and sustainable synthesis of shaped iron oxide nanoparticles: Effect of iron precursor salts on the shapes of iron oxides, *Sci. Rep.* **5**, 9733 (2015).

31. M. Tadic, M. Panjan, V. Damnjanovic and I. Milosevic, Magnetic properties of hematite (α-Fe_2O_3) nanoparticles prepared by hydrothermal synthesis method, *Appl. Surf. Sci.* **320**, 183–187 (2014).

32. A. S. Teja and P. Y. Koh, Synthesis, properties, and applications of magnetic iron oxide nanoparticles. Progress in crystal growth and characterization of materials, *Prog. Cryst. Growth Charact. Mater.* **55**, 22–45 (2009).

33. A. Rufus, N. Sreeju and D. Philip, Synthesis of biogenic hematite (α-Fe_2O_3) nanoparticles for antibacterial and nanofluid applications, *RSC Adv.* **6**, 94206–94217 (2016).

34. A. Rufus, N. Sreeju, V. Vilas and D. Philip, Biosynthesis of hematite (α-Fe_2O_3) nanostructures: Size effects on applications in thermal conductivity, catalysis, and antibacterial activity, *J. Mol. Liq.* **242**, 537–549 (2017).

35. P. C. Nagajyothi, M. Pandurangan, D. H. Kim, T. V. M. Sreekanth and J. Shim, Green synthesis of iron oxide nanoparticles and their catalytic and *in vitro* anticancer activities, *J. Clust. Sci.* **28**, 245–257 (2017).

36. B. Iandolo, B. Wickman, E. Svensson, D. Paulsson, A. Hellman, Tailoring charge recombination in photoelectrodes using oxide nanostructures, *Nano Lett.* **16**, 2381–2386 (2016).

37. Y. Zhu, S. Cheng, W. Zhou, J. Jia, L. Yang, M. Yao, M. Wang, J. Zhou, P. Wu and M. Liu, Construction and performance characterization of α-Fe$_2$O$_3$/rGO composite for long-cycling-life supercapacitor anode, *ACS Sust. Chem. Eng.* **5**, 5067–5074 (2017).

38. H. Liu, K. Tian, J. Ning, Y. Zhong, Z. Zhang and Y. Hu, One-step solvothermal formation of Pt nanoparticles decorated Pt^{2+}-doped α-Fe$_2$O$_3$ nanoplates with enhanced photocatalytic O$_2$ evolution, *ACS Catal.* **9**, 1211–1219 (2019).

39. E. Yuan, G. Wu, W. Dai, N. Guan and L. Li, One-pot construction of Fe/ZSM-5 zeolites for the selective catalytic reduction of nitrogen oxides by ammonia, *Catal. Sci. Technol.* **7**, 3036–3044 (2017).

40. G. Magnacca, A. Allera, E. Montoneri, L. Celi, D. E. Benito, L. G. Gagliardi, M. C. Gonzalez, D. O. Martire and L. Carlos, Novel magnetite nanoparticles coated with waste-sourced biobased substances as sustainable and renewable adsorbing materials, *ACS Sust. Chem. Eng.* **2**, 1518 (2014).

41. F. Shabani and A. Khodayari, Structural, compositional, and biological characterization of Fe$_3$O$_4$ Nanoparticles synthesized by hydrothermal method, *Synth. React. Inorg. Met.-Org. Nano-Met. Chem.* **45**, 356–362 (2015).

42. L. Gutiérrez, L. de la Cueva, M. Moros, E. Mazarío, S. de Bernardo, J. M. de la Fuente, M. Puerto Morales and G. Salas, Aggregation effects on the magnetic properties of iron oxide colloids, *Nanotechnology* **30**, 112001 (2019).

43. Y. N. Chen, T. Liu, Q. Zhang, C. Shang and H. Wang, Nanostructured biogel templated synthesis of Fe$_3$O$_4$ nanoparticles and its application for catalytic degradation of xylenol orange, *RSC Adv.* **7**, 758–763 (2017).

44. S. Lian, E. Wang, Z. Kang, Y. Bai, L. Gao, M. Jiang, C. Hu and L. Xu, Synthesis of magnetite nanorods and porous hematite nanorods, *Solid State Commun.* **129**, 485–490 (2004).

45. E. Y. Jomma and S. N. Ding, One-pot hydrothermal synthesis of magnetite prussian blue nano-composites and their application to fabricate glucose biosensor, *Sensors* **16**, 243 (2016).

46. Y. Jiang, N. Song, C. Wang, N. Pinna and X. Lu, A facile synthesis of Fe$_3$O$_4$/nitrogen-doped carbon hybrid nanofibers as a robust

peroxidase-like catalyst for the sensitive colorimetric detection of ascorbic acid, *J. Mater. Chem. B* **5**, 5499 (2017).

47. W. Lei, Y. Liu, X. Si, J. Xu, W. Du, J. Yang, T. Zhou and J. Lin, Synthesis and magnetic properties of octahedral Fe_3O_4 via a one-pot hydrothermal route, *Phys. Lett. A* **381**, 314–318 (2017).

48. G. Fu, W. Liu, S. Feng and X. Yue, Prussian blue nanoparticles operate as a new generation of photothermal ablation agents for cancer therapy, *Chem. Commun.* **48**, 11567–11569 (2012).

49. S. Rajput, C. U. Pittman Jr. and D. Mohan, Magnetic magnetite (Fe_3O_4) nanoparticle synthesis and applications for lead (Pb^{2+}) and chromium (Cr^{6+}) removal from water, *J. Colloid Inter. Sci.* **468**, 334–346 (2016).

50. G. Nabiyouni, M. Julaee, D. Ghanbari, P. C. Aliabadi and N. Safaie, Room temperature synthesis and magnetic property studies of Fe_3O_4 nanoparticles prepared by a simple precipitation method, *J. Indus. Eng. Chem.* **21**, 599 (2015).

51. C. Han, D. Zhu, H. Wu, Y. Li, L. Cheng and K. Hu, Magnetic memory signals variation induced by applied magnetic field and static tensile stress in ferromagnetic steel, *J. Magnet. Magnetic Mat.* **408**, 213–219 (2016).

52. A. Demir, R. Topkaya and A. Baykal, Green synthesis of superparamagnetic Fe_3O_4 nanoparticles with maltose: Its magnetic investigation, *Polyhedron* **65**, 282–287 (2013).

53. J. García-Aguilar, J. Fernαndez-García, E. V. Rebrov, M. R. Lees, P. Gao, D. Cazorla-Amorós and A. Berenguer-Murcia, Magnetic zeolites: Novel nanoreactors through radiofrequency heating, *Chem. Commun.* **53**, 4262–4265 (2017).

54. J. Du, Y. Ding, L. Guo, L. Wang, Z. Fu, C. Qin, F. Wang and X. Tao, Micro-tube biotemplate synthesis of Fe_3O_4/C composite as anode material for lithium-ion batteries, *Appl. Sur. Sci.* **425**, 164–169 (2017).

55. F. Zhang, Y. Song, S. Song, R. Zhang and W. Hou, Synthesis of magnetite-graphene oxide-layered double hydroxide composites and applications for the removal of Pb(II) and 2,4-dichlorophenoxyacetic acid from aqueous solutions, *ACS Appl. Mater. Interfaces.* **7**, 7251–7263 (2015).

56. M. Zhang, L. Chen, J. Zheng, W. Li, T. Hayat, N. S. Alharbi, W. Gan and J. Xu, The fabrication and application of magnetite coated

N-doped carbon microtubes hybrid nanomaterials with sandwich structures, *Dalton Trans.* **46**, 9172–9179 (2017).

57. H. C. Vu, A. D. Dwivedi, T. T. Le, S. H. Seo, E. J. Kim and Y. S. Chang, Magnetite graphene oxide encapsulated in alginate beads for enhanced adsorption of Cr (VI) and As (V) from aqueous solutions: Role of crosslinking metal cations in pH control, *Chem. Eng. J.* **307**, 220–229 (2017).

58. S. Zhang, W. Li, B. Tan, S. Chou, Z. Li and S. Doua, One-pot synthesis of ultra-small magnetite nanoparticles on the surface of reduced graphene oxide nanosheets as anodes for sodium-ion batteries, *J. Mater. Chem. A* **3**, 4793–4798 (2015).

59. X. Zhang, G. Wang, M. Yang, Y. Luan, W. Dong, R. Dang, H. Gao and J. Yu, Synthesis of a Fe_3O_4–CuO@meso-SiO_2 nanostructure as a magnetically recyclable and efficient catalyst for styrene epoxidation, *Catal. Sci. Technol.* **4**, 3082–3089 (2014).

60. X. Zheng, Q. Zhu, H. Song, X. Zhao, T. Yi, H. Chen and X. Chen, *In situ* synthesis of self-assembled three-dimensional graphene–magnetic palladium nanohybrids with dual-enzyme activity through one-pot strategy and its application in glucose probe, *ACS Appl. Mater. Interfaces.* **7**, 3480–3491 (2015).

61. N. Mei, B. Liu, J. Zheng, K. Lv, D. Tang and Z. Zhang, A novel magnetic palladium catalyst for the mild aerobic oxidation of 5-hydroxymethylfurfural into 2,5-furandicarboxylic acid in water, *Catal. Sci. Technol.* **5**, 3194–3202 (2015).

62. X. Y. Li, X. Wang, S. Y. Song, D. P. Liu and H. J. Zhang, Selectively deposited noble metal nanoparticles on Fe_3O_4/graphene composites: Stable, recyclable, and magnetically separable catalysts, *Chem. Eur. J.* **18**, 7601–7607 (2012).

63. M. I. Kim, M. S. Kim, M. A. Woo, Y. J. Ye, K. S. Kang, J. Lee and H. G. Park, Highly efficient colorimetric detection of target cancer cells utilizing superior catalytic activity of graphene oxide–magnetic-platinum nanohybrids, *Nanoscale* **6**, 1529–1536 (2014).

64. N. Bossa, A. W. Carpenter, N. Kumar, C. F. De Lannoy and M. Wiesner, Cellulose nanocrystal zero-valent iron nanocomposites for groundwater remediation, *Environ. Sci. Nano.* **4**, 1294–1303 (2017).

65. T. Raychoudhury and T. Scheytt, Potential of zerovalent iron nanoparticles for remediation of environmental organic contaminants in water: a review, *Water Sci. Technol.* **68**, 1425–1439 (2013).

66. D. O'Carrol, B. Sleep, M. Krol, H. Boparai and C. Kocur, Nanoscale zero valent iron and bimetallic particles for contaminated site remediation, *Adv. Water Resour.* **51**, 104–122 (2013).

67. Y. Mu, F. Jia, Z. Ai and L. Zhang, Iron oxide shell mediated environmental remediation properties of nano zero-valent iron, *Environ. Sci. Nano* **4**, 27–45 (2017).

68. M. Dickinson and T. B. Scott, The effect of vacuum annealing on the remediation abilities of iron and iron–nickel nanoparticles, *J. Nanopart. Res.* **13**, 3699–3711 (2011).

69. M. R. Jamei, M. R. Khosravi and B. Anvaripour, A novel ultrasound assisted method in synthesis of NZVI particles, *Ultrason. Sonochem.* **21**, 226–233 (2014).

70. S. Li, W. Yan and W. X. Zhang, Solvent-free production of nanoscale zero-valent iron (nZVI) with precision milling, *Green Chem.* **11**, 1618–1626 (2009).

71. S. Machado, J. P. Grosso, H. P. A. Nouws, J. T. Albergariaand and C. Delerue-Matos, Utilization of food industry wastes for the production of zero-valent iron nanoparticles, *Sci. Total Environ.* **496**, 233 (2014).

72. S. Bae, S. Gim, H. Kim and K. Hanna, Effect of NaBH4 on properties of nanoscale zero-valent iron and its catalytic activity for reduction of p-nitrophenol, *App. Catal. B: Environ.* **182**, 541–549 (2016).

73. J.Tuček, Z. Sofer, D. Bouša, M. Pumera, K. Holá, A. Malá, K. Poláková, M. Havrdová, K. Čépe, O. Tomanec and R. Zbořil, Air-stable superparamagnetic metal nanoparticles entrapped in graphene oxide matrix, *Nat. Commun.* **7**, 12879 (2016).

74. W. Teng, J. Fan, W. Wang, N. Bai, R. Liu, Y. Liu, Y. Deng, B. Kong, J. Yang, D. Zhao and W. X. Zhang, Nanoscale zero-valent iron in mesoporous carbon (nZVI@C): Stable nanoparticles for metal extraction and catalysis, *J. Mat. Chem. A* **5**, 4478 (2017).

75. L. Li, C. Zeng, L. Ai and J. Jiang, Synthesis of reduced graphene oxide-iron nanoparticles with superior enzyme-mimetic activity for biosensing application, *J. Alloys Compd.* **639**, 470–477 (2015).

76. S. Kment, F. Riboni, S. Pausova, L. Wang, L. Wang, H. Han, Z. Hubicka, J. Krysa, P. Schmuki and R. Zboril, Photoanodes based on TiO_2 and α-Fe_2O_3 for solar water splitting- superior role of 1D

nanoarchitectures and of combined heterostructures, *Chem. Soc. Rev.* **46**, 3716–3769 (2017).

77. Q. Huang, M. Cao, Z. Ai and L. Zhang, Reactive oxygen species dependent degradation pathway of 4-chlorophenol with Fe@Fe$_2$O$_3$ core–shell nanowires, *Appl. Catalysis B: Environ.* **162**, 319–326 (2015).

78. Y. Wang, X. Guo, Z.Wang, M. Lü, B. Wu, Y. Wang, C. Yan, A. Yuan and H. Yang, Controlled pyrolysis of MIL-88A to Fe$_2$O$_3$@C nanocomposites with varied morphologies and phases for advanced lithium storage, *J. Mater. Chem. A.* **5**, 25562–25573 (2017).

79. R. Han, L. Wei, W. Pan , M. Zhu, D. Zhuo and L. Fa-shen, 1D magnetic materials of Fe$_3$O$_4$ and Fe with high performance of microwave absorption fabricated by electrospinning method, *Sci. Rep.* **4**, 7493 (2014).

80. C. Zhang, H. C. Yang, L. S. Wan, H. Q. Liang, H. Li and Z. K. Xu, Polydopamine-coated porous substrates as a platform for mineralized β-FeOOH nanorods with photocatalysis under sunlight, *ACS Appl Mater Inter.* **7**, 11567–11574 (2015).

81. T. Zhu, W. L. Ong, L. Zhu and G. W. Ho, TiO$_2$ fibers supported β-FeOOH nanostructures as efficient visible light photocatalyst and room temperature sensor, *Sci. Rep.* **5**, 10601 (2015).

82. N. Liu, R. Qu, Y. Chen, Y. Cao, W. Zhang, X. Lin, Y. Wei, L. Feng and L. Jiang, *In situ* dual-functional water purification with simultaneous oil removal and visible light catalysis, *Nanoscale.* **8**, 18558–18564 (2016).

83. K. Gandha, J. Mohapatra, M. K. Hossain, K. Elkins, N. Poudyal, K. Rajeshwar and J. P. Liu. Mesoporous iron oxide nanowires: Synthesis, magnetic and photocatalytic properties, *RSC Adv.* **6**, 90537–90546 (2016).

84. P. Shende, P. Kasture and R. S. Gaud, Nanoflowers: The future trend of nanotechno logy for multi-applications, *Artif. Cells Nanomed. Biotechnol.* **46**, 413–422 (2018).

85. A. Curcio, A. K. A. Silva, S. Cabana, A. Espinosa, B. Baptiste, N. Menguy, C. Wilhelm and A. Abou-Hassan, Iron oxide nanoflowers @ CuS hybrids for cancer tri-therapy: Interplay of photothermal therapy, magnetic hyperthermia and photodynamic therapy, *Theranostics* **9**, 1288–1302 (2019).

86. X. L. Liu, C. T. Ng, P. Chandrasekharan, H. T. Yang, L. Y. Zhao, E. Peng, Y. B. Lv, W. Xiao, J. Fang, J. B. Yi, H. Zhang, K.-H. Chuang, B. H. Bay, J. Ding and H. M. Fan, Synthesis of ferromagnetic $Fe_{0.6}Mn_{0.4}O$ nanoflowers as a new class of magnetic theranostic platform for in vivo T_1-T_2 dual-mode magnetic resonance imaging and magnetic hyperthermia therapy, *Adv. Healthcare Mater.* 5, 2092–2104 (2016).

87. H. Gavilán, E. H. Sánchez, M. E. F. Brollo, L.Asín, K. K. Moerner, C. Frandsen, F. J. Lázaro, C. J. Serna, S. Veintemillas-Verdaguer, M. P. Morales and L. Gutiérrez, Formation mechanism of maghemite nanoflowers synthesized by a polyol-mediated process, *ACS Omega* 2, 7172–7184 (2017).

https://doi.org/10.1142/9781786349620_0002

Chapter 2

Keys and New Trends of Iron-Based Catalysts in Selective Oxidation of Propylene in Gas Phase

Javier Fernández-Catalá, Jaime García-Aguilar,
Diego Cazorla-Amorós, and Ángel Berenguer-Murcia*

*Materials Science Institute and Inorganic Chemistry Department,
Alicante University, Ap. 99, E-03080 Alicante, Spain*

*a.berenguer@ua.es

In this chapter, we propose the use of iron as a potential candidate for its implementation in an industrial process of great relevance. The production of alkene oxides (in this case, propylene oxide) involves not only the polymer industry, but also fine chemicals. In this framework, iron emerges as a candidate among transition metals to substitute noble metals such as Gold or Silver, which despite performing better at present, possesses the severe drawback of high cost. Moreover, this chapter is focused on the interaction of the iron species with the support, the effect of the oxidant (NO_2 and O_2) in the catalytic performance and the crucial role of the iron species and their distribution in the selective oxidation of propylene and the mechanism of this reaction using iron catalyst.

2.1 Introduction

Iron (Fe) is one of the most abundant elements on Earth. Although the majority of iron is present in the nucleus of our planet, a large fraction of iron is located in its crust forming oxides (Fe represents about 5% in the Earth's crust).[1] This element has had a prominent role in human society, a fact which is emphasized with the occurrence of the Iron Age in mankind's timeline, as shown in Fig. 2.1.[2] The importance of this element in human society is due to its relatively easy manipulation, cheap production cost, and environmental friendliness.[2] Another important aspect of iron is the presence of this element in all living organisms in the form of proteins. This makes iron essential for the adequate function of biological systems. Iron is generally stored in complexes (metalloproteins), and this type of complexes can catalyze numerous redox reaction in living organisms. As a prime example, hemoglobin is the protein that is responsible for oxygen transport in the humans.[3–5]

In the last 50 years, the utilization of iron as a catalyst has experienced an unprecedented increase.[6,7] In this respect, the most important and economically relevant industry process uses iron as a

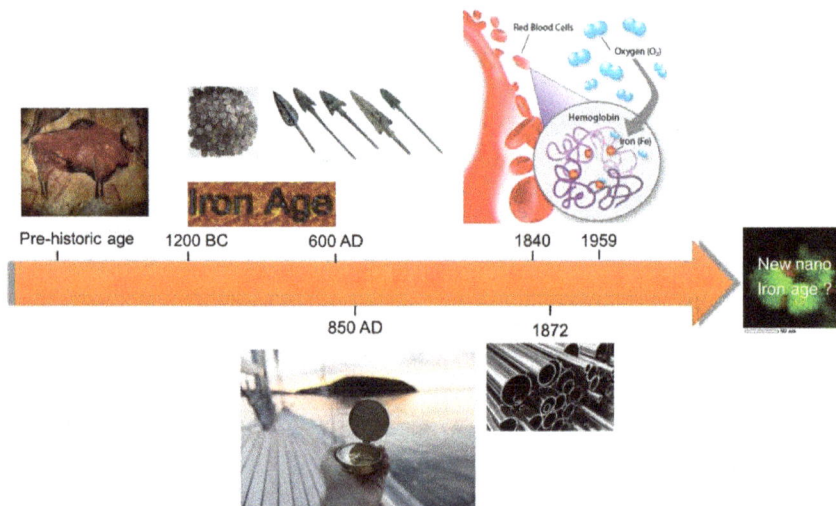

Fig. 2.1.   Use of iron/iron oxides in human society.[2]

catalyst, being the Haber–Bosch ammonia synthesis.[8,9] The interest of companies and the scientific community in the use of iron as a catalyst is due to the rise of catalysis (which usually brings forth more cost-effective and sustainable processes) and the need of industries for substituting catalysts based on noble metals (Pt, Au...) for cheaper, sustainable, and environmental friendly ones. In this sense, transition metals have emerged as excellent candidates for this challenge.[6–10] In this respect, iron has once again risen to the challenge.

Selective oxidation processes have modified the modern chemical industry, since many chemicals and intermediates are produced by selective oxidation of hydrocarbons using catalysts.[11] In this sense, one interesting selective oxidation process is the synthesis of propylene oxide (PO) from propylene. Since propylene epoxidation is considered an emerging technology due to the use of PO as a precursor of polymers and other added-value products. At present the most established chemical processes for PO synthesis do not use catalytic reactions.[12]

With this in mind, the aim of this chapter is to provide the reader with the new advances in the use of Fe-based catalysts for the gas-phase catalytic epoxidation of propylene. We shall put the focus on the interaction of the iron species with the support, the effect of the oxidant (NO_2 and O_2) in the catalytic performance and the crucial role of the iron species and their speciation in the selective oxidation of propylene and the mechanism of this reaction using iron catalyst.

2.2 Propylene Epoxidation

Chemical companies and the scientific community both have a significant interest in PO due to its high reactivity and versatility in the formation of a large variety of products, including polymers. These properties allow the use of PO as a pre polymer for polyurethanes for the automobile and housing industries, polyester for the textile and construction sectors, and propylene glycol as additives in drugs or cosmetics, among others. This fact has caused a continuous

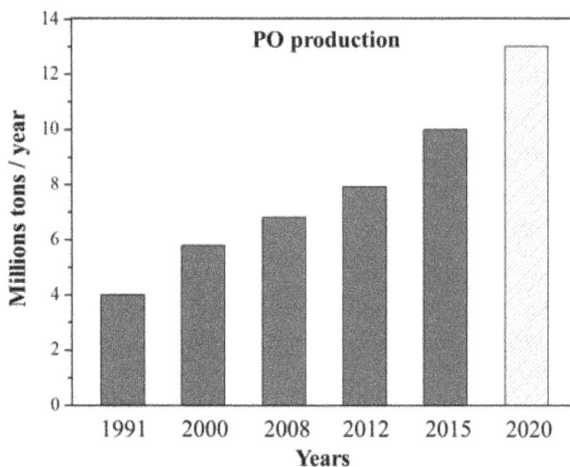

Fig. 2.2. Evolution of the propylene oxidation production.

increase in the worldwide production of PO over the years by the main chemical companies, such as DOW, LyondellBasell, or BASF, with an annual estimated production of 13 million tons of PO in 2020 (see Fig. 2.2).[12–14]

At present, PO is mainly produced by non catalytic processes in liquid phase using the chlorohydrin process (40% of PO production) or the organic hydroperoxide processes (see Table 2.1).[12,14,15] While they do present benefits, these industrial processes are not devoid of some problems. For example, the use of dangerous and highly oxidizing reagents, such as Cl_2 or organic peroxides (R-OOH), and the generation of by-products, which are difficult to separate, are two key elements that would be economically advantageous should they be minimized or completely suppressed. For this reason, an alternative to solve this problem is to produce PO by selective oxidation of propylene via heterogeneous catalysis, since this methodology can avoid the use of Cl_2 or organic peroxides enabling the use of other more sustainable and environmentally friendly reagents like H_2/O_2 or O_2.[16,17]

In the last 20 years, the scientific community has strongly focused on the study of PO production in gas and liquid phase using

Table 2.1.   Commercial PO routes and production distribution.[12,14]

Process	Feedstock	By-product	Companies	Production
Chlorohydrin process	Propylene, chlorine	Calcium chloride	Dow, Asahi Glass, Tokuyama	43%
Hydroperoxidation	Isobutane	t-butyl alcohol	Lyondell, Huntsman	33%
	Ethylbenzene		Lyondell, Shell,	15%
	Cumene	Styrene	Repsol Sumitomo	4%
Hydrogen peroxide-based process (HPPO)	Propylene and hydrogen peroxide	Only water	BASF/Dow, Evonik/ Uhde	5%

catalysis for this process. In this respect, epoxidation studies in the gas phase have the upper hand due to the use of less dangerous oxidants (O_2 or H_2/O_2 mixtures).[14] Ever since Haruta *et al.*[18] showed that their Au/TiO_2 catalysts (gold particles dispersed on TiO_2) could perform the propylene epoxidation reaction using H_2/O_2 mixtures with very high PO selectivity (>95%) at low propylene conversions (1%), gold-based catalysts supported on different oxide catalysts have attracted a great deal of attention for this purpose due to the almost negligible production of unwanted by-products such as acrolein (Acr) of carbon oxides (COx). This type of catalyst has been studied by the scientific community until reaching a PO selectivity about 90% with a propene conversion around 10% at mild temperatures using H_2/O_2 mixture. These results show that Au-based titanosilicate (Ti-SiO$_2$) catalysts are very interesting for propylene epoxidation.[19–22] Under these conditions, it is mandatory to reach and maintain a PO selectivity around 90% since these types of catalysts present a relatively low conversion (10%).

Au-based Ti-SiO$_2$ catalyst is a good candidate for its use in propylene epoxidation using a H_2/O_2 mixture as oxidant. However, the use of a noble metal, in this case Au, presents some drawbacks, such as high catalyst price (due largely to the use of gold), low H_2 efficiency, and also relatively low propylene conversion. For these reasons, it is crucial to develop new catalysts based on transition metals. Once again, iron comes to mind as a potential candidate.

2.3 Iron-Based Catalysts as Candidates for Catalysts in the Selective Propylene Oxidation

The scientific community has dedicated significant efforts in developing different catalysts for carrying out the generation of PO in the gas phase without the use of noble metal-based catalysts due to the aforementioned drawbacks. One alternative to solve this problem is the use of silver-based catalysts. This catalyst can be interesting since silver has been used for decades for ethylene epoxidation using molecular oxygen to produce ethylene oxide (EO).[23,24] Nevertheless, Ag-based catalysts are not adequate for propylene epoxidation because the allylic hydrogen atoms (which have a relatively strong acidity) in propylene can be attacked by the oxygen and the main product is the total combustion products (CO_x).[25–27]

Another tentative solution is the use of catalysts with transition metals (Fe,[28–32] Cu,[33–36] V[37,38] and Mo,[39] among others) using different oxidants (NO_2, O_2, and O_2/H_2). However, these catalysts do not reach as high a selectivity as the gold-based catalysts, but these catalysts may have higher propylene conversions/yields or similar PO generation values.

In the last 20 years, the scientific community has paid attention to the use of iron as a catalyst in the synthesis of propylene epoxidation, since in nature some enzymes possess iron as an active center. Furthermore, these active centers can perform this type of selective oxidation.[40] For example, it is the case of the enzyme soluble methane monooxygenase (sMMO).[41] The interest of this enzyme is an intermediate, namely compound Q, that presents active oxygen species. This species can hydroxylate the C–H bonds of methane or attack the C = C bonds of propylene to form PO selectively. For this reason, the synthesis of new transition metal-based catalysts that mimic the process of this enzyme is a topic of great interest.[30]

Moreover, transition metal-based catalysts (e.g., Fe-based) are very interesting for this proposed reaction due to the benefits of iron such as its low cost and its being environmentally friendly. In this sense, Wang *et al.* tested different transition metal oxides (loading 1 wt%), such as V, Cr, Fe, Co, Ni, among others, supported on mesoporous silica (SBA-15) with and without KCl modification for this reaction using NO_2 as oxidant. As shown in Fig. 2.3, iron is especially interesting for the epoxidation of C_3H_6 by N_2O since it presents the best conversion of propylene and selectivity toward PO in these conditions.[42] These results highlight Fe-based catalysts as a good candidate for the challenge of switching from Au-based

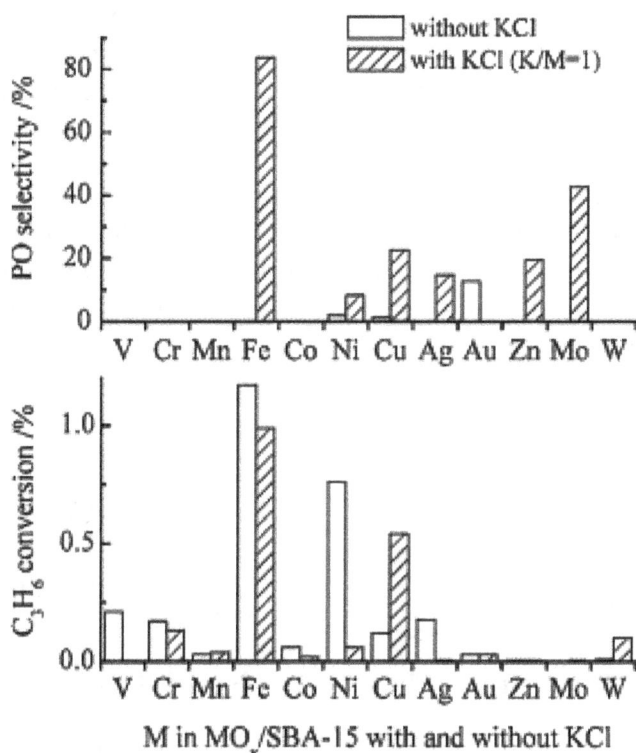

Fig. 2.3. Catalytic performances of various SBA-15-supported transition metal oxides (MO_x/SBA-15) with and without KCl modification for C_3H_6 oxidation by N_2O. The loading of each metal oxide was 1 wt%, and the K/M (molar ratio) was 1.0. Reaction conditions: T = 325 °C; W = 0.2 g; F = 60 mL min^{-1}; P(C_3H_6) = 2.53 kPa; P(N_2O) = 25.3 kPa.[42]

Ti-SiO$_2$ catalysts in the propylene epoxidation, using different oxidants such as NO$_2$ or O$_2$.

2.3.1 Effect of the oxidant in the iron-based catalyst for the selective propylene oxidation

One interesting effect to consider in the study, preparation, and use of iron-based catalysts for the selective propylene oxidation is the interaction between molecular oxygen and the metal(s) (in this case iron) of the catalyst, since this interaction produces electrophilic and nucleophilic oxygen species necessary for the reaction with propylene and the subsequent formation of PO. Duma *et al.* showed three possible reaction paths between the oxygen species formed and propylene (Fig. 2.4)[43]:

1. Reaction between mild electrophilic oxygen species and propene. In this case, the products of this reaction are propene oxide, propanal, and acetone. However, high temperatures and acidic or basic conditions favor deformation of propanal and acetone, among others.

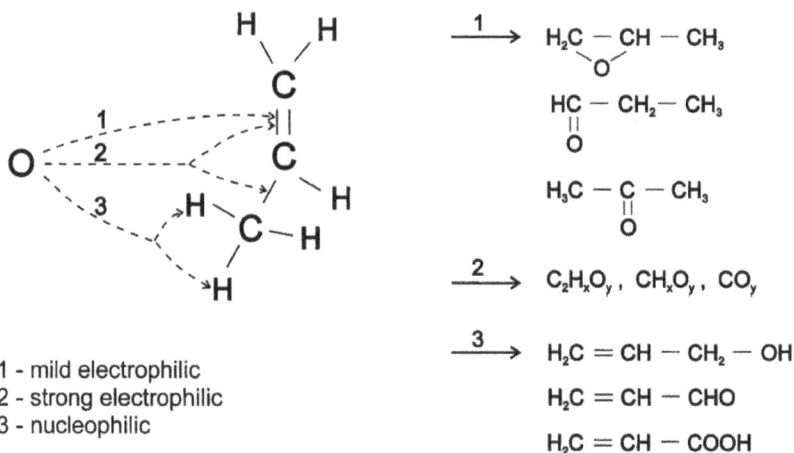

Fig. 2.4. Possible reaction paths of the reaction between propene and oxygen species.[43]

2. Strong electrophilic oxygen species will attack the C–C bonds. The reaction products are the total combustion of the molecule (carbon oxides).
3. Nucleophilic oxygen species have affinity to react with H atoms in the allylic position. The products formed in this condition are allyl alcohol, acrolein, and acrylic acid.

In order to obtain a good selectivity toward PO using Fe-based catalysts, it is necessary to use one oxidant that only generates mild electrophilic oxygen species when the oxidant is activated on the iron-based catalyst.

One interesting oxidant for this reaction can be molecular oxygen (O_2) due to the great advantages of this compound. Moreover, some proteins in nature (enzyme sMMO) can use this reagent (O_2) to perform selective oxidation reactions. However, in the oxidation of propylene, Fe-based catalysts do not achieve great conversions or selectivities.[42] This fact is due to the rapid formation of nucleophilic oxygen species (lattice oxygen) by the interaction with the iron species and the oxygen. These generated nucleophilic oxygen species react with H atoms in the allylic position forming allyl alcohol, acrolein, and acrylic acid. At present, the scientific community is turning toward the possibility of studying the use of O_2 as oxidant. In order to use molecular O_2 as oxidant, it becomes necessary to decrease or block the acidity of the Fe-based catalyst since its interaction with molecular oxygen generates nucleophilic oxygen species that favor propylene combustion.[44,45] In this respect, the scientific community has not hitherto met with satisfactory results in this topic.

Another alternative is the use of nitrous oxide. As described by Duma *et al.* this oxidant compound presents properties interesting for its use as oxidant in the formation of PO, since nitrogen oxide has a low reactivity with iron and can generate electrophilic species selective toward PO formation. The study of Duma *et al.* was the first where a PO selectivity of 55% and a propylene conversion of 5% were achieved using N_2O as an oxidant with an Fe-based catalyst

(Fe-SiO$_2$ catalyst).[43] Recently, Horvárth *et al.* have made some studies along this line and they present a better yield of about 20% with similar selectivity 60% using the same oxidant (NO$_2$) and similar catalyst (Fe-SiO$_2$ catalyst).[46,47]

2.3.2 *Effect of support in the iron-based catalyst for the selective propylene oxidation*

Beyond its cost-effectiveness, iron is an interesting catalyst for meeting the challenge of epoxidizing propylene without the use of noble metal catalysts. However, the iron species present in the catalyst must be well dispersed and in small quantities (preferably as single-site species) in a support in order to selectively epoxidize propylene, since iron in the form of FeO$_x$ aggregates favors the combustion of propylene (CO$_x$). This fact is due to the Lewis acidity of this material.[42,43,48] For this reason it is key to understand the effect of both iron and support so as to reach a suitable preparation which renders an adequate Fe-based catalyst with a performance capable of delivering a high selectivity toward PO and a high propylene conversion.

In this sense, Horvárth *et al.* tested different Fe-based catalysts on different supports (magnesium, thorium and titanium oxides, faujasite, ZSM-5, and silicalite S-1) using Fe(acac)$_3$ as a precursor of iron using N$_2$O as oxidant. The catalytic tests yielded acrolein, CO$_x$, and condensation products in the majority of the supports, as shown in Table 2.2.[46] These results were attributed to the acidity or basicity present in the different supports tested. Before this work, Zhang *et al.* showed that the use of zeolites with medium-sized pores (e.g., Fe-MFI): as a support for iron applied to this selective oxidation is not adequate due to the acidity of the catalysts.[48]

In this sense, Duma *et al.*[43] previously indicated that silica gel is a good candidate as a support for iron species in the reaction of interest. The advantages of using silica as support are the high surface area, low acidity, and non intrinsic activity in N$_2$O decomposition. Furthermore, in this study the authors demonstrated that the alkali promotion and high temperature calcination reduced the acidity of the catalyst obtaining a good selectivity toward PO (55%).[43]

Table 2.2.   Effect of the various supports on product distribution in iron-based catalyst described by Horvárth *et al.* Data were measured for the authors after 6 min TOS, using 1 g of catalyst (1000 ppm Fe deposited from Fe(acac)$_3$ modified with ethylenediamine), at 723 K, feed: 3% propylene in N$_2$O. S$_{PO}$, S$_{Ac}$, S$_{COx}$–selectivities toward PO, acroleine and CO$_x$, respectively, X is the conversion of propylene.[46]

Support	S$_{PO}$ (%)	S$_{Ac}$ (%)	S$_{COx}$ (%)	X (%)
Rutile, thoria	Traces	Traces	Traces	0.1
Anatase	Traces	1	70	0.2
ZSM-5	Traces	Traces	90	0.5
S-1	8	15	60	0.5
MgO	Traces	10	60	0.2

Another important factor in the preparation of the catalysts is the porosity of the support and its effect in catalytic activity. In this sense, Duma *et al.* also studied in their work three silica gel supports with different textural properties (narrow, intermediate, and large mesopores for ranges between 2–5, 5–10, and 10–100, respectively). The authors observed that those catalysts with a major proportion of narrow pores showed the lowest activity and selectivity due to the appearance of desorption and/or diffusional limitations. As a result, the catalysts with large pores presented better catalytic activity than those with narrow pores. This is due to large pores permitting a good diffusion of the reactants and products. Nevertheless, these porous materials did not have a good adsorption capacity. For this reason, the catalyst with intermediate mesopores had the best catalytic activity since it combined the advantages of narrow pores and large pores. The slim mesopores presented good diffusion of the reagents and products together with a good adsorption capacity. These properties of the slim porous silica make iron supported on the surface of mesoporous silica an interesting catalyst for this reaction.

Another important factor to consider in this type of catalyst is the residence time of the reagent in the pores of the catalyst. Horvárth *et al.* observed that it is "is necessary to maintain short residence time of the substrate inside the pores to prevent

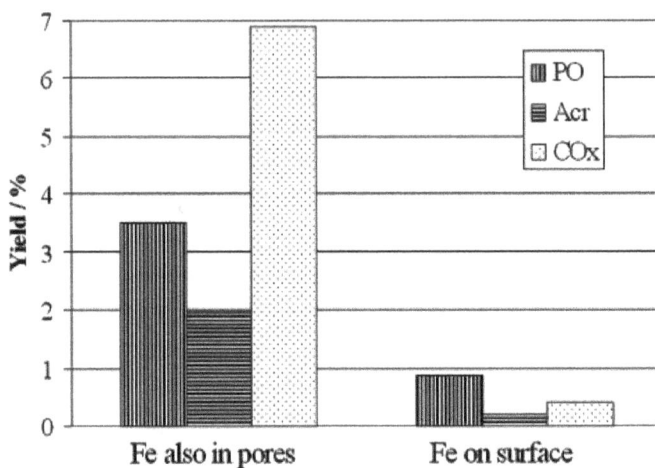

Fig. 2.5. Effect of the iron localization on the product distribution at 723 K. In all experiments 1 g of AE-2 supported catalyst was used (1000 ppm Fe), prepared from the Fe(acac)$_3$ precursor. Feed: 3% propylene in N$_2$O, at 723 K.[46]

overoxidation of PO." This is shown in Fig. 2.5, where Fe on the surface is more selective toward PO than Fe in the pores. Nevertheless, the conversion of propylene was still very low.[46]

As a general rule, the catalysts used in this reaction are based on the impregnation of the iron precursor on the support, generally silica due to its interesting properties as described previously. Another alternative was described by our research group, where García-Aguilar *et al.* described a well-dispersed iron catalyst synthesized in silica (Fe$_{0.0X}$SiO$_2$) by a one-step synthesis procedure, studied previously in the literature.[49] The prepared catalysts were tested in the propylene epoxidation reaction using gaseous O$_2$ as only oxidant. This iron-dispersed catalyst performs more efficiently in the propylene epoxidation reaction than the same catalyst prepared by impregnation.[45]

2.3.3 *Importance of the iron species in the PO production and possible mechanism of reaction*

One of the most important aspects in the study of iron-based catalysts in the propylene epoxidation reaction in gas phase is understanding the active sites in the reaction when using different oxidants (e.g., NO_2 or O_2). In this sense, Duma *et al.* had already observed an effect of the loading of the iron supported on silica. An increase of the iron loading improved propylene conversion but decreased the selectivity toward the desired product using nitrogen dioxide as oxidant. Moreover, the authors observed that it was essential to impregnate the iron precursor with alkali metal salts (sodium) since this impregnation reduced the acidity and the reduction temperature of the catalyst. This fact showed that not all iron species are active.[43]

Most recently, the scientific community has focused on the study of the iron location (framework or extra-framework), coordination (tetrahedral or octahedral), or its distribution (single-sites, clusters, or small particles) on the different supports used. Among those, silica stands out due to its good properties, which were described in Section 2.3.2 "Effect of support in the iron-based catalyst for the selective propylene oxidation." The scientific community has also focused on the effect that alkali metal salts exert on the iron species present in the catalyst. In this sense, Wang *et al.* observed that these salts are key to obtain highly dispersed iron species in the catalyst in order to reach a partial oxidation of propylene using N_2O. Furthermore, they showed that the iron active species in this catalyst is tetrahedrally coordinated iron (Fe^{3+}). Their work showed that the incorporation of alkali metal salts (being KCl the most appropriate) in the iron-based catalyst shifts the reaction route from allylic oxidation to propylene epoxidation as shown in Table 2.3. This effect is due to the decrease of the reactivity of lattice oxygen associated with iron species (FeO_x), the improvement of the dispersion of the iron species, and the local coordination of iron, which changes into a surface tetrahedral configuration, as shown in Fig. 2.6.[50]

In another recent work, it was described that the framework iron cations or extra-framework FeO_x clusters in iron-based catalysts (Fe-MFI or Fe-MCM-41) are selective toward the allylic oxidation of C_3H_6 by N_2O.[48] However, the catalyst prepared with mononuclear iron exhibited an increase in propylene conversion with respect to the catalyst containing FeO_x clusters. As shown in their previous work,[50] with the incorporation of KCl in the iron catalyst it was

Table 2.3. Selective oxidation of propylene by N_2O over FeOx/SBA-15 and KCl-FeOx/SBA-15,[a] adapted from the work of Wang *et al.*[50]

Catalyst	Conv. (%)	TOF[b] (h^{-1})	Sel. (%) PO	Sel. (%) acrolein	Sel. (%) allyl alcohol	Sel. (%) others[c]	Sel. (%) COx
1 wt% FeOx/ SBA-15	1.2	1.9	0	42	4.5	7.5	46
KCl-1 wt% FeOx/ SBA-15[d,e]	4.5	7.6	72	1.6	1.1	7.3	18
α-Fe$_2$O$_3$	0.5	0.008	0	10	0	5.0	85
KCl-α-Fe$_2$O$_3$[f]	trace	trace	0	0	0	0	100

[a] Reaction conditions: $P(C_3H_6) = 2.5$ kPa, $P(N_2O) = 25$ kPa; T = 598 K; catalyst, 0.2 g; total flow rate, 60 mL min^{-1}.
[b] Turnover frequency (TOF) was evaluated from the moles of C_3H_6 converted per mole of Fe in the catalyst per hour.
[c] Other products mainly include acetone, propionaldehyde, and acetaldehyde.
[d] K/Si (molar ratio) = 0.04/1.
[e] K/Fe (molar ratio) = 5/1.
[f] K/Fe (molar ratio) = 1/1.

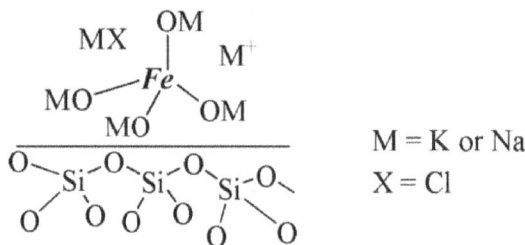

Fig. 2.6. Possible iron site proposed for KCl-modified FeO_x/SBA-15 described by Wang *et al.*[50]

possible to achieve a PO selectivity of 80%. Under these conditions (after KCl incorporation), the catalysts that presented extra-framework iron species exhibited an improvement of PO formation with respect to the catalyst containing framework iron.

In the same research line, Horváth *et al.* studied the addition of an alkali promotor (such as KCl) on the catalyst, resulting in an improvement of the catalytic properties of the iron-based catalyst (PO selectivity 75%) using N_2O as oxidant as described previously. However, the catalyst deactivated due to the formation of carbon deposits on the catalyst surface, even after the incorporation of the promoting agent (KCl).[46,47,51] This research group also observed that it may be possible to discriminate between three different forms of iron in the catalysts. These species are Fe species strongly bound in silicates being this iron inactive in propylene oxidation, Fe species in the form of crystalline hematite (this species not being selective for propylene epoxidation), and Fe species formed after an oxidative pretreatment in the presence of alkali metal salts, these species being selective for propylene epoxidation.[47]

Another important factor to understand the active Fe species in the iron-based catalyst for the selective oxidation of propylene is to understand the possible reaction mechanism of propylene epoxidation through the use of iron-based catalysts. Some authors have proposed that the epoxidation reaction takes place via the oxygen atoms abstracted from the N_2O (oxidant) by the iron (Fe^{3+}) active species for this reaction. Afterward, these abstracted oxygen atoms $(O)_x$ may be transferred to the propylene and complete the reaction assisted by alkali metal salts as shown in Scheme 2.1.[51]

Along these lines, but using molecular O_2 as oxidant, our research group has observed that highly dispersed Fe species incorporated within the SiO_2 framework present an improved selectivity

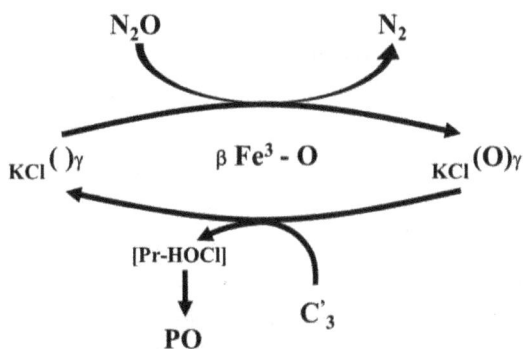

Scheme 2.1.   Propylene epoxidation reaction scheme proposed by Horváth *et al.*[51]

Table 2.4. Catalytic performance of the samples prepared in propylene epoxidation by O_2 molecule under steady-state conditions at 350°C, adapted from the work
of García-Aguilar *et al.*[44]

Catalyst	C_3H_6 conversion (%)	PO generation (%)	PO selectivity (%)	Selectivity (%) others	Selectivity (%) CO_2	TOF (h^{-1})
$Fe_{0.005}SiO_2$	5.5	1.8	33.6	<0.5	66	10.8
$Fe_{0.01}SiO_2$	7.8	1.9	24.6	<0.5	75	4.8
$Fe_{0.02}SiO_2$	10.3	2.4	23.1	<1.0	76	3.6
$Fe_{0.03}SiO_2$	15.4	3.0	19.5	<1.0	80	2.7
$Fe_{impreg}SiO_2$	5.4	1.2	22.9	<1.0	76	3.5

toward PO. This fact indicated that a good distribution of isolated
iron species (Fe^{3+}) into the silica framework presents a higher selectivity to PO production with respect to small iron surface clusters
and larger particles of iron oxide supported on the silica by an
impregnation method (Table 2.4).[44]

In this study, it was also observed that the presence of isolated
iron species in tetrahedral or pseudo-tetrahedral coordination
(Fe^{3+}) into the silica framework might be responsible for the adsorption of both molecular oxygen on the isolated iron and propylene
through an interaction with the neighboring acidic proton favoring
the reaction between these reagents (Fig. 2.7). These results were
obtained by means of molecular simulation studies.[44]

Our research group has also studied the effect of alkali metal
salts incorporated on ferrosilicates, as previously described in the
literature.[45] However, in the study of García-Aguilar *et al.* O_2 was used
as the oxidant. In our work, it was evidenced that the addition of
alkali metal-based promoters (K and Ca) led to the removal of the
small particles/clusters of iron oxide present in the catalyst by the
generation of superficial K- or Ca-Fe species as well as the blocking
or substitution of the hydroxyl groups (Brønsted acid character) by
K or Ca promoters (Scheme 2.2). This fact favors selectivity toward
PO reducing propylene conversion toward CO_2. Nevertheless, propylene conversion dropped due to the elimination of Fe_2O_3. The

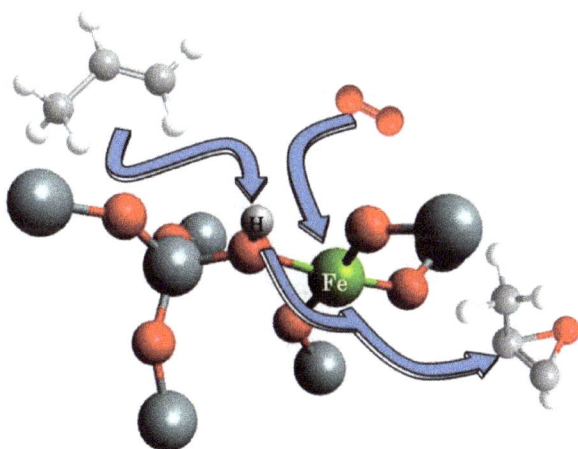

Fig. 2.7. Possible mechanism of the propylene epoxidation by adsorption of both molecular oxygen in the isolated iron and propylene in the acidic proton.[44]

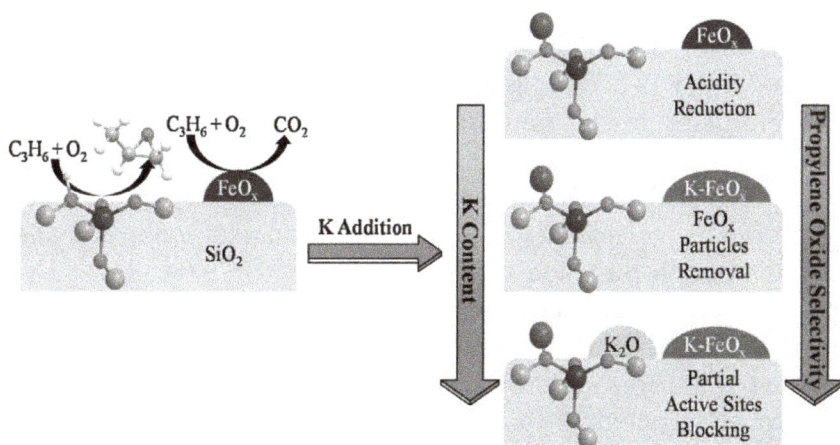

Scheme 2.2. Possible interpretation of the alkali promoter effects on the Fe-SiO$_2$ surface and its influence on the PO selectivity.[45]

cracking of propylene on the catalyst surface was also reduced due to the reduction in the surface acidic sites of the catalyst.

The results presented in this section show that controlling the chemical nature of the iron species present in the catalyst is key in the development of a catalyst with good performance. In this sense, the most active iron species are those in tetrahedral or pseudo-tetrahedral coordination species. We have also shown that the incorporation of alkali metal salts (KCl) can be significant for the improvement of the selectivity toward PO but may negatively affect propene conversion.

2.4 Conclusion and New Perspectives

While there is undoubtedly a substantial amount of ground to cover, in this chapter, we have hinted at the use of iron as a potential candidate for its implementation in an industrial process of both great volume and great value, since the production of alkene oxides (in this case, PO) involves not only the polymer industry, but also fine chemicals. In this sense, iron emerges as a substitute for "conventional" noble metals such as Gold or Silver, which despite performing better at present, possess the severe drawback of high cost. While having bulk iron or any of its oxides is inviable for this kind of processes, isomorphic iron incorporated into a suitable inorganic oxides network such as silica provides us with a viable alternative, which serves as a proof-of-concept of what this transition metal can do if prepared in the right way. Nevertheless, its overall performance still lags behind that of noble metals. Efforts need to be made towards increasing the iron loading in oxide networks without compromising its dispersion. In this sense, the use of metal chelates or fine control over sol-gel and/or hydrothermal synthesis should bring forth catalysts with a high loading in iron in tetrahedral coordination incorporated into an oxide framework, which can perform similarly to its noble metal counterparts.

Acknowledgments

The authors thank Ministerio de Ciencia, Innovación y Universidades and FEDER (Project RTI2018-095291-B-I00) and Generalitat Valenciana (PROMETEO/2018/076) for financial support.

References

1. P. A. Frey and G. H. Reed, The ubiquity of iron, *ACS Chem. Biol.* **7**, 1477–1481 (2012).
2. S. A. Theofanidis, V. V. Galvita, C. Konstantopoulos, H. Poelman and G. B. Marin, Fe-based nano-materials in catalysis, *Materials* **11**, 831 (2018).
3. O. Montellano and R. Paul, *Cytochrome P450 Structure, Mechanism, and Biochemistry*, Springer, US (2005).
4. D. H. R. Barton, A. E. Martell and D. T. Sawyer, *The Activation of Dioxygen and Homogeneous Catalytic Oxidation*, Srpinger, US (1993).
5. L. Banci, *Metallomics and the Cell*, Springer, the Netherlands (2013).
6. I. Bauer and H. J. Knölker, Iron catalysis in organic synthesis, *Chem. Rev.* **115**, 3170–3387 (2015).
7. E. Bauer, *Iron Catalysis II*, Springer International Publishing, Switzerland, (2015).
8. M. North, *Sustainable Catalysis: With Non-endangered Metals.* Royal Society of Chemistry (2016).
9. J. R. Jennings, *Catalytic Ammonia Synthesis*, Springer, US (1991).
10. D. Lopez-Tejedor, R. Benavente and J. M. Palomo, Iron nanostructured catalysts: Design and applications, *Catal. Sci. Technol.* **8**, 1754–1776 (2018).
11. Y. Wang, Selective oxidation of hydrocarbons catalyzed by iron-containing heterogeneous catalysts, *Res. Chem. Intermed.* **32**(3–4), 235–251 (2006).
12. M. Hoeven, Y. Kobayashi and R. Diercks, *Technology Roadmap Energy and GHG Reductions in the Chemical Industry via Catalytic Processes*, International Energy Agency, Germany (2013).
13. J. Huang and M. Haruta, Gas-phase propene epoxidation over coin-age metal catalysts, *Res. Chem. Intermediat.* **38**(1), 1–24 (2012).
14. S. J. Khatib and S. T. Oyama, Direct oxidation of propylene to propylene oxide with molecular oxygen: A review, *Cataly. Rev.* **57**(3), 306–344 (2015).
15. Ullmans, Propylene oxide, *Rep. Carcinog.* **12**, 367–369 (2011).
16. V. Russo, R. Tesser, E. Santacesaria and M. Di Serio, Chemical and technical aspects of propene oxide production via hydrogen peroxide (HPPO process), *Ind. Eng. Chem. Res.* **52**(3), 1168–1178 (2013).
17. T. A. Nijhuis, M. Makkee, J. A. Moulijn and B. M. Weckhuysen, The production of propene oxide: Catalytic processes and recent developments, *Ind. Eng. Chem. Res.* **45**(10), 3447–345 (2006).

18. T. Hayashi, K. Tanaka and M. Haruta, Selective vapor-phase epoxidation of propylene over Au/TiO_2 catalysts in the presence of oxygen and hydrogen, *J. Catal.* **178**(2), 566–575 (1998).

19. S. T. Oyama, *Mechanisms in Homogeneous and Heterogeneous Epoxidation catalysis*, Elsevier, the Netherlands (2008).

20. A. K. Sinha, S. Seelan, S. Tsubota and M. Haruta, A three-dimensional mesoporous titanosilicate support for gold nanoparticles: Vapor-phase epoxidation of propene with high conversion, *Angew. Chem. Int. Ed.* **43**, 1546–1548 (2004).

21. B. Chowdhury, J. J. Bravo-Suárez, M. Daté, S. Tsubota and M. Haruta, Trimethylamine as a gas-phase promoter: Highly efficient epoxidation of propylene over supported gold catalysts, *Angew. Chem. Int. Ed.* **45**, 412–415 (2006).

22. J. Lu, X. Zhang, J. J. Bravo-Suárez, T. Fujitani and S. T. Oyama, Effect of composition and promoters in Au/TS-1 catalysts for direct propylene epoxidation using H_2 and O_2, *Catal. Today* **147**, 186–195 (2009).

23. R. A. Van Santen and H. P. Kuipers, The mechanism of ethylene epoxidation, *Adv. Catal.* **35**, 265–321 (1987).

24. J. G. Serafin, A. C. Liu and S. R. Seyedmonir, Surface science and the silver-catalyzed epoxidation of ethylene: An industrial perspective, *J. Mol. Catal. A-Chem.*, **131**(1–3), 157–168 (1998).

25. M. A. Barteau and R. J. Madix, Low-pressure oxidation mechanism and reactivity of propylene on silver (110) and relation to gas-phase acidity, *J. Am. Chem. Soc.* **105**(3), 344–349 (1983).

26. M. Akimoto, K. Ichikawa and E. Echigoya, Kinetic and adsorption studies on vapor-phase catalytic oxidation of olefins over silver, *J. Catal.* **76**(2), 333–344 (1982).

27. J. T. Roberts, R. J. Madix and W. W. Crew, The rate-limiting step for olefin combustion on silver: experiment compared to theory, *J. Catal.* **141**(1), 300–307 (1993).

28. E. Ananieva and A. Reitzmann, Direct gas-phase epoxidation of propene with nitrous oxide over modified silica supported FeO_x catalysts, *Chem. Eng. Sci.* **59**(22–23), 5509–5517 (2004).

29. X. Wang, Q. Zhang, Q. Guo, Y. Lou, L. Yanga and Y. Wang, Iron-catalysed propylene epoxidation by nitrous oxide: Dramatic shift of

allylic oxidation to epoxidation by the modification with alkali metal salts, *Chem. Commun.* **12**, 1396–1397 (2004).

30. G. I. Panov, E. V. Starokon, M. V. Parfenov and L. V. Pirutko, Single turnover epoxidation of propylene by α-complexes (Fe^{III}–O•)α on the surface of FeZSM-5 zeolite, *ACS Catal.* **6**(6), 3875–3879 (2016).

31. M. A. Sainna, S. Kumar, D. Kumar, S. Fornarini, M. E. Crestoni and S. P. de Visser, A comprehensive test set of epoxidation rate constants for iron(IV)–oxo porphyrin cation radical complexes, *Chem. Sci.* **6**, 1516–1529 (2015).

32. Y. Wang, Q. Zhang, T. Shishido, K. Takehira, Characterizations of Iron-containing MCM-41 and its catalytic properties in epoxidation of styrene with hydrogen peroxide, *J. Catal.* **209**(1), 186–196 (2002).

33. Y. Wang, H. Chu, W. Zhu and Q. Zhang, Copper-based efficient catalysts for propylene epoxidation by molecular oxygen, *Catal. Today* **131**(1–4), 496–504 (2008).

34. A. Seubsai, M. Kahn, B. Zohour, D. Noon, M. Charoenpanich, and S. Senkan, Copper–Manganese mixed metal oxide catalysts for the direct epoxidation of propylene by molecular oxygen, *Ind. Eng. Chem. Res.* **54**(10), 2638–2645 (2015).

35. A. Marimuthu, J. Zhang and S. Linic, Tuning selectivity in propylene epoxidation by plasmon mediated photo-switching of Cu oxidation state, *Science* **339**(6127), 1590–1593 (2013).

36. X. Yang, S. Kattel, K. Xiong, K. Mudiyanselage, S. Rykov, S. D. Senanayake, J. A. Rodriguez, P. Liu, D. J. Stacchiola and J. G. Chen, Direct epoxidation of propylene over stabilized Cu$^+$ surface sites on titanium-modified Cu$_2$O, *Angew. Chem. Int. Ed.* **54**(41), 11946–11951 (2015).

37. C. Zhao and I. E. Wachs, Selective oxidation of propylene over model supported V$_2$O$_5$ catalysts: Influence of surface vanadia coverage and oxide support, *J. Catal.* **257**(1), 181–189 (2008).

38. A. Held, J. Kowalska, A. Łapiński and K. Nowińska, Vanadium species supported on inorganic oxides as catalysts for propene epoxidation in the presence of N$_2$O as an oxidant, *J. Catal.* **306**, 1–10 (2013).

39. B. Horváth, M. Hronec, I. Vávra, M. Šustek, Z. Križanová, J. Dérer and E. Dobročk, Direct gas-phase epoxidation of propylene over nanostructured molybdenum oxide film catalysts, *Catal. Comm.* **34**, 16–21 (2013).

40. A. Sigel and H. Sigel, *Metal Ions in Life Sciences.* Springer International Publishing, Switzerland (2015).

41. H. W. Liu and L. Mander, *Comprehensive Natural Products II: Chemistry and Biology,* Elsevier, the Netherlands (2010).

42. Y. Wang, W. Yang, L. Yang, X. Wang and Q. Zhang, Iron-containing heterogeneous catalysts for partial oxidation of methane and epoxidation of propylene, *Catal. Today* **117**, 156–162 (2006).

43. V. Duma and D. Hönicke, Gas phase epoxidation of propene by nitrous oxide over silica-supported iron oxide catalysts, *J. Catal.* **191**(1), 93–104 (2000).

44. J. García-Aguilar, I. Miguel-García , J. Juan-Juan, I. Such-Basáñez, E. San Fabián, D. Cazorla-Amorós and Á. Berenguer-Murcia, One step-synthesis of highly dispersed iron species into silica for propylene epoxidation with dioxygen, *J. Catal.* **338**, 154–167 (2016).

45. J.García-Aguilar, D. Cazorla-Amorós and Á. Berenguer-Murcia, K- and Ca-promoted ferrosilicates for the gas-phase epoxidation of propylene with O_2, *Appl. Catal. A.* **538**, 139–147 (2017).

46. B. Horváth, M. Hronec, and R. Glaum, Epoxidation of propylene in the gas phase, *Top. Catal.* **46**(1–2), 129–135 (2007).

47. B. Horváth, M. Šustek, I. Vávra, M. Mičušík, M. Gál and M. Hronec, Gas-phase epoxidation of propylene over iron-containing catalysts: The effect of iron incorporation in the support matrix, *Catal. Sci. Technol.* **4**, 2664–2673 (2014).

48. Q. Zhang, Q. Guo, X. Wang, T. Shishido and Y. Wang, Iron-catalyzed propylene epoxidation by nitrous oxide: Toward understanding the nature of active iron sites with modified Fe-MFI and Fe-MCM-41 catalysts, *J. Catal.* **239**(1), 105–116 (2006).

49. C. Nozaki, C. G. Lugmair, A. T. Bell and T. D. Tilley, Synthesis, characterization, and catalytic performance of single-site iron(III) centers on the surface of SBA-15 silica, *J. Am. Chem. Soc.* **124**(44), 13194–13203 (2002).

50. X. Wang, Q. Zhang, S. Yang and Y. Wang, Iron-catalyzed propylene epoxidation by nitrous oxide: studies on the effects of alkali metal salts, *J. Phys. Chem. B* **109**(49), 23500–23508 (2005).

51. B. Horváth and M. Hronec, Iron-catalyzed gas-phase epoxidation of propylene by N_2O via halide-assisted oxygen transfer, *Appl. Catal. A: General* **347**, 72–80 (2008).

Chapter 3

Iron Oxide Nanoenzymes for the Treatment of Polluted Water

N. Pariona[a], F. Mondaca[b], and A. I. Mtz-Enriquez[b,*]

[a]*Red de Estudios Moleculares Avanzados, Instituto de Ecología, A.C., Carretera Antigua a Coatepec 351, El Haya, 91070 Xalapa, Veracruz, Mexico*

[b]*Cinvestav Unidad Saltillo, Av. Industria Metalúrgica 1062, Parque Industrial Ramos Arizpe, Coahuila, 25900, Mexico*

*arturo.martinez@cinvestav.edu.mx

The ubiquitous distribution of the iron oxides makes them interesting materials for different fields. Iron oxides occur at the nanoscale in the environment such as in loess, peat bog, and paddy fields and in living organisms such as bacteria, animals, and plants. On the other hand, the engineered iron oxide nanoparticles (NPs) are materials of great interest in different technological applications, such as cancer diagnosis and therapy, drug delivery vehicles, resonance imaging, fabrication of biocompatible magnetic fluids, tunneling magnetoresistance, magnetic components, electrochromic devices, lithium–ion batteries, photoelectrochemical systems, and solar filters. In addition, the iron oxides can be used as nanoenzymes, which are nanostructured iron oxides mimicking the activity of diverse biological

57

catalysts (enzymes). This chapter describes the peroxidase-like catalytic activity of different iron oxides and their application for water
remediation. This chapter is organized as follows: a brief description
of different iron oxides is given in the first part. Next, the surface
properties of iron oxides are explained. Finally, the peroxidase-like
catalytic activity of different iron oxides is described, focusing on the
Fenton-like and photo-Fenton processes, which are of great value
for the treatment of polluted water. Additionally, the improvements
for the peroxidase-like catalytic activity of iron oxides are presented.

3.1 Introduction

Iron is the fourth most abundant element in the Earth's crust.[1] In
nature, it can be found as ferrous (Fe^{2+}) and ferric (Fe^{3+}) ions, which
are transformed into iron oxides through biogeochemical processes.[2] In the environment, one can find a wide distribution of iron
oxides in air, soils, rocks, lakes, rivers, and at the bottom of the sea,
but they can also be synthesized in the laboratory.[3] Iron oxides are
of great importance because of their properties and processes that
take place in different ecosystems. In soils and sediments, iron
oxides are essential and crucial for microorganisms and plants.[4]
Since it regulates the concentration of phosphate, inorganic compounds, and organic compounds.[1] The iron oxides, which includes
the iron hydroxides and oxyhydroxides, are 16 crystal phases, the
most important are as follows: goethite (α-FeOOH), hematite
(α-Fe$_2$O$_3$), akaganeite (β-FeOOH), β-Fe$_2$O$_3$, lepidocrocite
(γ-FeOOH), maghemite (γ-Fe$_2$O$_3$), ferrihydrite (5 Fe$_2$O$_3$ ·9H$_2$O),
ϵ-Fe$_2$O$_3$, feroxyhyte (δ'-FeOOH), magnetite (Fe$_3$O$_4$), wüstite (FeO),
and bernalite Fe(OH)$_3$. The iron oxides are mainly composed of
iron (Fe), oxygen (O), and hydroxides (OH). Since the O^{2+} and OH^-
ions are in greater quantity than the Fe ions, the crystalline structure
of the oxides is controlled by the arrangement of the O^{2+} and OH^-.[1]
Ferrihydrite, goethite, hematite, and magnetite are the most import
iron oxides since they are part of the soil, sediments, and drainages
of mines.[5] Almost all iron oxides are crystalline, except for ferrihydrite and feroxyhyte that are poorly crystalline.

Each iron oxide has different stability, reactivity, and specific surface area, which will also depend on the particle size. Some of the most common iron oxides such as hematite, goethite, ferrihydrite, magnetite, and maghemite are nanosized in the environment. On the other hand, many iron oxides can be synthesized in the laboratory in the form of NPs.[1,6] Furthermore, nanosized iron oxides can be found in living systems (animals, plants, and microorganisms), the concentration of free iron in living beings is regulated by proteins, which control the growth of the iron oxide crystals. The protein/ nanosized iron oxide complexes found in living organisms are known as ferritin in mammals, phytoferritin in plants, and bacterioferritin in bacteria.[7] These proteins have a common function, which consists of storing iron when there is an excess of iron and release it when there is a deficiency.[8] Ferritin has the capacity to sequester and store iron in the form of iron oxides with a crystalline structure similar to ferrihydrite, magnetite, and hematite.[9,10]

Iron oxide NPs have attracted attention for their potential applications in the remediation of polluted water with a range of pollutants, such as heavy metals and organic substances. In addition, the low cost of iron oxides makes these nanomaterials considered for large-scale water treatment. It is important to mention that iron oxide nanoenzymes can be used in a range of areas, such as sensing/detection of specific chemicals, antioxidative protection, and detection/treatment of cancer. However, given the wide range of applications, these nanomaterials can be released to the environment, which may affect the ecosystem, particularly microorganisms.[11] Although that is an important issue, the effects of iron oxide NPs on living organisms will not be treated in this chapter.

3.2 Surface Properties of Iron Oxides

For water treatment, the iron oxide nanoenzymes are in direct contact with water, given this, the surface properties of iron oxides should be reviewed. The surface charge of the iron oxides depends on the pH of aqueous media. In addition, natural particles have a surface charge due to structural substitutions and reactions with ionic species.[12] The iron oxides in water can adsorb the following species: i) the protons and hydroxyl groups that come from the dissociation of water and ii) other ions or ligands present in solution.[3] Under dry conditions, the Fe atoms on the surface may be unsaturated and act as Lewis acid (a Lewis acid is the acceptor of a pair of electrons) and react with a Lewis base (electron pair donor). Therefore, in aqueous media, the Fe atoms on the surface act as Lewis acids and coordinate with Lewis bases (which can be OH⁻ or H_2O), which share a pair of electrons with Fe.[3]

During adsorption, water molecules usually dissociate and result in a surface covered with OH⁻ groups coordinated to the Fe atoms.[3] After the hydroxylation of the iron oxides, the adsorption of water molecules increases and hydrogen bonds occur with the OH groups on the surface.[3] Therefore, the hydroxyl surface groups are the functional groups of the iron oxides. Thus, in the presence of water, the surface of the iron oxides is hydroxylated and/or hydrated completely, giving rise to reactive surfaces. Surface OH groups can be coordinated by one, two, or three underlying iron atoms, which are defined as single, double, and triple coordinate groups, see Fig. 3.1. If a charge of +½ is assigned to each FeO bond, which comes from six-fold coordination of the Fe atom, these three types of coordinated groups will have charges of –½, 0, and +½, respectively.

The reactivity of the different types of OH groups varies according to the number of underlying Fe atoms coordinated to the functional groups on the surface. The overall density of these groups

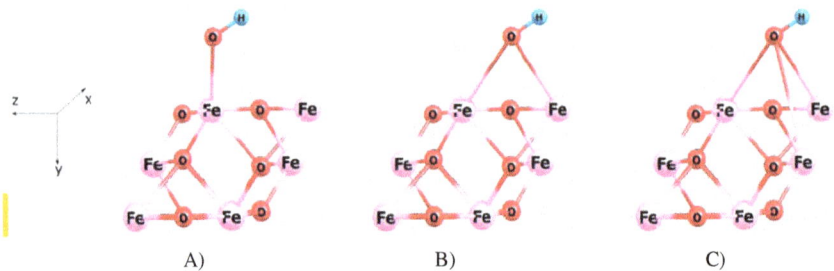

Fig. 3.1. Schematic diagram for A) single, B) double, and C) triple coordinated OH⁻ groups on the surface of hematite.

varies with the crystal structure and the crystal face; likewise, it is influenced by the crystal morphology.[3] Due to the presence of a pair of electrons and a dissociable hydrogen atom of the OH surface group, the iron oxides can react with acids and bases, therefore, they are amphoteric.[3] The charge on the surface of the iron oxides depends on the dissociation of the hydroxyl groups on the surface; i.e. the adsorption and desorption of the protons will depend on the pH of the solution; this dissociation reaction can be represented by the following equations:[3]

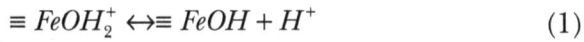

$$\equiv FeOH_2^+ \longleftrightarrow \equiv FeOH + H^+ \tag{1}$$

$$\equiv FeOH \longleftrightarrow \equiv FeO^- + H^+ \tag{2}$$

The charge of the NPs is a result of the ionization of the surface, which is balanced by a layer of counterions found in the aqueous solution, this ensures a neutral charge at interfacial region.[3] The surface charge of the particles together with the counterion layer in solution is defined as the electric double layer (EDL). The separation of charges in the EDL generates an electric potential between the surface and the solution, which depends on the pH and the temperature of the solution.[3] The point of zero charge (PZC) is defined as the pH at which the electrical charge density on the surface is zero. The sign of the surface charge and its magnitude depend on the pH and the concentration of the electrolyte in the solution. Although the iron oxides have little or no permanent charge, they can exchange cations and anions due to the adsorption of ions that determine the potential, usually H$^+$ and OH$^-$.[13] The total charge (σ_{tot}) of the surface of the iron oxides is described as follows:[3]

$$\sigma_{tot} = \sigma_{H^+} + \sigma_{is} + \sigma_{oS} \tag{3}$$

where σ_{H+} is the total charge of the proton due to the bond of the proton or OH$^-$ ions, σ_{is} is the charge of the internal sphere complex, and σ_{os} is the charge of the external sphere complex. Each term in Eq. (3) can be positive or negative, but in general, σ_{tot} is not zero despite the possibility of cancellation. Therefore, the total charge of the surface will be neutralized by ions from the solution that have not complexed with the functional groups on the surface. These ions, whether positive or negative, are dissociated from the surface of the iron oxides and are free to move in the solution beyond the interfacial region.[3] Most iron oxides have a PZC in the pH range of 8–9.[1] Thus, the positive, negative, and neutral functional groups can coexist on the FeO surface. For a pH below the PZC, the surface charge of the iron oxides is positive and for a pH above the PZC the surface charge is negative. The study of PZC is important to understand the role of the surface charge of insoluble iron oxides NPs for remediation of wastewater.

Adsorption is a traditional process for the separation of solutes from liquids or gases. An adsorbing material is called adsorbent and an adsorbed substance is called solute or adsorbate. An adsorption process involves a quantitative equilibrium distribution between adsorbent and adsorbate. The two main adsorption processes are governed by chemical or physical forces. Chemical adsorption or chemisorption results from a chemical (covalent) bond between the adsorbate and the adsorbent. The adsorbed molecules are located at specific sites and, therefore, are not free to migrate on the surface and are usually an irreversible and exothermic process. In contrast, physical adsorption (physisorption) is due to the attraction of weak forces (van der Waals) between adsorbate and adsorbent.[14] Therefore, in physisorption, the chemical structure of adsorbate and adsorbent does not undergo large chemical changes as a result of adsorption.

Chemical adsorption is usually associated with a greater enthalpy of adsorption and slower kinetics than physisorption.[15] Since chemisorption implies that adsorbates react with specific sites of adsorption of the adsorbent, there is an upper limit of the amount that can be adsorbed on the surface. That is, when all the available adsorption sites are occupied, a monolayer on the surface is formed. While in physisorption, multilayers of adsorbate molecules are often formed; this is because the molecules can be adsorbed on another layer by van der Waals forces.[15] Adsorption plays an important role in the transport, bioavailability, and fate of different water pollutants.

When the size of a particle is reduced, the ratio of surface atoms to interior atoms changes drastically, with a greater proportion of atoms on the surface than in the interior. This makes the NPs have a large area/volume ratio and therefore they have a disproportionately large area of reactive surfaces for their volume fraction.[5] The adsorption process involves the interaction of adsorbate/adsorbent. The Fe ions act as Lewis acids and react with OH⁻ groups of the oxyanions to form a surface complex.[3] In the case of chemisorption, the adsorption takes place by exchange, which involves the replacement of hydroxyl groups on the surface by the new adsorbate,[3] this process can be schematized as follows:

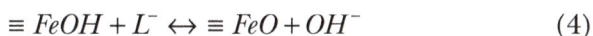

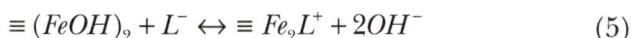

$$\equiv FeOH + L^- \leftrightarrow \equiv FeO + OH^- \tag{4}$$

$$\equiv (FeOH)_2 + L^- \leftrightarrow \equiv Fe_2L^+ + 2OH^- \tag{5}$$

The formation of the surface complexes takes place on the EDL formed on the surface of the oxides. The EDL comprises the surface of the adsorbent, an inner layer of cations immediately adjacent to the surface of the adsorbent (Stern layer), and the diffuse layer (a collection of cations and anions), see Fig. 3.2.

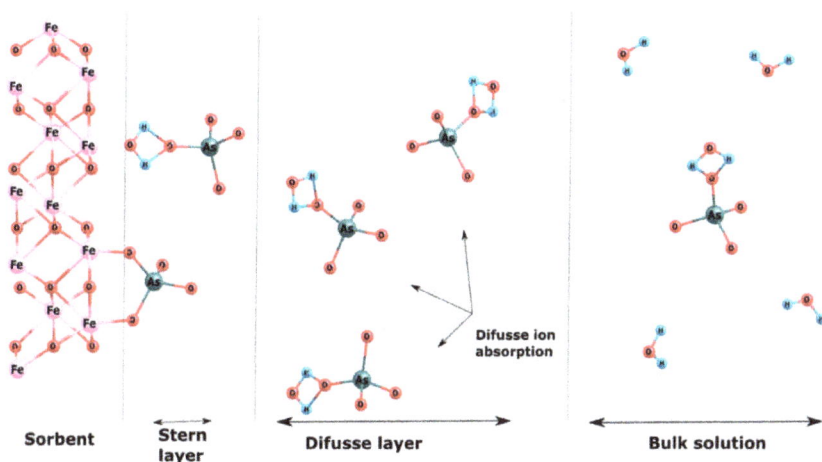

Fig. 3.2. Formation of a surface complex of different species on the surface of hematite.

3.3 Peroxidase-like Catalytic Activity of Iron Oxides: Degradation of Organic Pollutants in Water

Contamination of surface and groundwater by organic compounds is a serious problem that causes toxicological effects on microorganisms, plants, animals, and humans.[16] For example, pesticides that are widely used in agriculture pollute both surface water and groundwater.[17] Likewise, wastewater from textile industries contains toxic substances such as nonylphenol and trichloroaniline, which affect aquatic organisms.[18] Advanced oxidation processes (AOPs) have attracted attention in the treatment of polluted waters, it is because AOPs degrade organic compounds through the generation of reactive oxygen species (ROS) such as hydroxyl radicals (HO•), superoxide radicals (O_2^-•), and hydroperoxyl radicals (HO$_2$•).[19] Due to its efficiency and affordable cost, hydrogen peroxide (H_2O_2) is one of the most commonly used oxidants. However, H_2O_2 alone does not easily oxidize most organic substances. Iron ions have usually been used as catalysts for the decomposition of H_2O_2, which generates ROS, this process is known as the Fenton reaction.[19]

The Fenton process belongs to the AOPs and has been successfully used for the degradation of organic pollutants. This process involves the reduction of H_2O_2 by Fe^{2+} ions, generating ROS, which may oxidize almost any organic molecule,[20] the Fenton process involves the following steps:

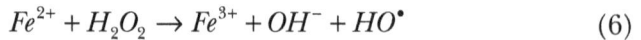

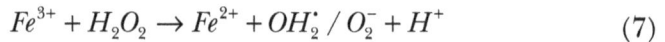

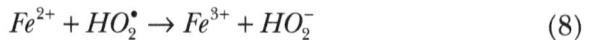

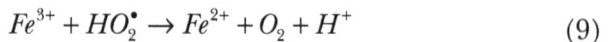

$$Fe^{2+} + H_2O_2 \rightarrow Fe^{3+} + OH^- + HO^\bullet \tag{6}$$

$$Fe^{3+} + H_2O_2 \rightarrow Fe^{2+} + OH_2^\bullet / O_2^- + H^+ \tag{7}$$

$$Fe^{2+} + HO_2^\bullet \rightarrow Fe^{3+} + HO_2^- \tag{8}$$

$$Fe^{3+} + HO_2^\bullet \rightarrow Fe^{2+} + O_2 + H^+ \tag{9}$$

$$H_2O_2 + HO^\bullet \rightarrow HO_2^\bullet + H_2O \tag{10}$$

Despite its effectiveness, there are some drawbacks in this process, such as the dependence of the pH of the media (optimum range from 2.5 to 4.0), iron precipitation at pH above 4.0, and it is hard to recover the catalyst (dissolved iron ions).[20,21] Moreover, the cost associated with acidification and subsequent neutralization can limit the application of the homogeneous Fenton process at large scale.[22] Likewise, a group of enzymes known as peroxidases may activate H_2O_2. Among them, the horseradish peroxidase (*Armoracia rusticana*) has been used for the degradation of organic compounds.[23] It has been reported that horseradish may degrade phenol and substituted phenols, however, this enzyme can be inhibited by some commercial dyes.[24] Additional drawbacks of peroxidases are the instability and high cost, which limit the application in the treatment of wastewater, especially dye effluents.[23]

It has been reported that Fe_3O_4 NPs exhibit peroxidase-like catalytic activity, which may be used for the development of spectrophotometric methods for the detection of H_2O_2 and glucose.[25] Furthermore, the H_2O_2/Fe_3O_4 system has been used for the degradation of phenol with an 85% efficiency in only 3 h of reaction.[26] As well, other iron oxides such as hematite, goethite, and ferrihydrite have attracted great attention for the activation of H_2O_2.[27] The reactions that take place between H_2O_2 and the iron oxides are called Fenton-like reactions or the heterogeneous Fenton process.

The Fenton-like process is also known as heterogeneous Fenton process or peroxidase-like catalysis.[28] Since iron oxides are insoluble in water, they can be used for the wastewater treatment, recovered, and reused for at least three cycles.[28] Contrasting the Fenton reaction, in the heterogeneous Fenton process, the hydroxyl radicals are generated in the presence of H_2O_2 through redox reactions initiated on the surface of the NPs.[27] The use of iron oxide NPs in the decontamination of wastewater has many advantages, such as the recovering and reuse; in addition, the pH of the media does not need a strict control. For example, it has been reported that iron oxide NPs/H_2O_2 can catalyze the oxidation of organic pollutants over a wide pH range (from 3 to 7).[27,28]

The mechanism of activation of H_2O_2 by iron oxides is still not clear, however, the following mechanism has been proposed.[27] First, H_2O_2 is absorbed on the surface of the NPs (Eq. 11), which react and slowly generate species of Fe(II) and hydroxyl radicals (Eqs. 12

and 13). Then H_2O_2 reacts with the absorbed Fe(II) and forms highly reactive HO• radicals (Eq. 14). These radicals can react with another H_2O_2 molecule to produce HO_2• radicals (Eq. 15). In addition, Fe(II) reacts with O_2^- to generate identical Fe(III) sites as those present in Eq. 11 (Eq. 16). Finally, the radicals HO_2• react with Fe(III) on the surface and generate Fe(II) (Eq. 17) and ending with the recovery of Fe (III) (Eq. 18).

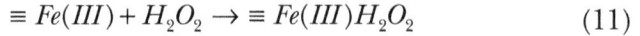

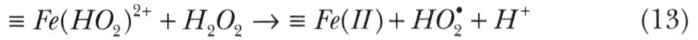

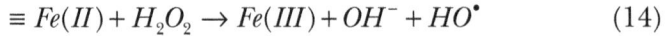

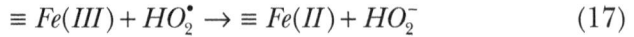

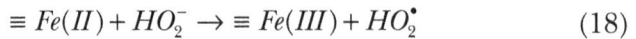

$$\equiv Fe(III) + H_2O_2 \rightarrow \equiv Fe(III)H_2O_2 \tag{11}$$

$$\equiv Fe(III) + H_2O_2 \rightarrow \equiv Fe(HO_2)^{2+} + H^+ \tag{12}$$

$$\equiv Fe(HO_2)^{2+} + H_2O_2 \rightarrow \equiv Fe(II) + HO_2^{\bullet} + H^+ \tag{13}$$

$$\equiv Fe(II) + H_2O_2 \rightarrow Fe(III) + OH^- + HO^{\bullet} \tag{14}$$

$$HO^{\bullet} + H_2O_2 \rightarrow H_2O + HO_2^{\bullet} \tag{15}$$

$$\equiv Fe(III) + O_2^- \rightarrow \equiv Fe(III) + O_2 \tag{16}$$

$$\equiv Fe(III) + HO_2^{\bullet} \rightarrow \equiv Fe(II) + HO_2^- \tag{17}$$

$$\equiv Fe(II) + HO_2^- \rightarrow \equiv Fe(III) + HO_2^{\bullet} \tag{18}$$

The efficiency of the Fenton-like reaction for the oxidation of organic pollutants is influenced by several parameters, such as the type of iron oxide, the size of the NPs, the concentration of H_2O_2, pH of the solution, and the kind of pollutants. Among the different iron oxide NPs, the most studied is magnetite,[29] goethite,[30] hematite,[27,31] and ferrihydrite.[27,32] However, the most effective iron oxide for the oxidation of organic pollutants is goethite (α-FeOOH).[27,32] Given that during the Fenton process, the H_2O_2 concentration is directly related to the amount of the produced HO•, this parameter strongly influences the degradation of organic compounds. It has been shown that higher concentrations of H_2O_2 increase their catalytic activity.[28]

In the study of the oxidation of dimethyl sulfoxide (DMSO) by H_2O_2/goethite, it was found that when the concentration of H_2O_2 was increased from 2.5 to 10 g/L, more hydroxyl radicals were generated, and the degradation rate was increased. However, when the

H_2O_2 dose was increased from 10 to 15 g/L, the decomposition rate decreased.[30] This behavior has been observed by several authors,[28] which consist on the overproduction of hydroperoxyl radicals that are less reactive and do not contribute to the oxidation of DMSO.[30] On the other hand, it was reported that the generation rate of HO• is proportional to the H_2O_2 consumption rate and the surface area of the iron oxide particles.[27] Therefore, the consumption rate of H_2O_2 will depend on the surface area and the type of iron oxide.

Another factor that influences the degradation of organic compounds by iron oxides is pH. In many cases, wastewater has an almost neutral to alkaline pH, in a Fenton process, it is necessary to acidify the medium for the oxidation process. However, in the Fenton-like reaction process, the oxidation of organic pollutants would take place over a wide range of pH, from acidic to neutral or alkaline.[33] For example, for the degradation of rhodamine B (RhB) using Fe_3O_4 NPs in the presence of H_2O_2, it was found that the NPs catalyze the decomposition of H_2O_2 to eliminate RhB in a wide range of pH (3.0–9.0).[33]

3.4 Light-Assisted Peroxidase-Like Catalysis

Numerous modifications to the classic Fenton reaction have been proposed, which include the electrochemical control of pH, the *in-situ* generation of H_2O_2, and the acceleration for the regeneration of Fe(II) species.[34] If the Fenton reaction is assisted by UV and/ or visible light, it is termed as photo-Fenton, which accelerate the regeneration of Fe(II) species and improve the Fenton process. However, it is a challenge to develop photocatalyst for the photo-Fenton process because some limitations must be overcome. The most common limitations are the reuse of the photocatalyst, mechanisms to increase the active sites, the total mineralization of the organic pollutants, the application in a wide range of pH, and to increase the efficiency of the photo generated charge carriers.[35]

Most of the iron oxide NPs have been used for the photo-Fenton process for the degradation of organic pollutants from water, which includes magnetite, maghemite, hematite, and goethite, see Table 3.1. To improve the efficiency and to avoid the recombination of the

Table 3.1. Different iron oxide NPs and their composites used for the photo-Fenton process.

Iron oxide NPs or their composites	Light source (pH of reaction)	Size of the NPs (nm)	Degraded pollutant	Notes
α-Fe$_2$O$_3$@ diatomite composites	UV (2–5)	200 (length)	Methylene blue	The composites were hydrothermally prepared. It was shown that the composites enhanced the formation of •OH.[39]
Geothite NPs (α-FeOOH)	UV (3)	12–83	Simulated petrochemical wastes	Geothite recovered from acid mine drainage, which generated singlet oxygen as active species.[41]
FeOx NPs supported on diamond NPs	Visible (4)	2–4	Phenol	The composites presented active sites that generate free •OH.[42]
Fe$_3$O$_4$@GO@ MIL-100(Fe) NPs	Visible (5.5)	380	2,4-dichlorophenol	The composites generated mainly •OH, O$_2^-$, and holes, which oxidize the pollutant.[40]
Fe$_3$O$_4$, γ-Fe$_2$O$_3$, and α-Fe$_2$O$_3$	UV-vis (2.8–11)	50–100	Diphen-hydramine	The NPS prepared by the solvothermal method shown good stability with low iron loss and reusability.[43]
α-Fe$_2$O$_3$@ g-C$_3$N$_4$ hetero structures	Visible (5.5)	~100	Tetra-cycline	The heterostructures were prepared by co calcination of melamine and Fe-based MOF. The heterostructures raised the formation of •OH and separate the photo-generated charge carriers.[35]
α-Fe$_2$O$_3$	Visible (–)	10–30	Methylene blue	The NPs obtained by calcination show the higher generation of •OH.[44]

Table 3.1. (*Continued*)

Iron oxide NPs or their composites	Light source (pH of reaction)	Size of the NPs (nm)	Degraded pollutant	Notes
Biosurfactant coated Fe_3O_4	Sunlight (3, 7.5, and 12)	23	Methyl violet	The coprecipitated NPs were coated with rhamnolipids glycolipids. The use of SDS improved the adsorption of dye and increased its degradation.[38]
rGO/Fe_3O_4 composite	UV and sunlight (7)	7.3	2-methyl-isoborneol	The coprecipitated composite improved the separation of the photo-generated charge pairs. Good reusability was found.[45]
TiO_2/GO/Fe_3O_4 composite	UV and visible (3, 5, and 7)	30–50	Amoxicillin	The separation of the photo-generated carriers was improved. Some degradation intermediates were identified as aromatic compounds and organic acids.[37]
rGO-Fe_3O_4 composites	Non mentioned (–)	<100	Methyl-orange	The coprecipitated composites separate the photo generated charge carriers and formed •OH mainly.[36]

photo generated charge carriers, some composites including conducting materials such as graphene, graphene oxide (GO), and reduced GO (rGO) have been proposed.[36,37] On the other hand, to increase the active sites for adsorption of the pollutants, some materials have been proposed, such as the synthesis of functionalized iron oxides with surfactants such as sodium dodecyl sulfate (SDS),[38] and the preparation of composites using porous materials such as diatomite and metal-organic frameworks such as MIL-100(Fe).[39,40]

It is hard to recuperate the photocatalysts from treated water, which limits the application of the Fenton process at large scales. Thus, the use of magnetic separation as a recovery method has been proposed, which employ composites including ferrimagnetic oxides such as magnetite and maghemite.[37,43,45] A list of the different materials used for the photo-Fenton process is shown in Table 3.1, which include some parameters used for the Fenton reaction, such as the pH of the polluted water, the kind of light used for the reaction, the size of the iron oxide NPs, and the degraded pollutant. Table 3.1 shows that iron oxide NPs and their composites work very well in a wide range of pH, use different light sources such as visible lamps, UV lamps, or solar light.

A wide range of pollutants such as dyes, pharmaceutical drugs, and petrochemicals have been oxidized by means of light-assisted peroxidase-like catalysts. Additionally, some notes are listed in the last column of Table 3.1, which include the preparation techniques, some improvements for the photo-Fenton process and the kind of the photo-generated ROS such as •OH, O_2^-, holes, and electrons. Finally, the photo-Fenton process can be used after other photocatalytic processes, such as the case of photo generation of hydrogen using the Ru-doped $LaFeO_3$ coupled to γ-Fe_2O_3 photocatalysts.[46] An additional advantage of that process is that after the photogeneration of hydrogen, the same photocatalyst can be used for the total mineralization of pollutants using the photo-Fenton process.

3.5 Conclusions

The development of very active iron oxide nanozymes will facilitate the creation of state-of-the-art technologies, which may improve human health. Future research can be focused on exploring nanozymes for a wide range of applications. To develop new and cost-effective nanozymes, it is important to correlate other characteristics of the iron oxide nanomaterials such as the structural, optical, magnetic, and electrical properties. It is very important because their catalytic activity can be improved if an external field is induced during their function.

References

1. U. Schwertmann and R. M. Cornell, *Iron Oxides in the Laboratory Preparation and Characterization*, Wiley-VCH, Weinheim (2000).
2. U. Schwertmann, *Occurrence and Formation of Iron Oxides in Various Pedoenvironments*, Springer, Dordrecht (1988).
3. R. M. Cornell and U. Schwertmann, *The Iron Oxides: Structure, Properties, Reactions, Occurences and Uses*, Wiley-VCH, Weinheim (2003).
4. C. Colombo, G. Palumbo, J. Z. He, R. Pinton and S. Cesco, Review on iron availability in soil: Interaction of Fe minerals, plants, and microbes, *J. Soils Sediments* **14**(3), 538–548 (2014).
5. G. A. Waychunas, C. S. Kim and J. F. Banfield, Nanoparticulate iron oxide minerals in soils and sediments: Unique properties and contaminant scavenging mechanisms, *J. Nanoparticle Res.* **7**(4), 409–433 (2005).
6. H. Hernαndez-Flores, N. Pariona, M. Herrera-Trejo, H. M. Hdz-García and A. I. Mtz-Enriquez, Concrete/maghemite nanocomposites as novel adsorbents for arsenic removal, *J. Mol. Struct.* **1171**, 9–16 (2018).
7. S. C. Andrews, J. M. A. Smith, S. J. Yewdall, J. R. Guest and P. M. Harrison, Bacterioferritins and ferritins are distantly related in evolution conservation of ferroxidase-centre residues, *FEBS Lett.* **293**(1–2), 164–168 (1991).
8. A. Galatro and S. Puntarulo, Mitochondrial ferritin in animals and plants, *Front. Biosci.* **12**(3), 1063–1071 (2007).
9. N. Gálvez, B. Fernández, P. Sánchez, R. Cuesta, M. Ceolín, M. Clemente-León, S. Trasobares, M. López-Haro, J. J. Calvino, O. Stéphan and J. M. Dominguez-Vera, Comparative structural and chemical studies of ferritin cores with gradual removal of their iron contents, *J. Am. Chem. Soc.* **130**(25), 8062–8068 (2008).
10. N. D. Chasteen and P. M. Harrison, Mineralization in Ferritin: An efficient means of iron storage, *J. Struct. Biol.* **126**(3), 182–194 (1999).
11. S. Rana and P. T. Kalaichelvan, Ecotoxicity of nanoparticles, *ISRN Toxicol.* **2013**, 574–648 (2013).
12. G. Sposito, *The Chemistry of Soils*, Oxford University Press, New York (2008).

13.   E. Tombácz, pH-dependent surface charging of metal oxides, *Period. Polytech. Chem. Eng.* **53**(2), 77–86 (2009).

14.   W. J. Thomas and B. Crittenden, *Adsorption Technology and Design*, Butterworth-Heinemann, Oxford (1998).

15.   J. M. Pettibone, J. Baltrusaitis and V. H. Grassian, *Chemical Properties of Oxide Nanoparticles: Surface Adsorption Studies from Gas- and Liquid-phase Environments*, John Wiley & Sons, Inc., Hoboken (2007).

16.   M. S. El-Shahawi, A. Hamza, A. S. Bashammakh and W. T. Al-Saggaf, An overview on the accumulation, distribution, transformations, toxicity and analytical methods for the monitoring of persistent organic pollutants, *Talanta* **80**(5), 1587–1597 (2010).

17.   M. Arias-Estévez, E. López-Periago, E. Martínez-Carballo, J. Simal-Gándara, J. C. Mejuto and L. García-Río, The mobility and degradation of pesticides in soils and the pollution of groundwater resources, *Agric. Ecosyst. Environ.* **123**(4), 247–260 (2008).

18.   K. Brigden, A. Cadena, K. Casper, M. Cobbing, M. Colin, T. Crawford, A. Dawe, S. Erwood, R. Estrada, N. Haiama, M. Hojsik, C. Rojas, T. Sadownichik, A. Simón and P. Terras, *Toxic Threads : Under Wraps Exposing the Textile Industry's Role in Polluting Mexico's Rivers*, Greenpeace International, Amsterdam (2012).

19.   H. Wu, J. J. Yin, W. G. Wamer, M. Zeng and Y. M. Lo, Reactive oxygen species-related activities of nano-iron metal and nano-iron oxides, *J. Food Drug Anal.* **22**(1), 86–94 (2014).

20.   S. Wang, A Comparative study of Fenton and Fenton-like reaction kinetics in decolourisation of wastewater, *Dye. Pigment.* **76**(3), 714–720 (2008).

21.   R. Andreozzi, V. Caprio, A. Insola and R. Marotta, Advanced oxidation processes (AOP) for water purification and recovery, *Catal. Today* **53**(1), 51–59 (1999).

22.   S. Rahim, A. A. A. Raman and W. M. A. Wan Daud, Review on the application of modified iron oxides as heterogeneous catalysts in Fenton reactions, *J. Clean. Prod.* **64**, 24–35 (2014).

23.   M. Hamid and Khalil-ur-Rehman, Potential applications of peroxidases, *Food Chem.* **115**(4), 1177–1186 (2009).

24. A. Bhunia, S. Durani and P. P. Wangikar, Horseradish peroxidase catalyzed degradation of industrially important dyes, *Biotechnol. Bioeng.* **72**(5), 562–567 (2001).

25. F. Yu, Y. Huang, A. J. Cole and V. C. Yang, The artificial peroxidase activity of magnetic iron oxide nanoparticles and its application to glucose detection, *Biomaterials* **30**(27), 4716–4722 (2009).

26. J. Zhang, J. Zhuang, L. Gao, Y. Zhang, N. Gu, J. Feng, D. Yang, J. Zhu and X. Yan, Decomposing phenol by the hidden talent of ferromagnetic nanoparticles, *Chemosphere* **73**(9), 1524–1528 (2008).

27. W. P. Kwan and B. M. Voelker, Rates of hydroxyl radical generation and organic compound oxidation in mineral-catalyzed fenton-like systems, *Environ. Sci. Technol.* **37**(6), 1150–1158 (2003).

28. K. Rusevova, F. D. Kopinke and A. Georgi, Nano-sized magnetic iron oxides as catalysts for heterogeneous Fenton-like reactions-Influence of Fe(II)/Fe(III) ratio on catalytic performance, *J. Hazard. Mater.* **241–242**, 433–440 (2012).

29. S. Liu, F. Lu, R. Xing and J. J. Zhu, Structural effects of Fe_3O_4 nanocrystals on peroxidase-like activity, *Chem. Eur. J.* **17**(2), 620–625 (2011).

30. J. J. Wu, M. Muruganandham, J. S. Yang and S. S. Lin, Oxidation of DMSO on goethite catalyst in the presence of H_2O_2 at neutral pH, *Catal. Commun.* **7**(11), 901–906 (2006).

31. K. N. Chaudhari, N. K. Chaudhari and J. S. Yu, Peroxidase mimic activity of hematiteiron oxides (α-Fe_2O_3) with different nanostructures, *Catal. Sci. Technol.* **2**(1), 119–124 (2012).

32. R. Matta, K. Hanna and S. Chiron, Fenton-like oxidation of 2,4,6-trinitrotoluene using different iron minerals, *Sci. Total Environ.* **385**(1–3), 242–251 (2007).

33. N. Wang, L. Zhu, M. Wang, D. Wang and H. Tang, Sono-enhanced degradation of dye pollutants with the use of H_2O_2 activated by Fe_3O_4 magnetic nanoparticles as peroxidase mimetic, *Ultrason. Sonochem.* **17**(1), 78–83 (2010).

34. C. Wang, H. Liu and Z. Sun, Heterogeneous photo-fenton reaction catalyzed by nanosized iron oxides for water treatment, *Int. J. Photoenergy* **2012**, 801–694 (2012).

35. T. Guo, K. Wang, G. Zhang and X. Wu, A novel α-Fe$_2$O$_3$@g-C$_3$N$_4$ catalyst: Synthesis derived from Fe-based MOF and its superior photo-Fenton performance, *Appl. Surf. Sci.* **469**, 331–339 (2019).

36. H. Y. Xu, B. Li, T. N. Shi, Y. Wang and S. Komarneni, Nanoparticles of magnetite anchored onto few-layer graphene: A highly efficient Fenton-like nanocomposite catalyst, *J. Colloid Interface Sci.* **532**, 161–170 (2018).

37. Q. Li, H. Kong, P. Li, J. Shao and Y. He, Photo-Fenton degradation of amoxicillin via magnetic TiO$_2$-graphene oxide-Fe$_3$O$_4$ composite with a submerged magnetic separation membrane photocatalytic reactor (SMSMPR), *J. Hazard. Mater.* **373**, 437–446 (2019).

38. S. S. Bhosale, S. S. Rohiwal, L. S. Chaudhary, K. D. Pawar, P. S. Patil and A. P. Tiwari, Photocatalytic decolorization of methyl violet dye using Rhamnolipid biosurfactant modified iron oxide nanoparticles for wastewater treatment, *J. Mater. Sci. Mater. Electron.* **30**(5), 4590–4598 (2019).

39. Y. He, D. Bin Jiang, D. Y. Jiang, J. Chen and Y. X. Zhang, Evaluation of MnO$_2$-templated iron oxide-coated diatomites for their catalytic performance in heterogeneous photo Fenton-like system, *J. Hazard. Mater.* **344**, 230–240 (2018).

40. Q. Gong, Y. Liu and Z. Dang, Core-shell structured Fe$_3$O$_4$@GO@MIL-100(Fe) magnetic nanoparticles as heterogeneous photo-Fenton catalyst for 2,4-dichlorophenol degradation under visible light, *J. Hazard. Mater.* **371**, 677–686 (2019).

41. G. Scaratti, T. G. Rauen, V. Z. Baldissarelli, H. J. José and R. D. F. P. M. Moreira, Residue-based iron oxide catalyst for the degradation of simulated petrochemical wastewater via heterogeneous photo-Fenton process, *Environ. Technol.* **39**(20), 2559–2567 (2018).

42. J. C. Espinosa, C. Catalá, S. Navalón, B. Ferrer, M. Alvaro and H. García, Iron oxide nanoparticles supported on diamond nanoparticles as efficient and stable catalyst for the visible light assisted Fenton reaction, *Appl. Catal. B Environ.* **226**, 242–251 (2018).

43. L. M. Pastrana-Martínez, N. Pereira, R. Lima, J. L. Faria, H. T. Gomes and A. M. T. Silva, Degradation of diphenhydramine by photo-Fenton using magnetically recoverable iron oxide nanoparticles as catalyst, *Chem. Eng. J.* **261**, 45–52 (2015).

44. J. Liu, B. Wang, Z. Li, Z. Wu, K. Zhu, J. Zhuang, Q. Xi, Y. Hou, J. Chen, M. Cong, J. Li, G. Qian and Z. Lin, Photo-Fenton reaction and H_2O_2 enhanced photocatalytic activity of α-Fe_2O_3 nanoparticles obtained by a simple decomposition route, *J. Alloys Compd.* **771**, 398–405 (2019).

45. M. Moztahida, M. Nawaz, J. Kim, A. Shahzad, S. Kim, J. Jang and D. S. Lee, Reduced graphene oxide-loaded-magnetite: A Fenton-like heterogeneous catalyst for photocatalytic degradation of 2-methyl-isoborneol, *Chem. Eng. J.* **370**, 855–865 (2019).

46. G. Iervolino, V. Vaiano, D. Sannino, L. Rizzo, A. Galluzzi, M. Polichetti, G. Pepe and P. Campiglia, Hydrogen production from glucose degradation in water and wastewater treated by Ru-LaFeO$_3$/Fe$_2$O$_3$ magnetic particles photocatalysis and heterogeneous photo-Fenton, *Int. J. Hydrogen Energy* **43**(4), 2184–2196 (2018).

Chapter 4

Design of Artificial Iron Metalloenzymes by Combining Proteins and Organometallics

Jose M. Palomo

*Department of Biocatalysis, Institute of Catalysis (CSIC),
Marie Curie 2, Campus UAM, Cantoblanco, 28049 Madrid, Spain*

josempalomo@icp.csic.es

This chapter describes the recent advances in the design of novel artificial iron metalloenzymes and their application as catalysts. The combination of enzymes and iron for the creation of new artificial metalloenzymes has represented a very exciting research line. In particular, the development of proteins with the synthetic ability to perform C–C functionalization represents a significant challenge. Novel synthetic artificial iron-enzymes prepared by protein engineering, organometallic-protein conjugation, or bionanohybrids are discussed.

4.1 Introduction

In the last 10 years, from the different metalloenzymes, iron-enzymes represent a very interesting class, with a huge number of important natural enzymes that present a core of iron metal or organometallic system involved in biological processes.[1-2]

77

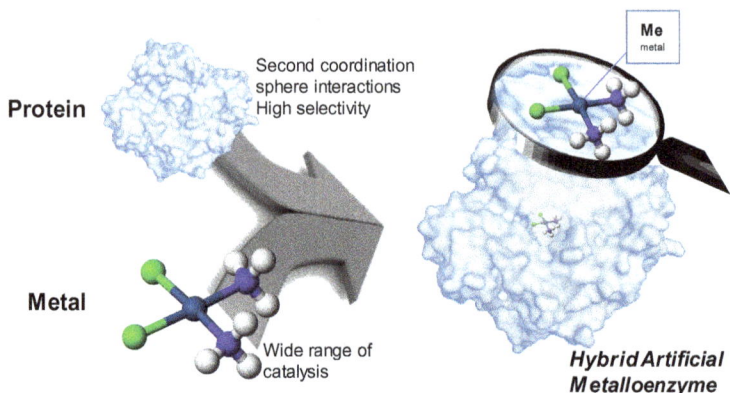

Fig. 4.1. Concept on the preparation of artificial metalloenzymes.

Thus, the use of iron catalysis as a future alternative to the precious metal being able to catalyze non natural chemical processes is a challenge.

In this way, the combination of proteins and metals or organometallic complexes has represented a very exciting line of research,[3–10] where it is possible to conjugate the potential of the three-dimensional structure of proteins, excellent for generating a coordination sphere and critical for high selectivity processes, and the broad range of chemical reactions that metals can provide (Fig. 4.1).

Indeed, in nature many different important biological process are controlled by metalloenzymes. So one of the most developed strategies to create artificial enzymes has been based on nature. Biological tools such as site-directed mutagenesis or directed evolution techniques have been successfully extended to modify the metal-coordinated aminoacids environment in natural metalloenzymes to produce enzymes with tuned or enhanced reactivity and selectivity toward non natural substrates.[11–12]

Following this strategy, the replacement of the existing metal ion or organometallic complex inside the natural metalloenzyme by engineered heme-containing proteins has demonstrated to be a very useful tool for creating metalloenzymes with new activities.[13–14]

Other successful strategies to create novel artificial metalloenzymes have been based on the use of different proteins as scaffold for a metal or organometallic complex, conferring enhanced selectivity by three-dimensional space of the protein cavity,[7,12,15–16] or a new concept based on the synthesis of metal nanoparticles (NPs) induced by a protein, generating the so-called nanobiohybrids.[17]

Remarkable results have been published to date of the catalytic performance of these different synthesized artificial metalloenzymes in reduction, oxidation, C–C bonding reactions, etc.

This chapter is focused on describing the examples that, up to now, exist in the literature about the creation of iron artificial metalloenzymes as novel synthetic catalysts.

4.2 Synthesis of Artificial Iron Metalloenzymes by Using Protein Engineering Tools

4.2.1 *Artificial iron metalloenzymes by directed evolution technology*

Directed evolution is one of the main research strategies actually applied for the creation of iron metalloenzymes with new synthetic activities. This technique requires a method for generating variations in a protein, followed by screening and selection of variants with improved characteristics. This methodology has allowed to create thousands of variants of the same enzyme in order to enhance the catalytic properties.[18–19]

A very interesting example is the bioengineering of cytochrome P450 monooxygenase and the application of the new enzymes (e.g., P411CHA) for C–H activation chemistry (Fig. 4.2).

This modification impacted positively on the selectivity and activity when compared with another variant that does not have them. This variant exhibited excellent enantioselectivity in the amination of (1) (Fig. 4.2) obtaining, in a particular set of conditions, the benzylic amine (2) at >99% enantiomeric excess (e.e.).

The mechanism proposed in Fig. 4.2 for the intermolecular amination catalyzed by the enzyme consists, first, in the reduction of the ferric state of the heme cofactor (Step A) with electrons derived from nicotinamide adenine dinucleotide phosphate (NADPH) gives the ferrous state (Step B). The reaction with tosyl azide (TsN_3),

J. M. Palomo

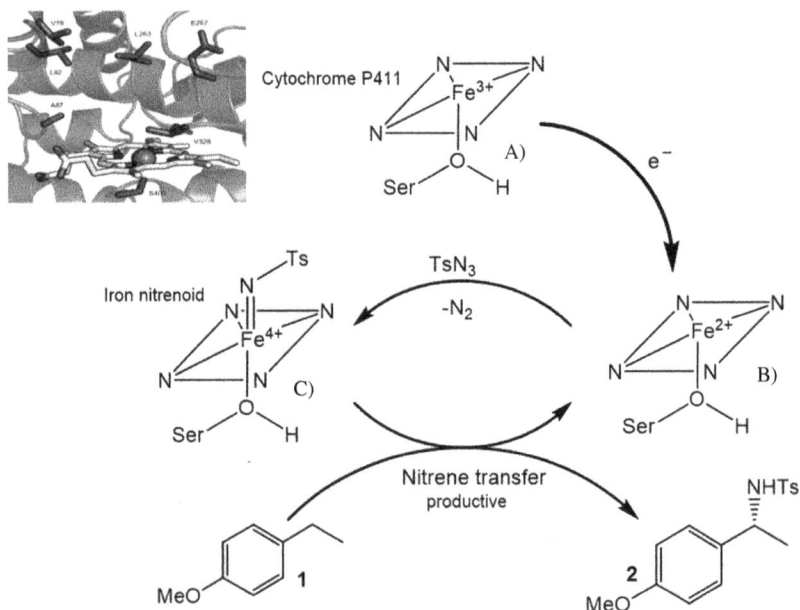

Fig. 4.2. Cytocrome P411 mutant (P-4 A82L A78V F263L)-catalyzed C–H amina-tion proposed mechanism. Nitrene transference is achieved thanks to the heme group reduction. Ts:p-toluenesulfonyl; Ser: serine.

a source of nitrene, provides the iron nitrenoid (Step C). Subsequent reaction of this intermediate with 4-ethylanisole (**1**) would yield the amination product C–H (**2**) and regenerate the ferrous state of the catalyst (Fig. 4.2, Step B).

Indeed, using 10 mM of both substrates, $P411_{CHA}$ artificial enzyme can support up to 1,300 turnovers, the highest turnover number (TON) reported with any chiral transition metal complex for intermolecular enantioselective C–H amination.[20] Optimization of the reaction allowed the synthesis of product (**2**) with a yield of 86%, with >99% e.e., and 670 turnovers in water at room tempera-ture in contrast to the chemical protocols reported. An extension of the application of this variant to other alkanes revealed the great synthetic interest of this artificial metalloenzyme.[20]

Furthermore, Arnold and coworkers recently demonstrated the high versatility of this evolved cytochrome P450 enzyme (Cytochrome

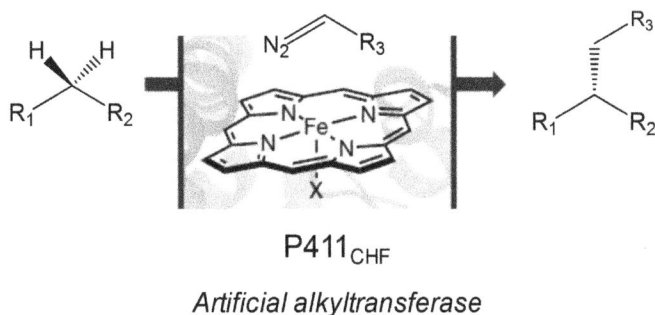

P411$_{CHF}$

Artificial alkyltransferase

Fig. 4.3. C–H alkylation reaction catalyzed by an artificial heme protein.

P411$_{CHF}$), which also showed an outstanding application in catalytic enantio-, region-, and chemoselective intermolecular alkylation of sp^3 C–H bonds through carbene C–H insertion with high turnover and excellent selectivity (Fig. 4.3).[21]

More than 70 variants from directed evolution were performed and applied in the C–H alkylation reaction at room temperature and anaerobic conditions using p-methoxybenzyl methyl ether (**3**) and ethyl diazoacetate (**4**) (Fig. 4.4A).[21]

The cytochrome P411$_{CHF}$ variant synthesized (**5**) with excellent stereoselectivity 96.7:3.3 enantiomeric ratio (e.r.) and 2,020 total turnover number (TTN). Even this enzymatic C–H activation was successfully extended to gram scale.[21]

Once more, the versatility of this artificial enzyme in C–H reaction was demonstrated giving excellent selectivity against different benzylic, allylic, propargylic, alkylamine, and diazo compounds (**6–11**) (Fig. 4.4B).[21]

A "lactam synthase" that can be tuned by directed evolution to convert individual substrates into different lactams through a catalyst-controlled C–H amidation process has been the recent application of an evolved cytochrome P450.[22]

A set of different variants with different catalytic properties was obtained in amidation reactions of divergent constructions of β-, γ- and δ-lactams.[22]

For example, the LS$_{sp3}$ variant was produced, which showed extraordinary catalytic properties, producing (**13**) with a TTN value

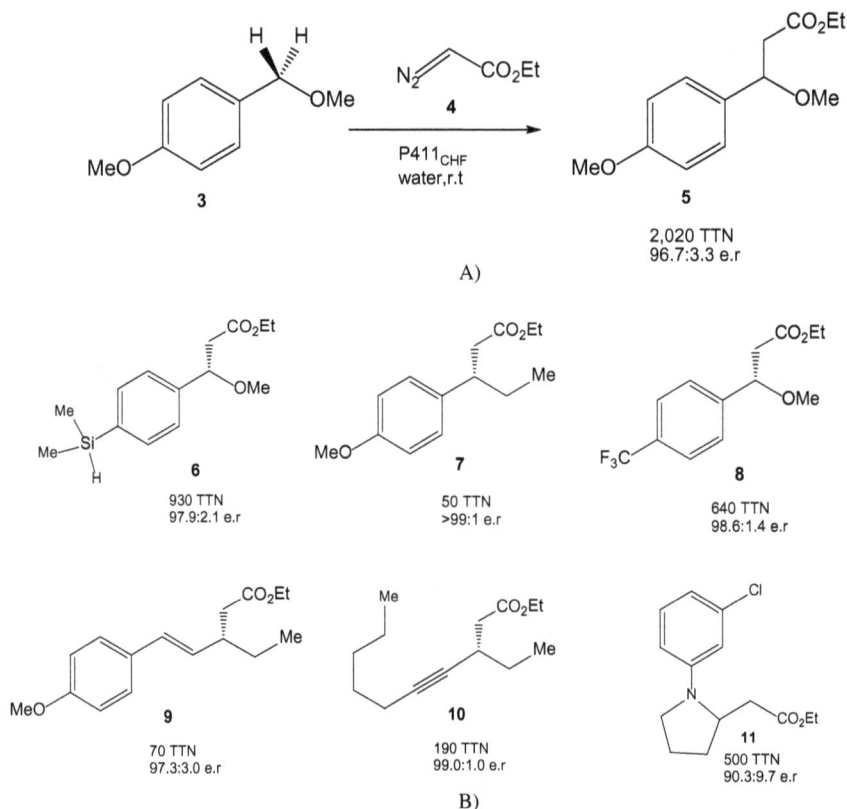

Fig. 4.4. C–H alkylation catalyzed by P411$_{CHF}$ enzyme variant. A) Alkylation of **3**. B) Selected examples of synthesized products by this strategy.

> 200,000 and an excellent selectivity (>95% e.e.) (Fig. 4.5). The application of this variant to other substrates once again demonstrated the potential of this technique.[22]

4.2.2 *Artificial iron metalloenzymes by site-specific genetic modification*

In addition to cytochrome, myoglobin (Mb) is one of the most used hemoproteins used as template for genetic modification. In particular, in order to create C–H synthases, recently, Fasan group has

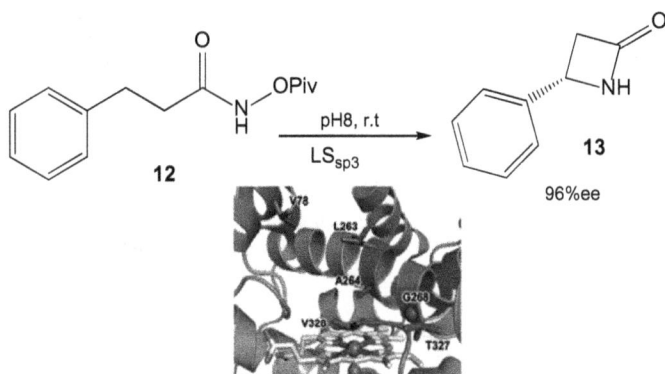

Fig. 4.5. Synthesis of β-lactams by enantioselective C–H reaction catalyzed by P411 variants.[22]

developed different evolved Mb variants for site-specific C–H cyclopropanation[23–25] or functionalization of unprotected indoles.[26]

The strategy to create new enzymes was based on the site-specific genetic modifications of particular amino acids at the active site of sperm whale Mb.[26]

A set of Mb variants was tested in the C–H functionalization of indol (**14**) in aqueous media at room temperature using ethyl α-diazoacetate (**6**) as electrophilic donor (Fig. 4.6). The Mb variant (H64V, V68A) showed the best results at pH 9, >80% conversion of (**15**). The catalytic efficiency of this enzyme was more than three times higher than that reported using metal-based catalysts for the C3 functionalization of N-protected indoles.[27]

An excellent result with a complete conversion of the C–H product was obtained using a whole cell biotransformation, using *Escherichia coli* cells expressing Mb (H64V, V68A) variant (Fig. 4.6). The reaction was extended to other substituted substrates using the whole cells as catalysts, in many cases with excellent results (Fig. 4.6).

Sulfite reductase mimic represents major advancement in engineering synthetic enzymes. Recently, Lu and coworkers created the first artificial enzyme to reduce sulfite molecules.[28] This is quite important because sulfite in the environment inhibits bioremediation of

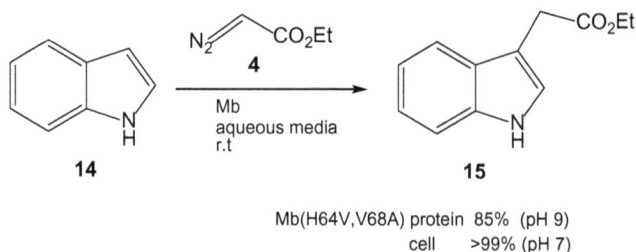

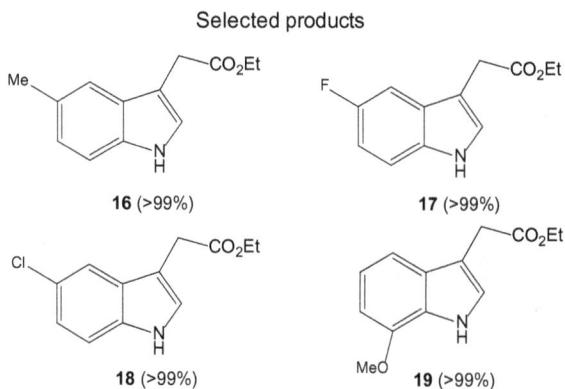

Fig. 4.6. C–H functionalization of indoles alkylation with ethyl α-diazoacetate cata-lyzed by Mb(H64V,V68A) variant.

other toxic pollutants such as chloride compounds or arsenate or nitrate. They were inspired in natural sulfite reductase structure cre-ated by rationally designing an artificial version using Cytochrome C as template.[28]

They found that its active site contains a cavity on the heme proximal face large enough to host a [4Fe-4S] cluster, critical for the sufite reduction activity. After bioinformatics analysis they found several positions for introducing the adequate mutations, founding a variant exhibiting sulfite reduction activity of 21.8 (±2.4) min^{-1}.

4.3 Artificial Iron Metalloenzymes by Organometallics-Protein Conjugation

A second approach to produce artificial metalloenzymes is based on anchoring a metal cofactor within a protein, combining characteristics that typically are from homogenous catalysts with those from enzyme. This strategy has been extensively used for the incorporation of typical transition metal catalysts (Pd, Rh, Ru, Ir) into a protein core for developing synthetic artificial enzymes.[7,12,15,29–31]

There are different strategies to achieve this. They differ in how the metal is incorporated into the protein (dative, covalent, etc.) and in how the purpose of the protein was redesigned.[12]

However, only few examples have been reported for the creation of artificial iron-enzymes by this way.[32–34]

The strong non covalent interaction between biotin and streptavidin has been excellently used for the design of an artificial metalloenzyme technology by the Ward group.[7]

In the case of iron metalloenzymes, this technology was applied to create a novel artificial hydrogenase.[33] For that, biotin molecule was chemically modified by insertion of different (cyclopentadienone) iron tricarbonyl complexes for the *in situ* generation of the corresponding streptavidin conjugates.[33] From the different artificial conjugates, the best results in hydrogenation test was the artificial streptavidin–Fe conjugate Strp–Fe (Fig. 4.7). This catalyst showed excellent hydrogenation conversion against different substrates but with very low selectivity.[33]

Another kind of artificial iron metalloenzymes has been focused on nitrogenases.[35–36]

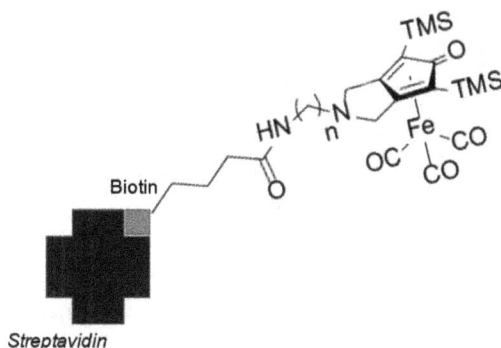

Fig. 4.7. Streptavidin–biotinylated iron complex.

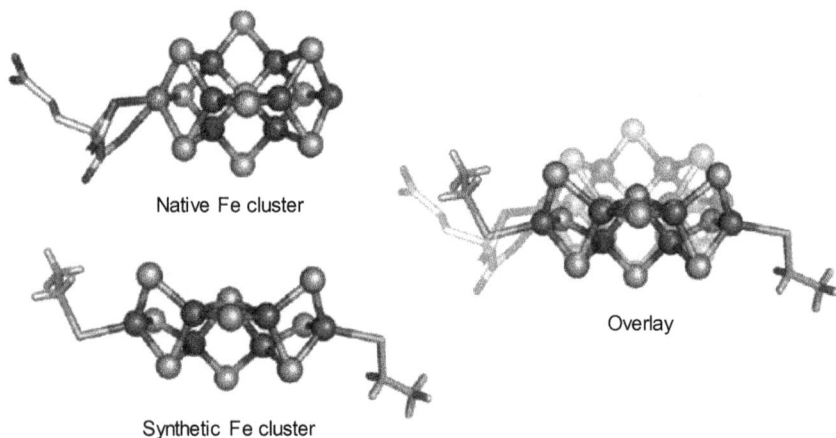

Fig. 4.8. Natural and artificial Fe cluster in nitrogenases.

These particular natural enzymes present a catalytic core with an iron complex cofactor center $(Cit)MoFe_7S_9C$, whose natural function is to transform N_2 into ammonia, however, this enzyme also is able to catalyze the reduction of nitrogen compounds (CN, N_3, S=C=N, etc.). Thus, two different strategies have been reported in order to create artificial nitrogenases by the insertion of a particular iron complex in the protein scaffold. [35–36]

In one case, Ribbe and coworkers[35] combined nitrogenase scaffold with a synthetic iron complex $[Fe_6S_9(SEt)_2]^{4-}$. Two Fe_6 clusters were inserted in the apo-nitrogenase where the synthetic metal complex resulted in a very similar conformation than the native M-cluster (Fig. 4.8).

This artificial enzyme showed activity against the reduction of C_2H_2 or CN^- to C_1–C_3 hydrocarbons.[35]

More recently, this group evaluated the effect of amount of sulfur group in the iron core, by the insertion of synthetic clusters $[Fe_4S_4] \times 2$ or $[Fe_8S_8C]$ in a protein template for reconstruction of FeS systems in the catalytic applications.[36]

4.4 Iron NP Biohybrids as Novel Artificial Metalloenzymes

Iron nanostructures with controlled size and composition have been recently developed with different interesting properties.[37–39] They are of great interest for heterogeneous catalysis (as metalloenzymes-like activity) because of their unique nanostructure-dependent properties, which drastically differentiated their catalytic performance from that of bulk metals.[40]

However, synthetic methods of iron nanoparticles (IPs) usually need the use of hard conditions (e.g. high temperatures or the presence of organic solvents) and the need of highly controllable conditions.[41–42]

Therefore, one of the main challenges is to design green and mild methodologies to obtain this kind of molecules. Because of this extremely fast capacity for oxidation, the control of the production of a particular iron species is always complex, and experimental conditions also have an enormous influence on the final iron species obtained. Morphology is also a versatile parameter for this metal, being possible to observe one-dimensional and two-dimensional structures depending on the experimental conditions.[43]

Iron oxides have been one of the most used iron species as artificial metalloenzymes because of the high stability, e.g., mainly as peroxidase in biological processes.[44–46]

Although, magnetite NPs have also been successfully used as artificial peroxidases.[47–49]

Of special interest are the synthesized IP on graphene derivatives as nanomaterials with metalloenzyme-like activities.

In this way, Wang and coworkers demonstrated the peroxidase activity of cubic Fe_3O_4 NPs (13 nm) loaded on graphene oxide carbon nanotubes (GCNT).[47]

Also Fe(0) NPs on reduced graphene oxide (GO) materials have been developed.[49] Nanonecklace-like IPs with diameters in the range of 20–50 nm and length in the micrometers scale were created. This nanocatalyst showed a remarkable peroxidase activity toward the oxidation of 3,3′,5,5′tetramethylbenzidine (TMB) in presence of H_2O_2 (Fig. 4.9) and showed a much lower Michaelis constant (Km) for TMB than natural horseradish peroxidase (HRP), which is indicative of its high affinity for that substrate.

Furthermore, some publications in the literature reported the application of Fe_2O_3 and Fe_3O_4 NPs with catalase-like activity, able to decompose H_2O_2 into water and oxygen.[46,50]

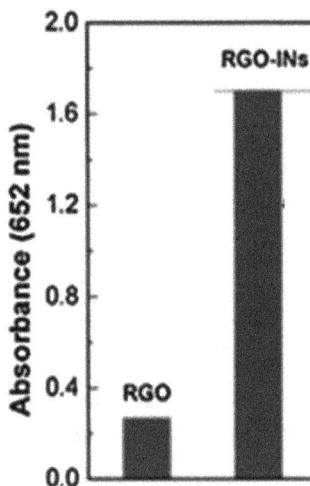

Fig. 4.9. Comparison of the peroxidase activity of the GO and iron IPs–GO nanomaterials.

However, one of the most interesting applications is focused—as in the previous strategies in this chapter—on the creation of artificial enzymes with non natural catalytic activities.

In this term, a new strategy to create artificial metalloenzymes was developed a few years ago, which consist in the synthesis of metal NPs using enzymes from simple metal salt in aqueous media at room temperature. This methodology allows the creation of, in a very fast and efficient way, of a heterogonous nanobiohybrid formed by metal NPs—directly synthesized *in situ* by the enzyme—in a protein network.[17,51]

This strategy has allowed to synthesize different, very stable, homogeneously dispersed, and functional metal NPs with excellent properties as heterogeneous catalysts.[17,51]

Between the different metals, it was possible to synthesize for the first time a nanostructured hybrid of iron (ii) carbonate nanorods (Fig. 4.10).[52] The size of the nanorods were the smallest than that described in the literature, which were embedded on the protein network creating a heterogeneous nanomaterial. The iron species was extremely stable compared with other synthetic methods.

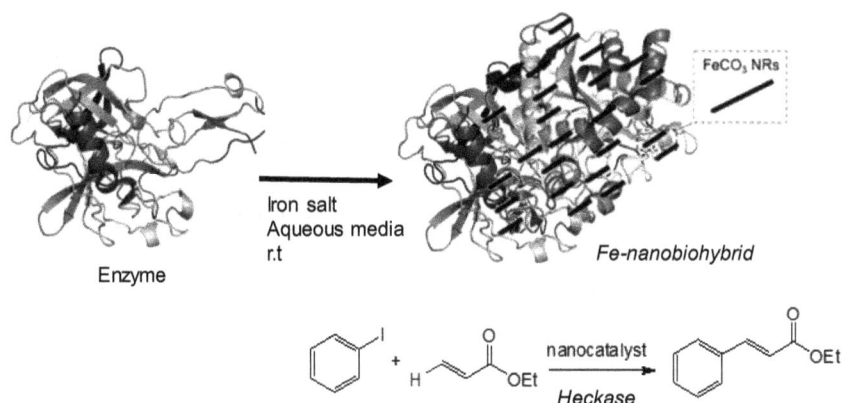

Fig. 4.10. Synthesis of iron-nanobiohybrid as Heckase.

This Fe-nanorods biohybrid showed excellent activities as artificial reductases and Heckase activity (Fig. 4.10).[52]

4.5 Conclusions

New kind of artificial iron catalysts based on the generation of newly created metalloenzymes has been developed by combination of metals or organometallic complexes and proteins.

Strategies such as protein engineering have allowed modified enzymes with unnatural synthetic activities, but also strategies for incorporating and complexing enzymes, it is a valid tool for creating metalloenzymes with natural and synthetic activities. In addition, both the strategies are being combined to achieve developments in the coupling of the compounds on enzymes for improving the selectivity (enantio, regio, etc.) of the metal complexes.

Nevertheless, the design of nanostructures with mimetic enzyme activities represents a very interesting and future alternative, especially that which combines the use of material surfaces, such as graphene. Finally, we can emphasize the strategy of creation of nanobiohybrids that combines the generation of nanostructures with the creation of metalloenzymes by insertion of metals into proteins. This method of nanohybrids allows to obtain heterogeneous artificial metalloenzymes more easily and efficiently and on a large scale, thus overcoming problems found in other methods, such as limiting the amount of enzyme, complicated organometallic complexes, etc. In addition, it has been demonstrated in other metals besides iron, and although it is a relatively new strategy, there are a lot of applications, especially considering both its metallic activity and enzyme activity (which retains its active site), as sustainable catalysts in cascade processes.

Acknowledgments

This work was supported by the Spanish Government the Spanish National Research Council (CSIC), Government of Spain, SAMSUNG Electronics by GRO PROGRAM 2017, the Ministry of Education, Youth and Sports of the Community of Madrid, and the European Social Fund (PEJD-2017PRE/SAL-3762). Authors thank the European Cooperation in Science and Technology (COST) program under the CA15106 grant (CHAOS: C–H Activation in Organic Synthesis).

References

1. L. Que Jr and W. B. Tolman, Biologically inspired oxidation catalysis, *Nature*. **455**, 333–340 (2008).
2. M. Costas, M. P. Mehn, M. P. Jensen and L. Que Jr, Dioxygen activation at mononuclear nonheme iron active sites: Enzymes, models, and intermediates, *Chem. Rev.* **104**(2), 939–986 (2004).
3. U. Markel, D. F. Sauer, J. Schiffels, J. Okuda and U. Schwaneberg, Towards the evolution of artificial metalloenzymes-a protein engineer's perspective, *Angew. Chemie Int. Ed.* **58**, 4454–4464 (2019).
4. J. C. Lewis, Beyond the second coordination sphere: Engineering dirhodium artificial metalloenzymes to enable protein control of transition metal catalysis, *Acc. Chem. Res.* **52**, 576–584. (2019).
5. F. Yu, V. M. Cangelosi, M. L. Zastrow, M. Tegoni, J. S. Plegaria, A. G. Tebo, C. S. Mocny, L. Ruckthong, H. Qayyum and V. L. Pecoraro, Protein design: Toward functional metalloenzymes, *Chem. Rev.* **114**, 3495–3578 (2014).

6. G. Roelfes, LmrR: A privileged scaffold for artificial metalloenzymes, *Acc. Chem. Res.* **52**, 545–556 (2019).

7. A. D. Liang, J. Serrano-Plana, R. L. Peterson and T. R. Ward, Artificial metalloenzymes based on the biotin–streptavidin technology: Enzymatic cascades and directed evolution, *Acc. Chem. Res.* **52**, 585–595 (2019).

8. J. C. Lewis, Artificial metalloenzymes and metallopeptide catalysts for organic synthesis, *ACS Catal.* **3**, 2954–2975 (2013).

9. K. M. Lancaster, Revving up an artificial metalloenzyme, *Science* **361**, 1071–1072 (2018).

10. F. Nastri, M. Chino, O. Maglio, A. Bhagi-Damodaran, Y. Lu and A. Lombardi, Design and engineering of artificial oxygen-activating metalloenzymes, *Chem. Soc. Rev.* **45**, 5020–5054 (2016).

11. R. K. Zhang, X. Huang and F. H. Arnold, Selective C–H bond functionalization with engineered Heme proteins: New tools to generate complexity, *Curr. Opin. Chem. Biol.* **49**, 67–75 (2019).

12. M. Diéguez, J.-E. Bäckvall and O. Pàmies, *Artificial Metalloenzymes and MetalloDNAzymes in Catalysis,* Eds., Weinheim, Germany: Wiley-VCH Verlag GmbH & Co. KGaA (2018).

13. C. L. Davies, E. L. Dux and A. K. Duhme-Klair, Supramolecular interactions between functional metal complexes and proteins, *Dalt. Trans.* **46**, 10141 (2009).

14. K. Oohora, Y. Kihira, E. Mizohata, T. Inoue and T. Hayashi, C(Sp 3)–H bond hydroxylation catalyzed by myoglobin reconstituted with manganese porphycene, *J. Am. Chem. Soc.* **135**, 17282–17285 (2013).

15. M. Filice, O. Romero, A. Aires, J. M. Guisan, A. Rumbero and J. M. Palomo, Preparation of an immobilized lipase-palladium artificial metalloenzyme as catalyst in the heck reaction: Role of the solid phase, *Adv. Synth. Catal.* **357**, 2687–2696 (2015).

16. M. Dürrenberger and T. R. Ward, Recent achievments in the design and engineering of artificial metalloenzymes, *Curr. Opin. Chem. Biol.* **19**, 99–106 (2014).

17. J. M. Palomo, Nanobiohybrids: A new concept for metal nanoparticles synthesis, *Chem. Commun.* **55**, 9583–9589 (2019).

18. R. K. Zhang, X. Huang and F. H. Arnold, Selective C–H bond functionalization with engineered Heme proteins: New tools to generate complexity, *Curr. Opin. Chem. Biol.* **49**, 67–75 (2019).

19. F. H. Arnold, Directed evolution: Bringing new chemistry to life, *Angew. Chem. Int. Ed.* **57**, 4143–4148 (2018).

20. C. K. Prier, R. K. Zhang, A. R. Buller, S. Brinkmann-Chen and F. H. Arnold, Enantioselective, intermolecular benzylic C–H amination catalysed by an engineered iron-Haem enzyme, *Nat. Chem.* **9**, 629–634 (2017).

21. R. K. Zhang, K. Chen, X. Huang, L. Wohlschlager, H. Renata and F. H. Arnold, Enzymatic assembly of carbon–carbon bonds via iron-catalysed sp^3 C–H functionalization, *Nature.* **565**, 67–194 (2019).

22. I. Cho, Z. Jia and F. H. Arnold, Site-selective enzymatic C–H amidation for synthesis of diverse lactams, *Science.* **364**, 575–578 (2019).

23. A. Tinoco, Y. Wei, J. P. Bacik, D. Carminati, E. Moore, N. Ando, Y. Zhang and R. Fasan, Origin of high stereocontrol in olefin cyclopropanation catalyzed by an engineered carbene transferase, *ACS Catal.* **9**, 1514–1524 (2019).

24. A. Chandgude and R. Fasan, Highly diastereo- and enantioselective synthesis of nitrile-substituted cyclopropanes by myoglobin-mediated carbene transfer catalysis, *Angew. Chem. Int. Ed.* **57**, 15852–15856 (2018).

25. E. Moore and R. Fasan, Effect of proximal ligand substitutions on the carbene and nitrene transferase activity of myoglobin, *Tetrahedron.* **75**, 2357–2363 (2019).

26. D. Vargas, A. Tinoco, V. Tyagi and R. Fasan, Myoglobin-catalyzed C–H functionalization of unprotected indoles, *Angew. Chem. Int. Ed.* **57**, 9911–9915 (2018).

27. X. Gao, B. Wu, Z. Yan and Y. G. Zhou, Copper-catalyzed enantioselective C–H functionalization of indoles with an axially chiral bipyridine ligand, *Org. Biomol. Chem.* **14**, 8237–8240 (2016).

28. E. N. Mirts, I. D. Petrik, P. Hosseinzadeh, M. J. Nilges, Y. Lu, A designed heme-[4Fe-4S] metalloenzyme catalyzes sulfite reduction like the native enzyme, *Science.* **361**, 1098–1101 (2018).

29. T. Heinisch and T. R. Ward, Artificial metalloenzymes based on the biotin-streptavidin technology: Challenges and opportunities, *Acc. Chem. Res.* **49**, 1711–1721 (2016).

30. P. O'Brien, D. Lopez-Tejedor, R. Benavente and J. M. Palomo, Pd nanoparticles-polyethylenemine-lipase bionanohybrids as heterogeneous

catalysts for selective oxidation of aromatic alcohols, *ChemCatChem.* **10**, 5006–5013 (2018).

31. M. Filice, O. Romero, J. Gutierrez-Fernandez, B. de las Rivas, J. A. Hermoso and J. M. Palomo, Synthesis of a heterogeneous artificial metallolipase with chimeric catalytic activity, *Chem. Commun.* **51**, 9324–9327 (2015).

32. T. Koshiyama, N. Yokoi, T. Ueno, S. Kanamaru, S. Nagano, Y. Shiro, F. Arisaka and Y. Watanabe, Molecular design of heteroprotein assemblies providing a bionanocup as a chemical reactor, *Small.* **4**, 50 (2008).

33. D. S. Merell, S. Gaillard, T. R. Ward and J.-L. Renaud, Achiral cyclopentadienone iron tricarbonyl complexes embedded in streptavidin: An access to artificial iron hydrogenases and application in asymmetric hydrogenation, *Catal. Lett.* **146**, 564–569 (2016).

34. F. Nastri, M. Chino, O. Maglio, A. Bhagi-Damodaran, Y. Lu and A. Lombardi, Design and engineering of artificial oxygen-activating metalloenzymes, *Chem. Soc. Rev.* **45**, 5020–5054 (2016).

35. K. Tanifuji, C. C. Lee, Y. Ohki, K. Tatsumi, Y. Hu and M. W. Ribbe, Combining a nitrogenase scaffold and a synthetic compound into an artificial enzyme, *Angew. Chem. Int. Ed.* **54**,14022–14025 (2015).

36. K. Tanifuji, C. C. Lee, N. S. Sickerman, K. Tatsumi, Y. Ohki, Y. Hu and M. W. Ribbe, tracing the 'ninth sulfur' of the nitrogenase cofactor via a semi-synthetic approach, *Nat. Chem.* **10**, 568–572 (2018).

37. Y. A. Nor, L. Zhou, A. K. Meka, C. Xu, Y. Niu, H. Zhang, N. Mitter, D. Mahony and C. Yu, Hollow nanospheres: Engineering iron oxide hollow nanospheres to enhance antimicrobial property: Understanding the cytotoxic origin in organic rich environment, *Adv. Funct. Mater.* **26**, 5408 (2016).

38. R. Gao, H. Zhang and D. Yan, Iron diselenide nanoplatelets: Stable and efficient water-electrolysis catalysts, *Nano Energy.* **31**, 90 (2017).

39. N. Lee, D. Yoo, D. Ling, M. H. Cho, T. Hyeon and J. Cheon, Iron oxide based nanoparticles for multimodal imaging and magnetoresponsive therapy, *Chem. Rev.* **115**, 10637–10689 (2015).

40. S. Shylesh, V. Schnemann and W. R. Thiel, Magnetically separable nanocatalysts: Bridges between homogeneous and heterogeneous catalysis, *Angew. Chem. Int. Ed.* **49**, 342–359 (2010).

41. G. Singh, P. A. Kumar, C. Lundgren, A. T. J. van Helvoort, R. Mathieu, E. Wahlstrom and W. R. Glomm, Tunability in crystallinity and magnetic properties of core–shell Fe nanoparticles, *Part. Part. Syst. Charact.* **31**, 1054 (2014).

42. L. H. Reddy, J. L. Arias, J. Nicolas and P. Couvreur, Magnetic nanoparticles: Design and characterization, toxicity and biocompatibility, pharmaceutical and biomedical applications, *Chem. Rev.* **112**, 5818–5878 (2012).

43. D. Lopez-Tejedor, R. Benavente and J. M. Palomo, Iron nanostructured catalysts: Design and applications, *Catal. Sci. Technol.* **8**, 1754–1776 (2018).

44. D. P. Cormode, L. Gao and H. Koo, Emerging biomedical applications of enzyme-like catalytic nanomaterials, *Trends Biotechnol.* **36**, 15–29 (2018).

45. M. Vázquez-González, R. M. Torrente-Rodríguez, A. Kozell, W. C. Liao, A. Cecconello, S. Campuzano, J. M. Pingarrón and I. Willner, Mimicking peroxidase activities with Prussian blue nanoparticles and their cyanometalate structural analogues, *Nano Lett.* **17**, 4958–4963 (2017).

46. K. Fan, H. Wang, J. Xi, Q. Liu, X. Meng, D. Duan, L. Gao and X. Yan, Optimization of Fe_3O_4 nanozyme activity *via* single amino acid modification mimicking an enzyme active site, *Chem. Commun.* **53**, 424–427 (2017).

47. H. Wang, S. Li, Y. Si, Z. Sun, S. Li and Y. Lin, Recyclable enzyme mimic of cubic Fe_3O_4 nanoparticles loaded on graphene oxide-dispersed carbon nanotubes with enhanced peroxidase-like catalysis and electrocatalysis, *J. Mater. Chem. B.* **2**, 4442–4448 (2014).

48. Y. Jiang, N. Song, C. Wang, N. Pinna and X. Lu, A facile synthesis of Fe_3O_4/nitrogen-doped carbon hybrid nanofibers as a robust peroxidase-like catalyst for the sensitive colorimetric detection of ascorbic acid, *J. Mater. Chem. B.* **5**, 5499–5505 (2017).

49. L. Li, C. Zeng, L. Ai and J. Jiang, Synthesis of reduced graphene oxide-iron nanoparticles with superior enzyme-mimetic activity for biosensing application, *J. Alloys Comp.* **639**, 470–477 (2015).

50. Z. Chen, J. J. Yin, Y. T. Zhou, Y. Zhang, L. Song, M. Song, S. Hu and N. Gu, Dual enzyme-like activities of iron oxide nanoparticles and

their implication for diminishing cytotoxicity, *ACS Nano.* **6**, 4001–4012 (2012).

51. M. Filice, M. Marciello, M. P. Morales and J. M. Palomo, Synthesis of heterogeneous enzyme-metal nanoparticle biohybrids in aqueous media and their applications in C–C bond formation and tandem catalysis, *Chem. Commun.* **49**, 6876–6878 (2013).

52. R. Benavente, D. Lopez-Tejedor and J. M. Palomo, Synthesis of a superparamagnetic ultrathin $FeCO_3$ nanorods–enzyme bionanohybrid as a novel heterogeneous catalyst, *Chem. Commun.* **54**, 6256–6259 (2018).

Chapter 5

Iron-Containing Enzyme Catalysts

Cesar Mateo

*Departamento de Biocátalisis, Instituto de Catálisis y
Petroleoquímica (CSIC), Marie Curie 2,
Cantoblanco, 28049 Madrid, Spain*

ce.mateo@icp.csic.es

Enzymes are highly active and selective catalysts that are capable of catalyzing reactions by different catalytic mechanisms (either through different amino acids located at the catalytic site or different trace elements or cofactors involved in catalysis). Among the different trace elements, iron is key both for the life of different living organisms, by catalyzing vital reactions for their survival, and on an industrial scale since it is capable of catalyzing different reactions of great interest. Therefore, when iron binds to the active site of an enzyme, it combines its catalytic activity with the properties of enzymes, such as their high selectivity. In this context, iron-dependent enzymes will be classified according to how they chelate iron. Thus, different examples of enzymes of each type, some significant reactions and some of the most important catalysis mechanisms will be discussed in more depth in this chapter.

5.1 Introduction

Enzymes are biocatalysts produced by living organisms being capable of functioning outside the cell. They have a high catalytic activity in very mild reaction conditions such as neutral pH, room temperature, and atmospheric pressure, and are highly selective and specific. This makes them potential candidates to catalyze reactions of industrial interest.[1]

Depending on their composition, they can be classified in two types:

- Enzymes: composed of one or several protein chains. In this case, the catalysis is produced exclusively by different interactions of the substrate with different amino acid residues present in the catalytic site yielding the product of interest and leaving the enzyme unchanged and ready to continue with other catalytic cycles.
- Holoenzymes: they have a protein part called apoenzyme and another protein part called cofactor. The cofactor can be an inorganic molecule such as metal ions (Fe, Cu, Zn, Mn, Mg, etc.) or it can be an organic molecule. If the connection to the apoenzyme is covalent, it is called the prosthetic group, and coenzyme, if it is not covalent. In these cases, residues of amino acids present in the catalytic site intervene for the fixation of the substrates and the real catalytic reaction is produced by the cofactor.[1]

Among the enzymes with trace elements, those containing iron are of great importance both for life and in different processes of industrial interest.

Iron is one of the essential elements in all living organisms because it is involved in the catalysis of many physiological processes. Iron in metalloenzymes has different functions such as oxygen transport, oxidation and reduction transformations, or transfer of electrons.[2]

Iron-containing proteins have traditionally been classified in function of the coordination of iron atom, thus it has been classified into two major groups: hemo- and non hemo-ironproteins depending on whether they contain a heme group at the catalytic site or not.

In this chapter, the different types of iron-containing proteins will be treated as well as the most representative proteins enclosed in each class.

5.2 Heme Proteins

Heme proteins also called hemeproteins or haemproteins are metalproteins with a heme as prosthetic group. This can be anchored through covalent or not covalent linkages to the protein chains. These proteins are located in different organisms. In humans and animals these are synthesized principally in the liver.[3]

They are formed by long protein chains linked to one or more heme groups. Most of the heme proteins are configured to catalyze red ox reactions through the change of the valence of the iron moiety. Heme proteins have been associated to three principal kinds of reactions: transport of dioxygen, activators of oxygen, and transport of electrons.

5.2.1 *Heme group*

The heme is a prosthetic group that is part of various proteins, among which hemoglobin is the most important. It consists of a Fe^{2+} ion coordinated to a porphyrin as a tetradentate ligand. Not all porphyrins contain iron, but a substantial fraction of the metalloproteins that contain the porphyrin nucleus possess the heme group as a prosthetic group; these proteins are known as hemoproteins. The heme group is mainly known to be part of hemoglobin, the red pigment in the blood, but it is also found in a large number of other biologically important hemoproteins such as myoglobin, cytochrome, catalase, or endothelial nitric oxide synthase.

Hemoproteins are key in diverse biological functions, including diatomic gas transport, chemical catalysis, diatomic gas detection, and electron transfer. The heme ion serves as a source of electrons during electron transfers or redox reactions. In the reactions of the peroxidases, the porphyrin molecule also serves as a source of electrons. In the transport or detection of diatomic gases, the gas binds to the heme ion. During the detection of diatomic gases, the binding of the gas to the heme group induces conformational changes in the surrounding protein.

It exists the theory that the original evolutionary function of heme proteins was the transfer of electrons in primitive photosynthesis based on the sulfur compounds that organisms similar to ancestral cyanobacteria made, before the appearance of molecular oxygen.

The hemoproteins have reached their remarkable functional diversity by modifying the immediate environment of the heme macrocycle within the protein matrix. For example, the ability of hemoglobin to deliver oxygen to tissues is due to specific amino acid residues located near the heme group of the protein. Hemoglobin binds oxygen in the pulmonary vasculature, where the pH is high and pCO_2 is low, and releases it into the tissues, where the situation reverses. This phenomenon is known as the Bohr effect. The molecular mechanism behind this effect is the steric organization of the globin chain; a histidine residue located in a position close to the hemegroup becomes positively charged when the pH becomes acidic (which is caused by the dissolution of carbon dioxide in tissues with high metabolic rate), sterically releasing the oxygen from the hemegroup.[3]

The heme group contains iron and a porphyrin ring, which corresponds to a cyclic tetrapyrrole, this macrocycle is composed of four pyrrole rings joined by bridges (=CH–). In the center of this ring is the iron atom (II) tetracoordinated by the four pairs of non shared electrons of the nitrogens of the porphyrin ring.[4]

The most common type of heme in nature is heme B (Fig. 5.1); the rest of the other heme groups are derivatizations of heme B. Other important types are heme A and heme C. Common heme

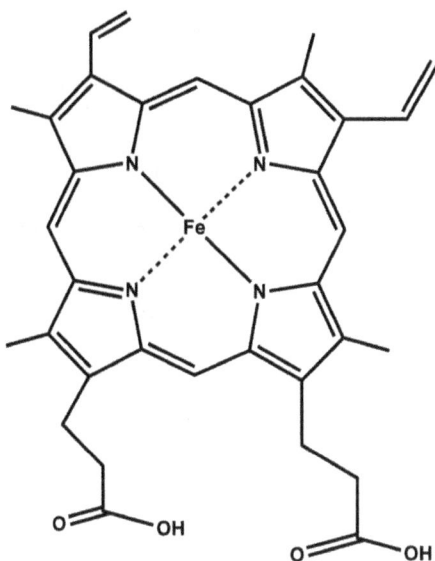

Fig. 5.1. Heme B structure.

groups are designated with capital letters, while heme groups attached to proteins are designated with lowercase letters.

5.2.2 *Hemoglobin and myoglobin*

Both the globular proteins are oxygen transporters in different living organisms. The heme group is located in a hydrophobic cleft of the tertiary structure of the protein in each subunit (Fig. 5.2).

5.2.2.1 *Myoglobin*

It is a monomeric transporter of oxygen, which is found inside cells, especially in muscle cells (it gives the reddish brown color to the muscle), and its function is to capture oxygen from the blood and give it to the mitochondria, where it is used in cellular respiration.[5]

Myoglobin also acts as a small storehouse of oxygen in cells. This function is especially important in aquatic mammals, such as the

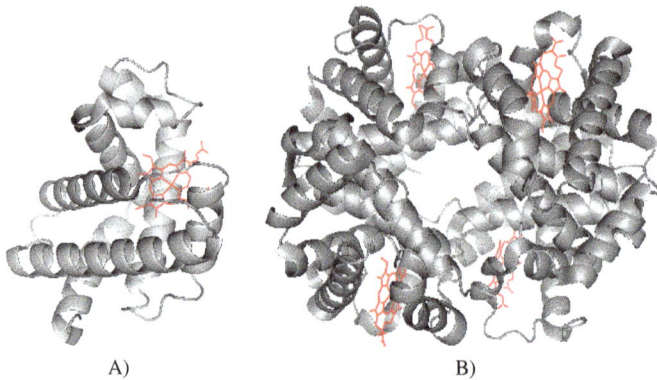

Fig. 5.2. Structure of both proteins. A) Myoglobin (protein data bank (PDB) entry 1VXA). B) Hemoglobin (PDB entry 1GZX).

sperm whale, which, due to the great depths in which they live, need to stay for long periods of time without breathing.

The protein part is constituted by a single peptide chain with secondary structure in alpha helix, formed by 153 amino acid residues, which are arranged in 8 segments of different size, with between 7 and 26 amino acid residues each. These segments are separated by loops formed by amino acids incompatible with the alpha helix structure.

These segments with secondary structure in alpha helix are folded in turn in space, giving rise to the tertiary structure of this protein, which is globular[6] (PDB entry 1VXA).

As usual in globular proteins, the amino acids with polar R group are located in the outer part of the protein, except two histidine residues that, being necessary for the activity of this protein, are located inside.

The prosthetic group consists of a heme group, formed by an organic part, protoporphyrin IX, and an iron (Fe) atom. The iron atom, in its ferrous form, can establish six coordination bonds that are arranged in an octahedral distribution, with the six links directed toward the six vertices of the octahedron. The iron atom is located in the center of the complex ring of protoporphyrin and establishes four of these bonds with the four central nitrogens of the four pyrrolic rings, two covalent and two coordinated. The two remaining

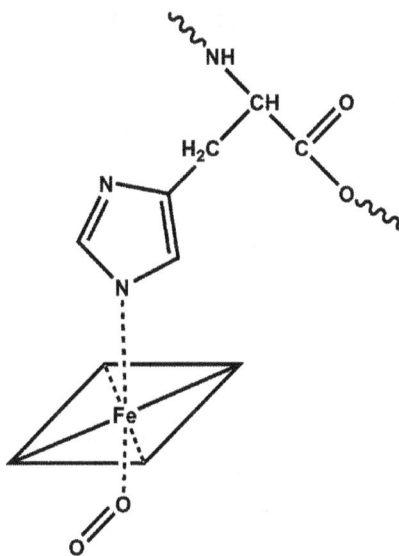

Fig. 5.3. Linkage of dioxygen to heme group.

bonds are oriented vertically to the plane of the protoporphyrin, each one on one side. One of these is established with a histidine residue of the peptide chain, specifically with the histidine F-8 (rest eight of the helix-alpha F), and this will be the anchor point of the heme group to the peptide chain; this histidine F-8 is also called proximal histidine. The other bond is the point of attachment for oxygen and, therefore, it will only form when oxygen binds to the heme group. Near this oxygen-binding link there is another histidine residue of the peptide chain, which does not form a bond with either the heme group or the oxygen, is the histidine E-7 or the distal histidine, which by its proximity generates a molecular environment that makes the union of oxygen with iron a weaker and non permanent union that can be formed and broken, which allows these proteins to transport oxygen (they capture it and then release it) (Fig. 5.3).

5.2.2.2 *Hemoglobin*

Hemoglobin is an oligomeric protein responsible for transporting oxygen from the lungs to different tissues through the blood. It is found inside specialized cells called erythrocytes. The protein part is formed by four peptide chains that are not equal to each other; there are two chains of alpha type (α) and two chains of beta type (β), being, therefore, a heterotetrameric protein (composed of four subunits that are not equal to each other with a tetrahedral arrangement) (PDB entry 1GZX). The alpha chains are formed by 141 amino acid residues and the beta chains by 146. The secondary and tertiary structures of these chains are very similar between them and similar to that of myoglobin. In the quaternary structure of hemoglobin, two blocks are formed, each consisting of the union of an alpha and a beta subunits, which are the alpha-1–beta-1 block and the alpha-2–beta-2 block; the two blocks are equal to each other. The union between the two blocks is not perfect, so there is a channel in the center of the hemoglobin molecule that is important for its function. The union between the subunits is mainly done by hydrophobic interactions and also by ionic (saline) bonds between groups with opposite charges. In this quaternary structure, the heme groups of the four subunits are also arranged symmetrically and with the greatest possible distance among them.[7]

The heme group is located in a cavity between two helices of the globin chain and in turn is protected by a residue of valine. The non polar vinyl groups of the heme group are found in the hydrophobic interior of the cavity, while the charged polar porphyrin groups are oriented toward the hydrophilic surface of the subunit (Fig. 5.2). Histidine residues of the polypeptide chains are also found, which bind to the iron atom and are designated as proximal histidines, since they are present near the heme group, while the distal histidine is located far from the heme group. The iron atom is in the center of the porphyrin ring and has six coordination bonds. The iron is coordinated to the nitrogen of the four rings of pyrol, to the nitrogen of the proximal histidine, and to the distal histidine or by oxygen.

The linkage of the oxygen to a binding site increases the likelihood that oxygen is linked to an empty binding site. Also, the release of oxygen from a binding site facilitates the release of dioxygen from other binding sites. This behavior is called cooperative, because the binding reactions at individual binding sites in each hemoglobin molecule are related and directly influence the binding reactions of the other binding sites of each molecule. The cooperative behavior of hemoglobin is essential for an efficient transport of the dioxygen within the body. In the lungs, hemoglobin is saturated in 98% oxygen. This means that 98% of the binding sites of each hemoglobin molecule are linked to one molecule of oxygen. When hemoglobin is mobilized by the blood, it releases the oxygen to the cells, and its level of saturation is reduced to 32%. This means that 66% of the hemoglobin binding sites contribute to the transport and discharge of oxygen. If a protein that does not have a cooperative binding behavior performs the same work as hemoglobin, its efficiency will be significantly reduced; e.g., myoglobin has an efficiency of 7%.[8]

5.2.2.3 *Industrial uses of myo- and hemo globin*

Both proteins are used in different fields such as in cosmetic industry for the preparation of tissue formulations; in agroindustry for preparation of nitrogen fertilizers; in biochemistry for preparation of biological reagents; in pharmaceutical industry as dietary supplements for children, the elderly and athletes; in therapeutics of nutrient deficiency and iron deficiency; in veterinary or human food as special formulations; in meat processing; or bakery industry.

5.2.3 *Cytochromes*

Cytochromes are proteins that play a vital role in the transport of chemical energy in all living cells. Animal cells obtain energy from food through a process called aerobic respiration; plants capture the energy of sunlight through photosynthesis. The cytochromes intervene in the two processes.

These are metalloporphyrins (proteins that have a ring composed of four pyrrol, called porphyrin, which encloses a metal atom through coordinated bonds) of the heme type, that is, it is iron, whose oxidation state varies from +3 to +2, that is part of the ring.

The metallic atom gives the cytochrome the characteristic dark color. There are three major types of cytochromes called a, b, and c, classified according to the absorption spectrum and the type of heme group.[9]

- Cytochrome a contains heme A, which is a modified form of protoporphyrin IX. It is important to mention that the cytochromes a of the transport chain are associated with copper atoms, which will be oxidized or reduced by the iron atoms of the porphyrins, but copper will never be part of the heme groups.
- Cytochrome b contains iron-protoporphyrin IX (there are 15 kinds of porphyrins although only nine appear in nature) known as heme B.
- Cytochrome c contains the heme C group, which, unlike the other we have, directly linked (and not by coordinated links) to the protein, specifically through two cysteines that form separate thiol bonds. Cytochromes are incorporated in the cell membrane of bacteria and in the inner membranes of mitochondria (organelles present in animal and plant cells) and chloroplasts (which are only found in plant cells). During respiration and photosynthesis, cytochrome molecules accept and release electrons alternately, which pass to another cytochrome in a chain of chemical reactions called electron transfer, which works with energy release. This energy is stored in the form of adenosine triphosphate (ATP). When the cell needs energy, it takes its reserves of ATP.[10]

5.2.3.1 Cytochrome P450

The cytochrome P450 is the main responsible for the oxidative metabolism of xenobiotics. It is not a single enzyme, but is actually a family of heme-proteins present in numerous species, from bacteria to mammals, and of which more than 2,000 different isoforms have already been identified.[11] Families 1, 2, and 3 are constituted by enzymes responsible for the biotransformation of xenobiotics, while the rest of families include P450s that are involved in the bio-synthesis and metabolism of endogenous compounds. The P450 system has an enormous functional versatility that is reflected both in the great variety of processes that it can catalyze, and in the high number of substrates that are capable of metabolizing. Although the P450 intervenes fundamentally in oxidation reactions, it is also capable of catalyzing reductions, hydrations, or hydrolysis. With a few exceptions, P450 requires molecular oxygen and nicotinamide ade-nine dinucleotide phosphate (NADPH) to oxidize the substrate. These are monooxygenation reactions in which only one of the oxy-gen atoms is incorporated in the substrate molecule, while the other is reduced to water. The enzymes that catalyze this type of oxidation are known as monooxygenases or mixed function oxidases.

The oxidations catalyzed by P450 include aromatic and aliphatic hydroxylations, N- and S-oxidations, epoxidations, O-, N-, and S-dealkylations, deaminations, desulfurizations, dehalogenations, and dehydrogenations. Their substrates include both small mole-cules and others much larger (e.g. ethanol and cyclosporin, with molecular weights of 40 and 1,203 D, respectively), aromatic or lin-ear, flat or globular, and containing or not containing heteroatoms. This broad substrate specificity is due to the existence of multiple forms of the enzyme, each of which has been adapted for the metab-olism of structurally related groups of compounds.

P450s are catalytic hemoproteins in which a thiol group of the amino acid cysteine serves as a fifth ligand to the iron atom of the heme group and the sixth ligand is a water molecule. The P450 enzymes catalyze the oxidative attack to compounds of an organic nature (hydrocarbons and their derivatives) not activated. These oxidation reactions are regio- and estero-specific and take place at physiological temperature. The details of the mechanisms by which the P450s catalyze such a high number of reactions are still unknown. The activation of oxygen, which seems to be similar in all P450s (or at least in all studied), is precisely the best known aspect.[12] The catalytic site of the P450 is the hexacoordinated iron atom (with the four rings of protoporphyrin IX, with the thiol group of a cysteine residue of the polypeptide chain and with the solvent, usually water). The first step of the catalytic process consists of the union of the substrate and the displacement of the solvent in the sixth position of coordination of the iron atom. As a consequence, changes occur in the state of spin, in the redox potential and in the maximum absorbance of the hemoprotein. In the second step occurs, the reduction of the hemoprotein–substrate complex to the ferrous state (the Fe^{3+} of the heme group passes to Fe^{2+}) thanks to the contribution of an electron and the increase in the redox potential originated in the previous step. The third step is the union of molecular oxygen to form a superoxide complex and in the fourth step the contribution of a second electron occurs with the formation of an activated oxygen species. From this point on, the mechanism is not known with certainty. The nature of the activated oxygen species is unknown (it is suggested that it could be a mixture of iron-peroxo or iron-oxo complexes with the hemoprotein). In any case, it would be a short-lived electrophilic oxidant formed by the protonation of oxygen $(O=O)$.[13]

The final result would be the release of one of the oxygen atoms in the form of a water molecule and the incorporation of the other into the substrate. Figure 5.4 represents the process that would lead to the formation of a hydroxylated metabolite. The result of the enzymatic activity of P450 is not always the insertion of oxygen into

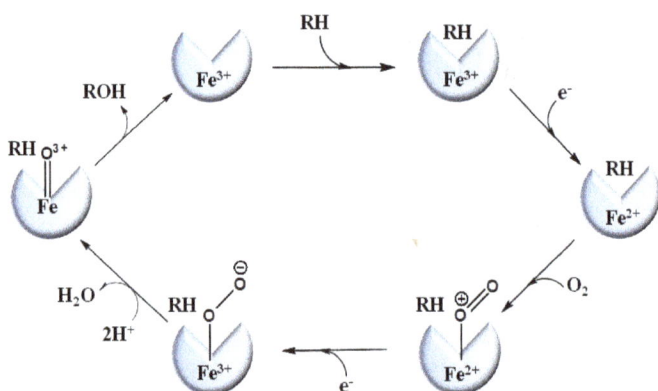

Fig. 5.4. Mechanism of hydroxylation alkyl compounds.

the substrate molecule; it can also catalyze reactions of dehydration, dehydrogenation, isomerization, dimerization, and even reduction.

The decoupling of the catalytic cycle of P450 occurs when the electrons of the NADPH cofactor are consumed without formation of oxidized metabolites. This occurs when i) the intermediate Fe^{2+}–O_2 is autooxidized releasing superoxide anion and regenerating the enzyme in the ferric state; ii) the intermediate Fe^{3+}–hydroperoxide dissociates into a molecule of H_2O_2 and ferric enzyme; or iii) the Fe=O species instead of oxidizing the substrate is reduced to one molecule of water by additional transfer of electrons.[12]

5.2.3.2 *Catalase*

Catalase is an enzyme belonging to the category of oxidoreductases that catalyzes the decomposition of hydrogen peroxide (H_2O_2) in oxygen and water. Hydrogen peroxide is a residue of the cellular metabolism of many living organisms and has, among others, a protective function against pathogenic microorganisms, mainly anaerobes, but given its toxicity it must be quickly transform into less dangerous compounds.[14] This function is carried out by this enzyme that catalyzes its decomposition in water and oxygen.

The chemical reaction occurs in two steps[15]:

1) $H_2O_2 + Fe(III)\text{-}E \rightarrow H_2O + O{=}Fe(IV)\text{-}E$
2) $H_2O_2 + O{=}Fe(IV)\text{-}E \rightarrow H_2O + Fe(III)\text{-}E + O_2$

The enzyme appears as a homotetramer and is located in peroxisomes. This enzyme can act as a peroxidase for many organic substances, especially for ethanol that acts as a hydrogen donor. The enzymes of many microorganisms, such as *Penicillium simplicissimum*, which exhibit catalase and peroxidase activity, are frequently called catalases-peroxidases.

This enzyme can be applied in different fields:

In general, the application of this enzyme is in preventing the oxidation in different fields. The use of this enzyme allows extending the useful life of citrus juices, beer, and wine since, by degrading hydrogen peroxide (an oxidizing agent) in non reactive substances (water and oxygen), oxidative reactions are inhibited without secondary problems.

In the textile industry, catalase is used to decompose the residual hydrogen peroxide (H_2O_2) in oxygen and water after the bleaching of cotton fibers. The removal of this product is necessary so that the fibers can then be dyed.

5.2.3.3 *Nitric oxide synthase*

Nitric oxide synthase or nitric oxide synthetase (NOS) is an enzyme that catalyzes the conversion of L-arginine (L-Arg) to L-citrulline by producing nitric oxide from the terminal nitrogen atom of the guanidine group of arginine. This reaction requires the presence of flavin mononucleotide, flavin adenine dinucleotide (FAD), NADPH, tetrahydrobiopterin, in addition to the participation of the heme group and calmodulin.[16] The reaction catalyzed by NOS is carried out without energy expenditure (without ATP expenditure).[17] NOS can be a dimer, calmodulin-dependent or cytochrome P450 heme protein containing calmodulin that combines catalytic domains reductase and oxygenase in a dimer, carrying FAD and flavin mononucleotide (FMN), and carries out an oxidation of an arginine with the help of tetrahydrobiopterin.[17]

The three isoforms (each of which are presumed to function as a homodimer) during activation share a carboxyl-terminal reductase domain homologous to cytochrome P450 reductase. They also share an amino-terminal oxygen domain, which includes a heme prosthetic group, which is linked in the center of the protein to a calmodulin-binding domain. The binding of calmodulin seems to act as a "molecular switch" to allow the electronic flow of prosthetic groups of flavin from the reductase domain to heme. This facilitates the conversion of O_2 and L-Arg to nitric oxide NO) and L-citrulline. The oxygenase domain of each NOS isoform also contains a BH_4 prosthetic group, which is required for the efficient generation of NO. Unlike other enzymes where BH_4 is used as a source of reduction and recycling equivalents by dihydrobiopterin reductase, BH_4

activates the heme bound to O_2 donate a single electron, which is then recaptured to allow the release of nitric oxide.

Nitric oxide can regulate the expression and activity of NOS by itself. In particular, it has been shown that NO plays an important negative regulatory role (negative feedback) on NOS3 and consequently in the function of vascular endothelial cells. This process, officially known as S-nitrosation, and mentioned by many in the field as S-nitrosylation, has been shown to reversibly inhibit NOS3 activity in vascular endothelial cells. This process can be important since it is regulated by the redox conditions of the cell and, therefore, can provide a mechanism for the association between "oxidative stress" and endothelial dysfunction. In addition to NOS3, both NOS1 and NOS2 have been found to be S-nitrosated, but the evidence for dynamic regulation of the NOS isoforms by this process is less complete. In addition, it has been shown that both NOS1 and NOS2 form nitrosyl ferrous complexes in the heme groups, which can partially act to auto inactivate these enzymes under certain conditions. The limiting step for the production of nitric oxide may also be the availability of L-Arg in some cell types. This may be particularly important after the induction of NOS2.

The synthase produces NO by catalyzing the oxidation of five electrons of the guanidine nitrogen of the amino acid L-Arg.

The oxidation of L-Arg to L-citrulline occurs via two successive monooxygenation reactions, producing Nω-hydroxy-L-arginine (NOHLA) as an intermediate. Two moles of O_2 and 1.5 moles of NADPH are consumed per mole of NO formed (Fig. 5.5).

5.3 Non Heme Iron Proteins

5.3.1 *Iron–sulfur proteins*

This type of iron-proteins is characterized by containing iron atoms linked with sulfur-forming clusters with different geometry. These clusters are capable not only of transferring electrons but also of linking oxygen and nitrogen atoms highly rich in electrons of different organic compounds. Fe–S clusters are forming four main types of scrutiny: [Fe–S] cluster consisting in a single Fe atom bound to

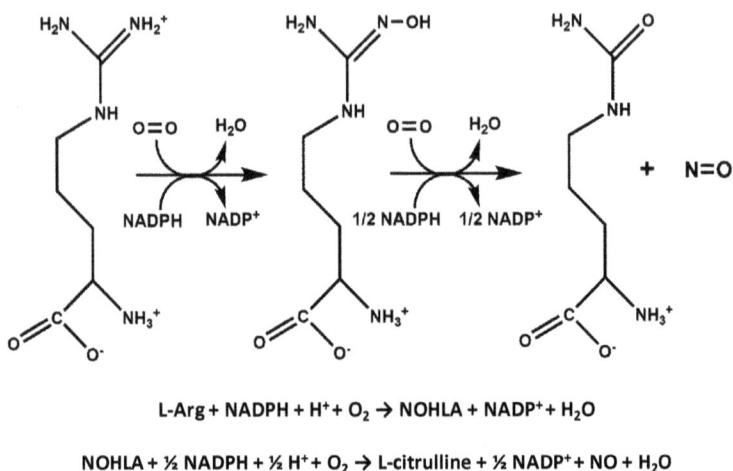

$$\text{L-Arg} + \text{NADPH} + \text{H}^+ + \text{O}_2 \rightarrow \text{NOHLA} + \text{NADP}^+ + \text{H}_2\text{O}$$

$$\text{NOHLA} + \tfrac{1}{2}\,\text{NADPH} + \tfrac{1}{2}\,\text{H}^+ + \text{O}_2 \rightarrow \text{L-citrulline} + \tfrac{1}{2}\,\text{NADP}^+ + \text{NO} + \text{H}_2\text{O}$$

Fig. 5.5. Reaction scheme of conversion of L-Arg to L-citrulline and NO catalyzed by NOS.

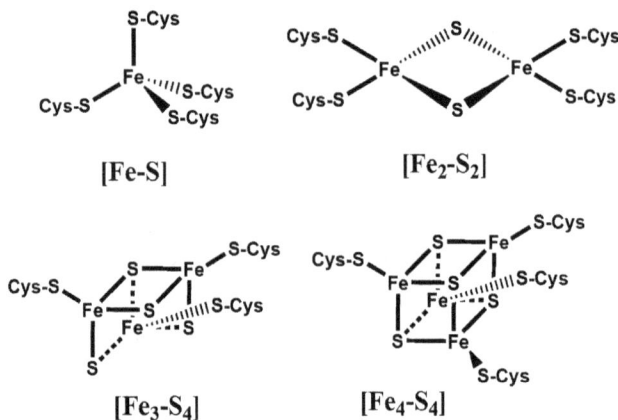

Fig. 5.6. Different iron–sulfur clusters.

four Cys residues. Typical examples are the rubredoxins that can be found in bacteria, $[\text{Fe}_2\text{-S}_2]$ rhombic are the ferredoxin-type clusters, cuboidal three-iron–four-sulfide $[\text{Fe}_3\text{-S}_4]$ clusters, and cubic four-iron–four-sulfide $[\text{Fe}_4\text{-S}_4]$ clusters (Fig. 5.6).[18]

5.3.1.1 *Rubredoxins*

The active site of rubredoxins consists in an iron coordinated with four sulfur of cysteines forming almost a perfect tetrahedron (Fig. 5.6 [Fe–S] cluster). Rubredoxins carry out the process of transferring an electron. The central iron atom changes its oxidation state between +2 and +3. In both oxidation states, the metal is maintained with a high spin, which helps minimize structural changes. The reduction potential of rubredoxin is in the range of +50 mV to –50 mV.

Many of the rubredoxins are biochemically characterized so far they are derived from sulfate-reducing bacteria and require electron carriers with reduction potentials between –400 and –200 mV.

However, other systems have been discovered where rubredoxin can act as an electron donor. The first was the role as the electron-donor enzyme rubredoxin-oxygen oxidoreductase, when it was discovered that the anaerobic bacterium *Desulfovibrio gigas* supposedly can actually use dioxygen as the final electron acceptor coupled to NADH oxidation. This system is catalyzed by different enzymes exchanging electrons as NADH-rubredoxin oxide-reductase (NRO), rubredoxin (Rd), and rubredoxin oxygen oxide-reductase (ROO) (Fig. 5.7).[19] Another system has also been reported that involves the detoxification of superoxide in anaerobic microorganisms by donating electrons for the enzyme superoxide reductase.[20]

5.3.1.2 *Ferredoxin*

They are usually small, soluble proteins, with low redox potentials and function as carriers of electrons in various metabolic pathways, both in bacteria and in plants, algae, and animals.[21–23]

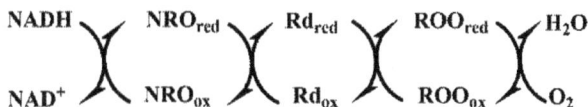

Fig. 5.7. Reduction of NADH catalyzed by rubredoxins.

Depending on the organization of the iron and sulfur atoms in the cluster, ferredoxins are organized into three types: [Fe–S], [Fe_2–S_2], [Fe_4–S_4], and [Fe_3–S_4] (see Fig. 5.6).

5.3.1.2.1 [Fe_2–S_2] type

Ferredoxins of this type are 11–15-kDa acidic proteins. This superfamily of ferredoxins, which contain a cluster [Fe_2–S_2] per molecule and a similar peptide folding, is organized into two subfamilies of ferredoxins: mitochondrial, halophilic archaebacteria, bacteria and vertebrates (called adrenoxins), and those of plants.[23] In all cases, the function is the transference of electrons from the reduced photosystem I to the ferredoxin-NADP reductase (FNR) for the assimilation of CO_2 in the case of plants; or transferring the electrons from the FNR to cytochrome P450 to synthetize steroids, bile, or vitamin D.

Mechanism of oxidation of ferredoxin[24]:

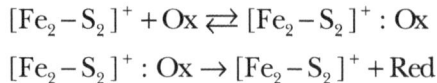

$$[Fe_2 - S_2]^+ + Ox \rightleftarrows [Fe_2 - S_2]^+ : Ox$$
$$[Fe_2 - S_2]^+ : Ox \rightarrow [Fe_2 - S_2]^+ + Red$$

5.3.1.2.2 [Fe_4–S_4] and [Fe_3–S_4] type

The ferredoxins of bacteria are different from those present in plants, not only in iron and sulfur clusters, they can have one or two clusters of [Fe_4–S_4] or [Fe_3–S_4], but also in the folding of polychain peptides. Clusters are atoms of iron and sulfur alternated at the vertices of a distorted cube. The reason for cysteine coordination in these types is given in a polypeptide chain of 60–100 aminoacids and is C-X2-C-X2-C, where the central cysteine is replaced by aspartic acid in the case of clusters with 3Fe.[25–27]

5.3.1.3 *Succinate dehydrogenase*

The enzyme succinate dehydrogenase (SDH), coenzyme Q reductase succinate, or mitochondrial complex II (EC number 1.3.5.1) is a protein

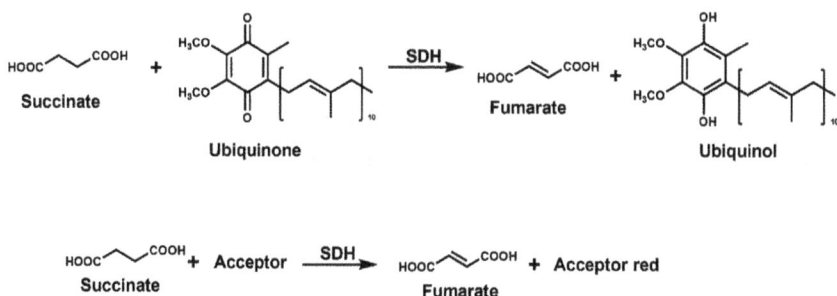

Fig. 5.8. Reaction catalyzed by succinate dehydrogenase.

complex linked to the mitochondrial membrane that is involved in the Krebs cycle and the electron transport chain, and containing FAD covalently bound. Catalyze the reaction described in Fig. 5.8.

The $FADH_2$ of SDH, unable to release the enzyme, must be oxidized again *in situ*. $FADH_2$ yields its two hydrogens to ubiquinone (coenzyme Q), which is reduced to ubiquinol (QH2) and leaves the enzyme, diffusing in the lipid bilayer until reaching the next enzymatic complex of the respiratory chain.

The FAD molecule is the electron acceptor of the reaction. In general, the biochemical function of FAD is to oxidize alkanes to alkenes, while NAD^+ oxidizes alcohols to aldehydes or ketones. This is because the oxidation of an alkane (such as succinate) to an alkene (such as fumarate) is sufficiently exergonic to reduce FAD to $FADH_2$, but not to reduce NAD^+ to NADH. It is unusual to find a covalent bond between FAD and a protein; in most cases, FAD is associated with enzymes in a non covalent way. SDH acts by separating two hydrogen atoms that are in a trans position from the methylene carbon atoms of succinate.[28]

This enzyme has some characteristics of an allosteric enzyme: it is activated by succinate, phosphate, ATP, and reduced coenzyme Q, and is inhibited by malonate, a structural analogue of succinate. On the other hand, this enzyme is constituted as one of the molecular targets in cyanide poisoning, a compound that inhibits its action and thus blocks the production of ATP, inducing cellular hypoxia.[28]

5.3.1.4 *Aconitase*

Aconitase (ACO) or aconitate hydratase (EC 4.2.1.3) is an enzyme that catalyzes the stereospecific isomerization of citrate to isocitrate through cis-aconitate in the Krebs cycle (Fig. 5.9). In humans, it can be found as cytosolic aconitase (ACO1) and mitochondrial aconitase (ACO2); these two proteins are isoenzymes. Both variants are moonlighting proteins, that is, proteins formed by a single polypeptide chain that have more than one function.[29]

Unlike most iron–sulfur proteins, which function as electron transporters, the iron–sulfur center of ACO reacts directly with the substrate of an enzyme. ACO has an active center $[Fe_4S_4]^{2+}$, which can be converted to an inactive form $[Fe_3S_4]^+$. It stabilizes the substrate through electrostatic interactions, placing it in the correct position to interact with the catalytic residues, and anchoring the substrate to the enzyme is guaranteed by the presence of amino acids such as serine (Ser), arginine (Arg), histidine (His), or aspartic acid (Asp). Three cysteine residues (Cys) have been shown to be center ligands $[Fe_4-S_4]$: Cys437, Cys503, and Cys506. Therefore, in the active site of the enzyme there are amino acids responsible for the binding and anchoring of the substrate to the enzyme, to which the iron-sulfurized center also contributes, and, on the other

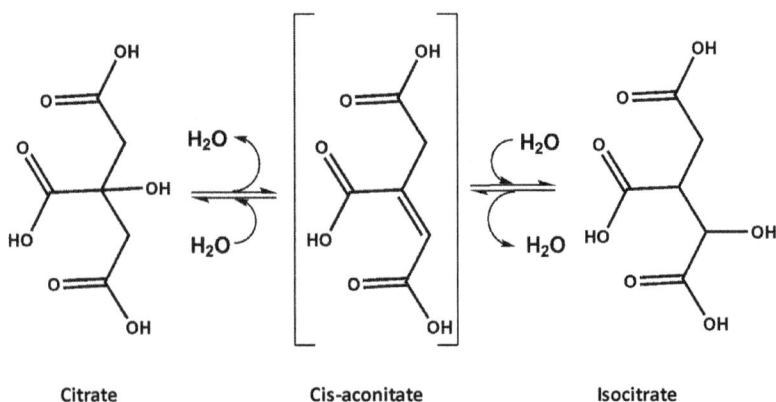

Citrate   Cis-aconitate   Isocitrate

Fig. 5.9. Conversion of citrate into isocitrate catalyzed by ACO.

hand, amino acids responsible for carrying out the catalysis properly. The closed conformation adopted by ACOs with the $[Fe_4-S_4]$ cluster is what causes a restriction of accessibility to this cluster. It is believed that access to this active center is through a movement of domain 4 of the protein with respect to the other domains (as occurs in mitochondrial ACO). The Ser642 is able to accept the proton that yields the substrate acting as a base, thanks to the chemical environment in which it is located. On the other hand, His101 acts as an acid, donating its proton for the reaction. In active state, the labile iron ions of the center $[Fe_4-S_4]$ are not coordinated by Cys, but by water molecules. ACO is competitively inhibited by fluoracetate, which is why it is poisonous. Thus, the fluoracetate in the form of fluoracetyl-CoA reacts with the oxaloacetate, by a condensation process, and produces fluorocitrate, which ACO does not recognize and paralyzes the cycle. This process inhibits oxidative phosphorylation, because no reduced equivalents are formed and, as a consequence, ATP synthase is inactivated, so ATP will not occur. It is also inhibited by trans-Aconite and is sensitive to peroxynitrium and carbonate radical.[30]

5.4 Non Heme Non-Sulfur Iron-Containing Proteins

Among the iron proteins that do not contain heme or sulfur clusters, the main ones are usually classified in mononuclear and dinuclear depending on whether they contain one or two iron atoms in the cluster.

5.4.1 *Mononuclear iron enzymes*

These iron proteins have in the catalytic site clusters with an only iron atom.[31]

Mononuclear non heme iron-dependent (NHI) enzymes catalyze an array of chemical transformations including hydroxylation, chlorination, and epimerization, as well as both cyclization and ring cleavage of organic substrates.[31,32] These enzymes can be oxygenases or oxidases depending or not of cofactors.

They have been classified into four principal groups: extradiol-cleaving catechol dioxygenases that catalyze the oxidative aromatic ring cleavage of catechol; Rieske dioxygenases that contain an extra $[Fe_2-S_2]$ ACO; α-ketoglutarate-dependent enzymes being this molecule the electron source; and aromatic amino acid hydrolases, which use tetrahydrobiopterin as a cofactor used for hydroxylation reactions.

5.4.1.1 Isopenicillin N-synthase

Isopenicillin synthase catalyzes the formation of isopenicillin N from the linear tripeptide δ-(L-α-aminoadipoyl)-L-cysteinyl-D-valine (LLD-ACV) (Fig. 5.10). Because of this, it is an enzyme with great potential for the design of new antibiotic compounds.

IPNS uses the four-electron oxidative power of O_2 to catalyze the transformation of this compound (ACV) to isopenicillin N (IPN). This is a bicyclic compound with a β-lactam ring and a five-member thiazolidine ring. This is a very special enzyme because the four electrons needed to reduce dioxygen come from the substrate. In addition, there is no atom of oxygen incorporated into the substrate. However, two water molecules are formed from two substrate hydrogens. The active site of the enzyme has a metal ion Fe(II) coordinated by at least two histidine residues, an aspartate residue, a glutamine, and two water molecules in the absence of bound substrate (Fig. 5.11).

When the ACV binds to the active site, glutamic 330 and a water molecule are replaced by ACV thiolate.[33,34] The ACV bonds with the iron complex resulting in an iron atom with five coordinates (**A**).

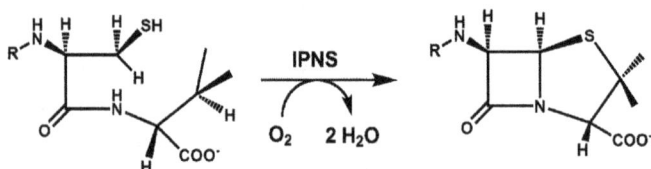

Fig. 5.10.   Conversion of LLD-ACV into isopenicillin N.

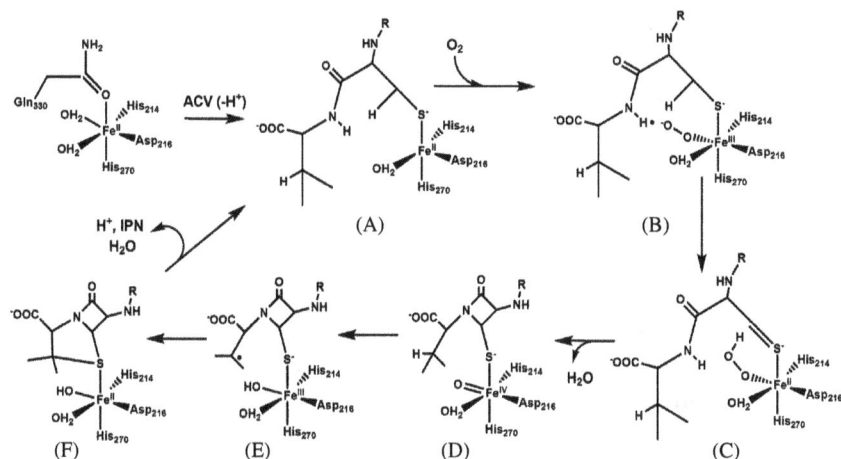

Fig. 5.11. Mechanism of IPNS.

The substrate binding increases the affinity of the enzyme for oxygen by decreasing the red ox potential of Fe(II) to Fe(III) (**B**), thus initiating the reaction cycle.[35] Subsequently, intramolecular transfer of hydrogen atoms occurs from C3 of the cysteine residue to the distal superoxide oxygen, converting Fe(III) to Fe(II). This produces a thioaldehyde and a hydroperoxy ligand (**C**). This ligand deprotonates the amide, which then closes the β-lactam ring by a nucleophilic attack on the thioaldehyde carbon[36] (**D**). This causes that hydrogen atom in the C3 of the valine residue to approach the Fe (IV) oxo ligand that is highly electrophilic. A second hydrogen transfer occurs, most likely producing an isopropyl radical that closes the thiazolidine ring by attacking the sulfur atom thiolate[36] (**E** and **F**).

5.4.2 *Dinuclear iron enzymes*

This family of proteins contains a carboxylate bridged di-iron center. They can catalyze different reactions but have the particularity of being able to bind dioxygen as part of the functional processes. Among the various functions can be found ferritins that make the storage of oxygen, hemerythrins that transport O_2 in some marine invertebrates, rubrerythrins as a peroxidase scavenger, or monooxygenases that are capable of hydroxylating different linear or aromatic hydrocarbons.[37]

5.4.2.1 *Methane monooxygenase*

This enzyme is possibly the most studied and representative enzyme within this group. It is an enzyme with very low specificity catalyzing different reactions as hydroxylation of different alkane compounds as well as the epoxidation of alkenes. It has studied two different methane monooxygenase (MMOs): soluble (sMMOs) and particulated (pMMOs). The sMMOs have a binuclear iron center bridged by μ-oxo and carboxylate bonds where oxygen reduction and methane hydroxylation are carried out.[38] A protein with iron sulfide and a flavoprotein transport the electrons from NADH through the FAD cofactors to the active hydroxylation site. The pMMOs use copper in their active site. Although some propose that pMMOs also use iron. The structures of both proteins have been determined by X-ray crystallography. The best characterized is hydroxylase (MMOH) that is part of the sMMO. When studied by X-ray crystallography it can be observed two different states, MMOHox and MMOHred, in the case of the first one each of the iron centers is hexacoordinated, bridged by a hydroxide ion, a bidentate ligand Glu γ-carboxylate ethyl, and a molecule of water. In the case of MMOred, iron centers change their coordination status to five and this is the way in which the dioxygen cluster is activated.[39–41]

The reaction of hydroxylation of methane is done by a complex mechanism (Fig. 5.12). The reaction is initiated with reduced complex H_{red}. This is capable of reacting with molecular oxygen by first giving a metastable species that is spontaneously converted to another intermediate called P. The binding of oxygen to iron atoms occurs through the transfer of metal-ligand electrons. This process concludes with the cleavage of the O–O bond to produce the reactive intermediate Q with the loss of a water molecule. Q can react with methane to insert oxygen into the C–H bond. A proposed mechanism is given by the abstraction of a hydrogen atom to form

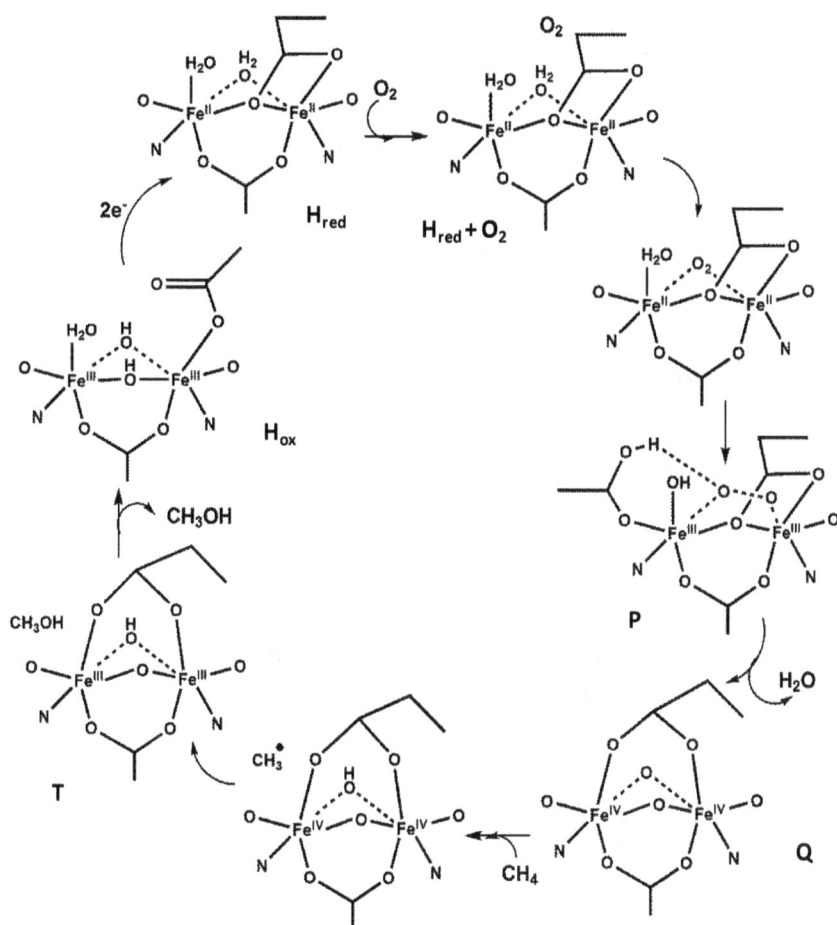

Fig. 5.12. Mechanism of action of MMO.

a cluster-bound hydroxyl radical and a species with a methyl radical to give a T complex. Finally, this complex produces methanol to form H_{ox}. As a final step, this H_{ox} complex is reduced by electrons donated by the NADH cofactor to give the reactive H_{red} complex.[42]

5.5 Conclusion and Perspective

This chapter has sought to give an overview of the state of the art in the field of iron-containing proteins in their active site. Currently, iron proteins are of great importance both in different living organisms and for the application as enzymes for industrial use. In the future, it is expected that many more iron proteins will be discovered and further development in this field will occur due to the great potential that this type of enzyme can have to catalyze different reactions in the industry. In the same way, it will be necessary to deepen into the study of reaction mechanisms that are currently unclear.

Acknowledgments

This work was supported by Spanish Government through Science, Innovation and Universities Ministry project (AGL2017-84614-C2-1-R).

References

1. J. M. Berg, J. L. Tymoczkoeremy M. and L. Stryer, *Biochemistry*, 6th ed. W. H. Freeman and Company, New York (2006).
2. R. Kipling, *Biological Inorganic Chemistry* 2nd ed. R. R. Crichton. Elsevier (2012).
3. D. L. Nelson and M. M. Cox, Lehninger, *Principles of Biochemistry*, 3rd ed. Worth Publishing: New York (2000).
4. T. Rae and T. H. Goff, The heme prosthetic group of lactoperoxidase. Structural characteristics of heme l and heme l-peptides, *J. Biol. Chem.* **273**, 27968–27977 (1998).
5. D. J. Garry, S. B. Kanatous and P. P. Mammen, Molecular insights into the functional role of myoglobin, *Adv. Exp. Med. Biol.* **618**, 181–193 (2007).
6. F. Yang and G. N. Philips Jr., Crystal structures of CO-, deoxy- and met-myoglobins at various pH values, *J. Mol. Biol.* **256**, 762–774 (1996).
7. M. F. Perutz, M. G. Rossmann, M. G. Cullis, H. Muirhead and G. Will, Structure of haemoglobin. A three-dimensional Fourier synthesis at 5.5Å resolution, obtained by X-ray analysis, *Nature* **185**, 416–422 (1960).
8. M. R. Mihailescu and I. M. Russu, A signature of the T → R transition in human hemoglobin, *Proc. Natl. Acad. Sci. U.S.A.* **98**, 3773–3777 (2001).

9. E. Margoliash, Primary structure and evolution of cytochrome C, *Proc. Natl. Acad. Sci. U.S.A.* **50**, 672–679 (1963).

10. C. J. Reedy and B. R. Gibney, Heme protein assemblies, *Chem. Rev.* **104**, 617–649 (2004).

11. D. Werck-Reichhart and R. Feyereisen, Cytochromes P-450: A success story, *Gen. Biol.* **1**, 1–8 (2001).

12. P. R. Ortiz de Montellano and J. J. Voss, Oxidizing species in the mechanism of cytochrome P-450, *Nat. Prod. Rep.* **19**, 477–493 (2002).

13. D. Mansuy, The great diversity of reactions catalized by cytochrome P-450, *Comp. Biochem. Physiol. Part. C Pharmacol. Toxicol. Endocrinol.* **121**, 5–14 (1998).

14. G. F. Gaetani, A. M. Ferraris, M. Rolfo, R. Mangerini, S. Arena and H. N. Kirkman, Predominant role of catalase in the disposal of hydrogen peroxide within human erythrocytes, *Blood* **87**, 1595–1599 (1996).

15. O. M. Lardinois, M. Mestdagh and P. G. Rouxhet, Reversible inhibition and irreversible inactivation of catalase in presence of hydrogen peroxide, *Biochim. Biophys. Acta*, **1295**, 222–238 (1996).

16. R. G. Knowles and S. Moncada, Nitric oxide synthases in mammals, *Biochem. J.* **298**, 249–258 (1994).

17. U. Förstermann and W. C. Sessa, Nitric oxide synthases: Regulation and function, *Eur Heart J.* **33**, 829–837 (2012).

18. V. P. Rao and R. H. Holm, Synthetic analogues of the active sites of iron-sulfur proteins, *Chem. Rev.* **104**, 527–559 (2004).

19. B. L. Victor, J. B. Vicente, R. Rodrigues, S. Oliveira, C. Rodrigues-Pousada, C. Frazão, C. M. Gomes, M. Teixeira and C. M. Soares, Docking and electron transfer studies between rubredoxin and rubredoxin: Oxygen oxidoreductase, *J. Biol. Inorg. Chem.* **8**, 475–488 (2003).

20. R. H. Calderon, J. G. García-Cerdán, A. Malnoë, R. Cook, J. J. Russell, C. Gaw, R. M. Dent, C. de Vitry and K. K. Niyogi, A conserved rubredoxin is necessary for photosystem II accumulation in diverse oxygenic photoautotrophs, *J. Biol. Chem.* **288**, 26688–26696 (2013).

21. H. Beinert, Iron-sulfur proteins: Ancient structures, still full of surprises, *J. Biol. Inorg. Chem.* **5**, 2–15, (2000).

22. J. M. Schmitter, J. P. Jacquot, F. De Lamotte-Guéry, C. Beau-vallet, S. Dutka, P. Gadal and P. Decottignies, Purification, properties and

complete amino acid sequence of the ferredoxin from a green alga, *Chlamydomonas rein-hardtii, Eur. J. Biochem.* **172**, 405–412 (1988).

23. K. Fukuyama, Structure and function of plant-type ferredoxins, *Photosynth. Res.* **81**, 289–301 (2004).

24. J. F. Gibson, D. O. Hall, J. H. Thornley and F. R. Whatley, The iron complex in spinach ferredoxin, *Proc. Natl. Acad. Sci. U.S.A.* **56**, 987–990 (1966).

25. I. K. Adzamli, A. Petrou, A. G. Sykes, K. K. Rao and D. O. Hall, Kinetic studies on reactions of iron-sulphur proteins. Oxidation of the reduced form of Spirulina platensis [2Fe-2S] ferredoxin with inorganic complexes, *Biochem. J.* **211**, 219–226 (1983).

26. R. Cammack, K. K. Rao, C. P. Bargeron, K. G. Hutson, P. W. Andrew and L. J. Rogers, Midpoint redox potentials of plant and algal ferredoxins, *Biochem. J.* **168**, 205–209 (1977).

27. T. Smith and A. Feinbergs, Redox properties of several bacte rial ferredoxins using square wave voltammetry, *J. Biol. Chem.* **265**, 14371–14376 (1990).

28. R. Horsefield, V. Yankovskaya, G. Sexton, W. Whittingham, K. Shiomi, S. Omura, B. Byrne, G. Cecchini and S. Iwata, Structural and computational analysis of the quinone-binding site of complex II (succinate-ubiquinone oxidoreductase): A mechanism of electron transfer and proton conduction during ubiquinone reduction, *J. Biol. Chem.* **281**, 7309–7316 (2006).

29. C. J. Jeffery, Moonlighting proteins-an update, *Mol. BioSyst.* **5**, 345–350 (2009).

30. P. R. Gardner, Aconitase: Sensitive target and measure of superoxide, *Methods Enzymol.* **349**, 9–23 (2002).

31. E. G. Kovaleva and J. D. Lipscomb, Versatility of biological non-heme Fe (II) centers in oxygen activation reactions, *Nat. Chem. Biol.* **4**, 186–193 (2008).

32. C. Loenarz and C. J. Schofield, Expanding chemical biology of 2-oxo-glutarate oxygenases, *Nat. Chem. Biol.* **4**, 152–156 (2008).

33. J. E. Baldwin and M. Bradley, Isopenicillin N synthase: Mechanistic studies, *Chem. Rev.* **90**, 1079–1088 (1990).

34. M. Lundberg, P. E. M. Siegbahn and K. Morokuma, The mechanism for isopenicillin N synthase from density-functional modeling

highlights the similarities with other enzymes in the 2-His-1-carboxylate family, *Biochemistry* **47**, 1031–1042 (2008).

35. P. L. Roach, I. J. Clifton, C. M. Hensgens, N. Shibata, C. J. Schofield, J. Hajdu and J. E. Baldwin, structure of isopenicillin N synthase complexed with substrate and the mechanism of penicillin formation, *Nature* **387**, 827–830 (1997).

36. W. A. Schenk, Isopenicillin N synthase: An enzyme at work, *Angew. Chem. Int. Ed. Engl.* **39**, 3409–3411 (2000).

37. W. J. Wei, P. E. M. Siegbahn and R. Z. Liao, Mechanism of the dinuclear iron enzyme *p*-aminobenzoate N-oxygenase from density functional calculations, *ChemCatChem.* **11**, 601–613 (2019).

38. R. Banerjee, J. C. Jones and J. D. Lipscomb, Soluble methane monooxygenase, *Annu. Rev. Biochem.* **88**, 409–431 (2019).

39. K. E. Liu, A. M. Valentine, D. Qiu, D. E. Edmondson, E. H. Appelman, T. G. Spiro and S. J. Lippard, Characterization of a Diiron(III) Peroxide intermediate in the reaction cycle of methane monooxygenase hydroxylase from *Methylococcus capsulatus* (Bath), *J. Am. Chem. Soc.* **117**, 4997–4998 (1995).

40. S. K. Lee, J. C. Nesheim and J. D. Lipscomb, Transient intermediates of the methane monooxygenase catalytic cycle, *J. Biol. Chem.* **268**, 21569–21577 (1993).

41. Y. Liu, J. C. Nesheim, S. K. Lee and J. D. Lipscomb, Gating effects of component B on oxygen activation by the methane monooxygenase hydroxylase component, *J. Biol. Chem.* **270**, 24662–24665 (1995).

42. R. Banerjee, J. C. Jones and J. D. Lipscomb, Soluble methane monooxygenase, *Annu. Rev. Biochem.* **88**, 409–430 (2019).

Chapter 6

Fe-Catalyzed C–H Activation/ Functionalization

Melania Gómez-Martínez and Olga García Mancheño*

*Organic Chemistry Institute, Münster University,
48149 Münster, Germany
olga.garcia@uni-muenster.de

6.1 Introduction

This chapter focuses on the application of iron complexes in C–H activation/functionalization reactions, which has recently become an interesting alternative approach respect to other transition metal-catalyzed C–H transformations.[1] Unlike to the most common catalysts based on Pd,[2] Ir,[3] Rh,[4] Ru,[5] or Pt,[6] iron is an abundant, benign, biocompatible, and greener metal.[7–9] Thus, the low toxicity of iron allows its use in pharmaceutical, cosmetic, and agricultural industries.[10]

C–H bond functionalization is a powerful synthetic tool for the formation of C–C and C-heteroatom bonds.[11] It has recently attracted a great interest among the scientific community since it avoids the classically required prefunctionalization of the corresponding starting materials, constituting an important breakthrough from an atom economy, sustainable and environmental point of view.[12] As a consequence, this approach is gaining weight in organic synthesis, constituting an appealing alternative to the traditional cross-coupling methodologies.[13] However, it is important to remark that transformations based on C–H bond activation/functionalization are very challenging due to i) the inert character of the C–H bonds, which posses high dissociation energies that difficult their activation, and ii) the selectivity issues of the process due to the ubiquitousity of C–H bonds in organic molecules. These points lead to the question of "*how to address selectively one C–H bond in the presence of the other C–H bonds in a target molecule?*" To overcome these fundamental issues in C–H activation/ functionalization, in the past 20 years several approaches have been revealed. According to the mechanism involving during the C–H activation step, the existing methods can be classified into two main groups (Scheme 6.1): i) *C–H activation* that involves the formation of a C-metal bond by insertion of the transition metal into the C–H bond and subsequent generation of new bond after reaction with the corresponding reagent (Scheme 6.1, top), ii) *Oxidative C–H functionalization* (not involving a C–H insertion), which relies on the selective oxidation of a C–H bond by a suitable oxidant, generating the corresponding cationic or radical intermediate species that can be then coupled with an appropriate nucleophile or radical reactant (Scheme 6.1, bottom).

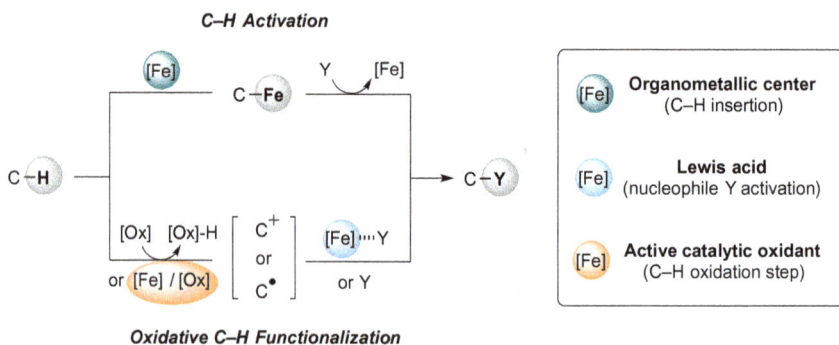

Scheme 6.1.   Fe-catalyzed C–H bond activation/functionalization approaches.

In the field of C–H bond functionalization, iron complexes can play a different role depending on the type of C–H activation mechanism involved in the reaction: i) in a "standard" C–H activation involving a metal C–H insertion, the iron species acts as the *active metal center that inserts into the C–H bond* forming an organometallic species; and ii) in oxidative C–H functionalization, the iron species can act as a *Lewis* **acid, activating the reaction partner** (usually a nucleophile, Y) or as the *active metallic oxidant* enabling the *oxidation step*.

In this chapter, the reports on homogeneous and heterogeneous iron-catalyzed C–H activation/functionalization until Summer 2019 will be summarized. Both types of C–H functionalization, i) C–H activation (C–H insertion) and ii) oxidative C–H functionalization, are included.[14] Furthermore, each part has been subdivided attending to the hybridation of the C-atom of the C–H bond involved.

6.2 Homogeneous Fe-Catalyzed C–H Activation/ Functionalization

The different homogeneous approaches in Fe-catalyzed C–H bond functionalization for the formation of C–C and C-heteroatom bonds are discussed in this section.

6.2.1 *C(sp^2)–H activation: C–C bond forming reactions*

Considering the intrinsic lower reactivity of iron species compared with the above cited most employed and studied transition metals such as Pd, many efforts have been carried out in the last decade toward the development of new and more efficient iron catalytic C–H activation approaches. Thus, initially stoichiometric studies were performed with this metal in order to explore the feasibility of such transformations. One of the first examples in Fe-mediated C–H functionalization was reported in 1968 by Miyake and coworkers, describing the synthesis of a cyclometallated iron complex (**2**) through C–H activation of one of the ligands of the $Fe(dppe)_2C_2H_5$

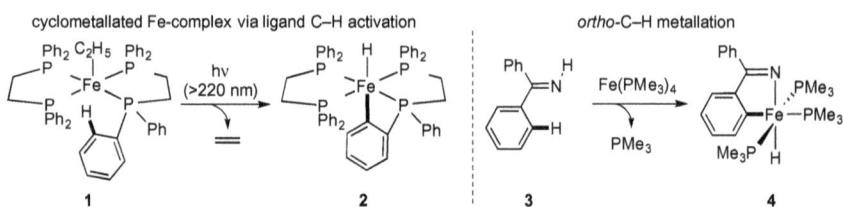

Scheme 6.2. Initial stoichiometric C–H bond activation with Fe-complexes.

(dppe = 1,2-bis(diphenylphosphino)ethane) complex (**1**) under UV irradiation (Scheme 6.2, left).[15] Since this work, a number of contributions appeared in this field, demonstrating the suitability of iron complexes to promote the activation of C–H bonds.[16] Among those, a representative example of the use of a directing group (DG) for the selective *ortho*-metalation of ketimines (**3**) catalyzed by the $Fe(PMe_3)_4$ complex was reported by Klein and coworkers obtaining the corresponding iron complex (**4**) (Scheme 6.2, right).[17]

These preliminary studies, especially the one using a DG, allowed the later development of several new catalytic C–H activation methodologies. In most of the cases, an *in situ* generated catalytic or stoichiometric organozinc species is used to promote the reaction with different organometallic compounds such as Grignard or organoboron reagents.[9b,14b] However, under the appropriate reaction conditions, the corresponding Zn-free transformations can also be achieved, as illustrated in the next sections.

6.2.1.1 *Fe-catalyzed C–H coupling with Grignard and organoboron reagents*

One of the first examples described in the literature on synthetically valuable Fe-catalyzed C–H activation dates from 2006, when the group of Nakamura "accidentally" discovered the formation of small amounts of a C–H activation product in the $FeCl_3$-catalyzed cross-coupling arylation reaction of 2-bromopyridine (**5**) with phenyl Grignard in the presence of the $ZnCl_2$•tetramethylethylenediamine (TMEDA) complex (Scheme 6.3, left).[18] Thus, besides the expected

First observation by Nakamura | Derived ortho-C–H arylation methodology

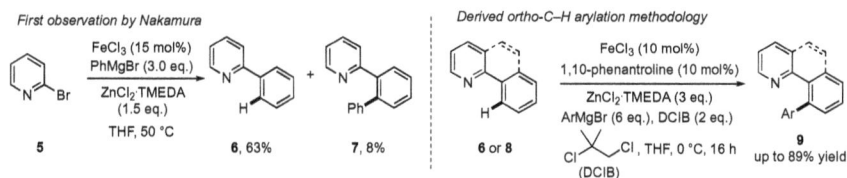

Scheme 6.3. Pioneer Fe-catalyzed directed *ortho*-C–H bond activation/arylations.

cross-coupling product (**6**, 2-phenylpyridine), the adduct formed by the C–H *ortho*-arylation of 2-phenylpyridine (**7**), was detected in 8% yield. This observation drove to the development in 2008 of a new and highly efficient methodology for the *ortho*-C–H arylation of 2-aryl pyridines (**6**) and α-benzoquinolines (**8**) catalyzed by iron(III) (Scheme 6.3, right).[18] The authors confirmed that the presence of ZnPh$_2$, generated *in situ* from PhMgBr, and ZnCl$_2$•TMEDA, and 1,2-dichloroisobutane (DCIB) as an oxidant were fundamental to carry out the C–H transformation in good to high yields (**9**), up to 89% yield).

Parallel to the development of the previous C–H activation reactions using Zn-salts as promoters, the related Zn-free coupling methodologies were also explored. Thus, an iron-catalyzed C–H arylation of 2-vinylic (**10**), 2-aryl pyridines (**11**), and imines (**12**) derivatives with aryl Grignard reagents was reported in 2011 (Scheme 6.4, left).[19] Remarkably, in the case of the vinyl substrates (**10**), the *E/Z* selectivity could be controlled depending on the solvent system used. Thus, the reaction in ethereal solvents such as tetrahydrofuran (THF) favored the formation of the (**13**) *E* adduct, while in the presence of a mixture of chlorobenzene and diethyl ether, the selectivity turns out being in favor of the (**13**) *Z* product. This effect was explained by coordination of the aromatic solvent to the *in situ* formed low-valent iron species to inhibit the iron-catalyzed olefin isomerization.[20]

Some years later, Deboef reported the C–H arylation of (hetero)arylimines with Grignard reagents (Scheme 6.4, right).[21] Thus, pyridines (**16**), furans, and thiophenes (**17**) could be then arylated with Grignard reagents in moderate to very good yields after 15 min at 0°C.

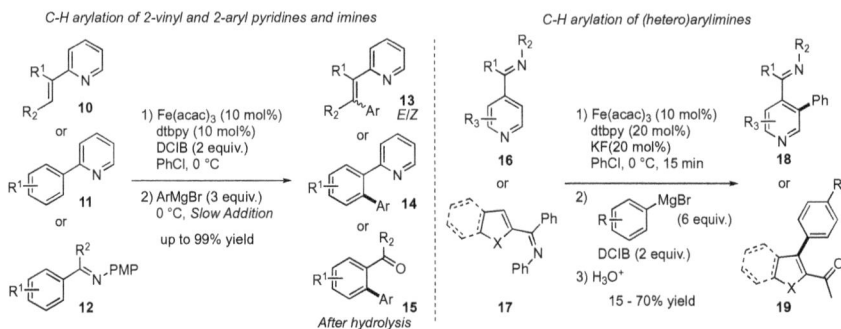

Scheme 6.4.   Fe-catalyzed Zn-free *ortho*-directed C–H arylations.

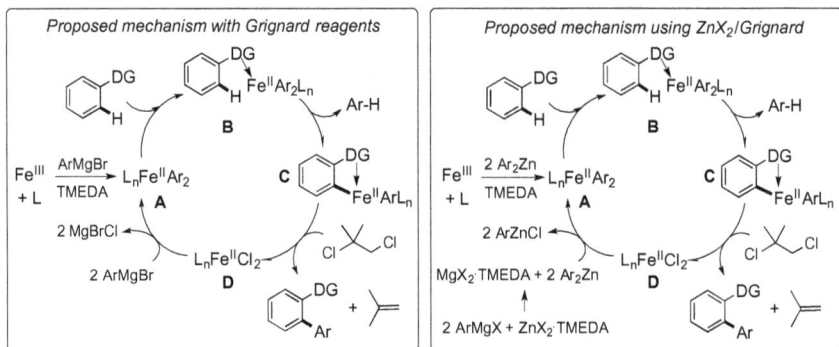

Scheme 6.5.   Simplified mechanisms involving a Fe(III)/Fe(II) catalytic cycle.

Concerning to the mechanism of these reactions, the most accepted catalytic cycle employing Grignard reagents is presented in Scheme 6.5 (left).[22] It implies a four-steps catalytic cycle, in which a first reversible coordination of the DG (**B**) to the diaryl iron(II) complex (**A**), formed *in situ* by reduction of the used Fe(III) precatalyst with the organometallic species, takes place. Then, a subsequent irreversible metalation into the *ortho*-C–H bond with the elimination of an arene leads to the species (**C**). This intermediate undergoes an oxidative-promoted reductive elimination (typically with DCIB), affording the arylated adduct, isobutene, and the dichloro-iron species (**D**). Finally, the active species (**A**) is regenerated by transmetalation of (**D**) with the aryl Grignard reagent, closing the catalytic cycle.

Regarding the reaction using Zn salts, the mechanism is more complex and less understood. Although a related Fe(III)/Fe(II) catalytic cycle can be envisioned (Scheme 6.5, right), based on the theoretical calculations carried out by the group of Chen, three possible mechanisms employing Zn salts as promoters are reasonable imploying either i) a Fe(III)/Fe(II), ii) a Fe(II)/Fe(III)/Fe(I), or iii) a Fe(II)/Fe(I) catalytic cycle.[23] In any case, the *in situ* formed Ar_2Zn from the ZnX_2 salt and Grignard (or other organometallic) reagent is key to regenerate the active Fe-catalytic species.

Inspired on the proposed mechanisms, the first breakthrough in this field was the introduction of the 8-quinolylamide as DG,[24] which was first employed by the group of Daugulis in 2005 for the palladium-catalyzed C–H functionalization of arenes.[25] Thus, assuming the generation of low-valence Fe-species by reduction with the Grignard or organozinc reagent under the standard reaction conditions, this bidentate DG was envisioned to stabilize the active Fe(II)-catalyst and minimized undesired side-reactions, while broadening the scope and functional group tolerance. Besides arylation processes (**22**), iron catalysis proved also to be capable to carry out Csp^2–H bond alkenylation (**23**), allylation (**24**), and alkylation (**25**) reactions of aryl and vinyl 8-quinolylamides (**20**) with Grignard and organoboron derivatives (Scheme 6.6).[14b]

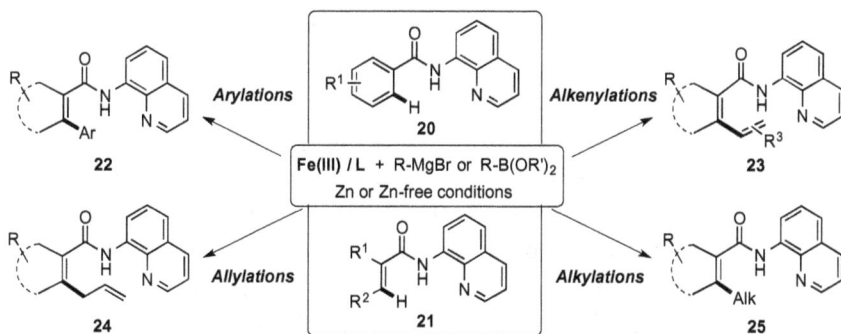

Scheme 6.6.   General overview of Fe-catalyzed C–H activation using 8-quinolylamide as DG.

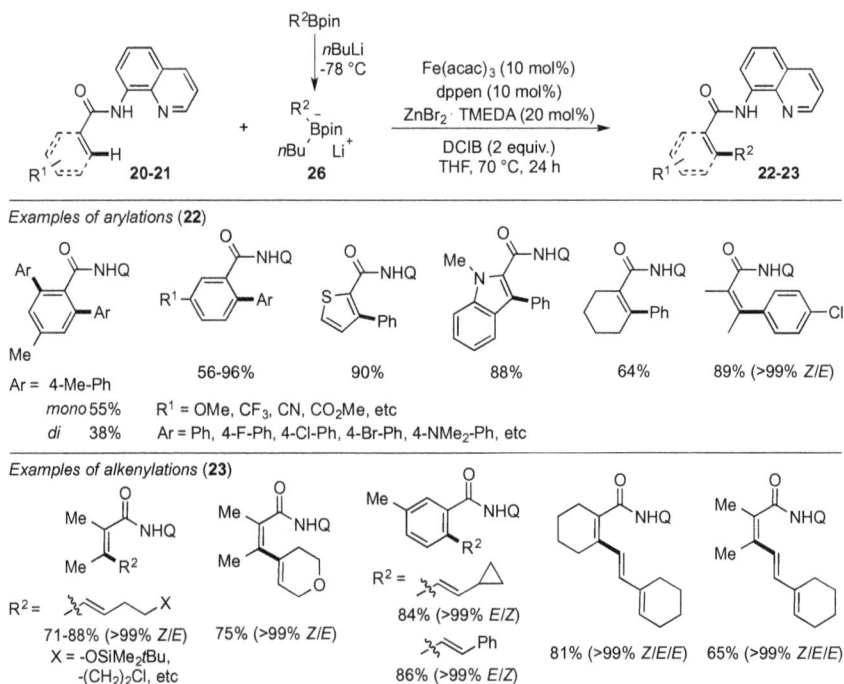

Scheme 6.7. Fe(III)-catalyzed arylations and alkenylations with boron compounds.

In the subject of arylation reactions, Nakamura, Ilies, and coworkers presented in 2014 a Fe(acac)$_3$-catalyzed C–H arylation of arenes (**20**) using aryl boronic acid pinacol esters in the presence of catalytic amounts of ZnCl$_2$•TMEDA (Scheme 6.7, top).[23b] *n*-Butyllithium was used to preactivate the boron reagent to its corresponding lithium borate salt (**26**).[26] In this catalytic system, the borate played a double role as both the coupling partner and the base to deprotonate the N–H of the amide and the C–H bond. Regarding the substrate scope, a wide range of (hetero)arenes (**20**) and boronic esters with different electronic and steric properties were well tolerated. Additionally, a highly selective monoarylation was obtained for *meta*-substituted aryl carboxamides, while a mixture of mono- and diarylation was observed for non- and *p*-substituted benzamides. For alkenyl quinolinyl amides derivatives (**21**), excellent regioselectivities (up to >99% *Z*) on the arylated products (**22**) were obtained. Alternatively, the heteroarylation of (**20**) was recently achieved under boron-free conditions with the $Fe(acac)_3$/dppen catalytic system in the presence of $Zn(CH_2SiMe_3)_2 \bullet 2MgCl_2$, Me_3SiCH_2MgCl, and dicumyl peroxide (DCP) in THF at 70 °C, obtaining up to 99% yield of the corresponding coupling adduct.[27]

The same group also described a $Fe(acac)_3$-catalyzed stereospecific alkenylation reaction of alkenyl and aryl amides (**20–21**) with alkenylborates (Scheme 6.7, bottom). As previously described, the boronic ester has to be firstly activated *in situ* with butyl lithium to form the active lithium borate species.[23b] Under those conditions, high yields and high levels of stereoselectivity were achieved, in which the stereochemistry of the alkene in the alkenylboronate is retained in the products (**23**). Additionally, the stereoselective synthesis of trienes could also be carried out in high selectivity.

A similar allylation reaction was further developed using electrophilic allylethers as coupling partners using the Grignard reagent also as the base to effectively deprotonate the C–H bond. In this regard, the use of 5 mol% of $Fe(acac)_3$ and dppen (dppen = 1,2-*cis*(diphenylphosphino)ethylene) as ligand in the presence of a Grignard and $ZnCl_2 \bullet TMEDA$ provided the C–H functionalization of *N*-(quinolin-8-yl)benzamides (**20**) with phenyl allyl ethers (**27**) in high yield (Scheme 6.8).[28] However, the reaction with α-substituted

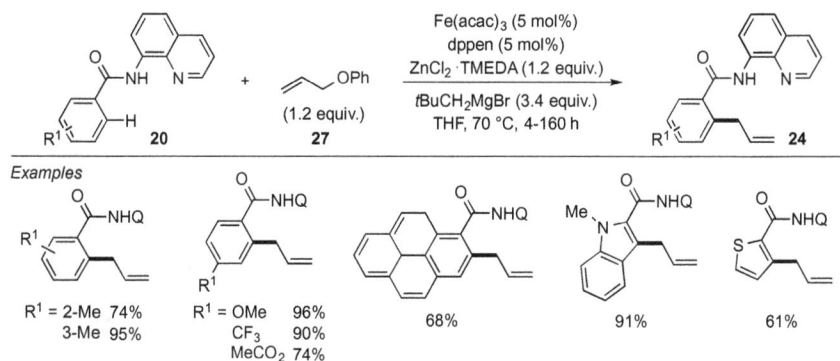

Scheme 6.8.   Fe(III)-catalyzed C–H bond allylation with phenyl allyl ethers.

allyl phenyl ethers gave the allylation products (**24**) in a remarkably low E/Z-stereoselectivity.

Although iron catalysis also allows the alkylation of Csp^2–H bonds, this transformation is more challenging than the previously discussed reactions due to the generally faster undesired β-hydride elimination reaction.[29] As a consequence, different research groups have been working on the development of new methodologies for the generation of C–C bonds avoiding the formation of the β-hydride elimination side products. In this regard, in 2014, Cook and coworkers reported the benzylation and alkylation reaction of (hetero)arene-substituted 8-quinolylamides (**20–21**) employing catalytic $Fe(acac)_3$/dppe (acac = acetylacetonate) and superstoichiometric amounts of PhMgBr as base in THF (Scheme 6.9).[30] Under the optimal reaction conditions, the alkylated products (**25**) were obtained in a range of 58%–93% yield, in which electron-rich benzyl halides provided the best results. For the coupling with secondary alkyl bromides, the radical scavenger 2,6-butylated hydroxytoluene (BHT) was needed as additive to suppress the formation of the undesired overalkylation products by trapping of the transient secondary alkyl radicals. However, in this case, a regioisomerization takes place leading to a mixture of linear and branched products.

Similarly, in 2015, Nakamura used the same DG for the reaction of alkenyl and (hetero)arene carboxamides (**20**) with alkylzinc reagents (Scheme 6.9).[31] In this case, the monoalkylzinc halides worked better as alkyl donors than dialkylzinc or alkyl magnesium halides. This observation might be explained by their weaker reducing

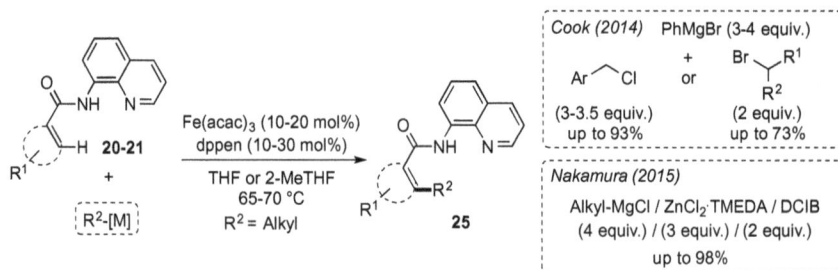

Scheme 6.9. Csp^2–H alkylations using 8-quinolylamide as DG.

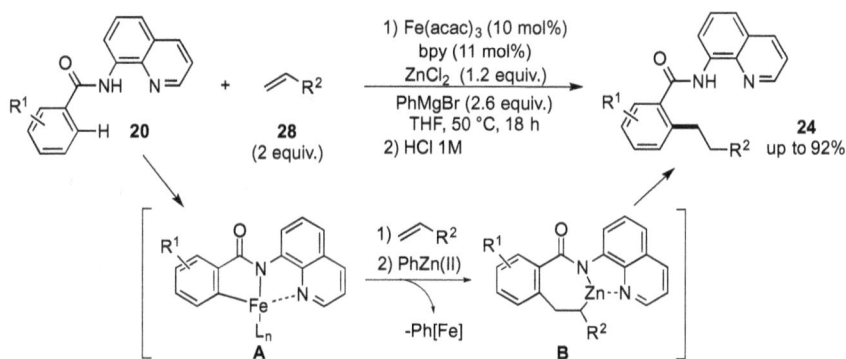

Scheme 6.10.   Fe(III)-catalyzed C–H alkylation of arenes with olefins.

ability, which reduce the side reactions and the risk of destroying the iron active species.

More recently, the direct C–H alkylation of aryl carboxamides (**20**) with olefins (**28**) involving a carbometallation step (**A → B**) was developed (Scheme 6.10).[32] The Fe(acac)$_3$/bipyridine catalytic system in the presence of an organozinc reagent, which acted as base activating the *ortho*-C–H bond of the carboxamide moiety, was used. Under these conditions, a variety of olefins reacted well with both electron-rich and electron-deficient substituted aryl carboxamides, leading selectively to the monoalkylated adducts (**23**) after treatment with HCl 1M.

Alternatively to the more widely employed 8-quinolylamide, Ackermann and coworkers introduced 1,2,3-triazole-based bidentate auxiliaries as a DG for the iron-catalyzed C–H arylation[33] and methylation or ethylation[34] reactions (Scheme 6.11). Thus, the functionalization of various (hetero)arenes (**29**) and alkenes (**30**) bearing a triazolyldimethylmethyl (TAM) DG were efficiently achieved under FeCl$_3$-catalysis, using dppe as ligand and *in situ* formed dimethylzinc as cocatalyst. The same group expanded the scope to the *ortho*-C–H allylation and alkylation of (hetero)arenes and alkenes under Zn-free conditions (Scheme 6.11).[35] The authors used allyl chloride or alkyl electrophiles such as primary and secondary alkyl bromides, as well as methyl iodide, in the presence of

R²-MgBr (7 equiv.)
FeCl₃ (20 mol%)/dppe (20 mol%)

R^2 = Ph, Me, Et
up to 96%

ZnCl₂·TMEDA (3 equiv.)
DCIB (2 equiv.)
THF, 25- 55 °C, 16 h

Fe(acac)₃ (10 mol%)
dppe (15 mol%), PhMgBr (3.75%)
THF, 65 °C, 0.5-16 h

R—CH=CH—Cl or Alkyl-Br
(3.75 equiv.)

29-30

TAM

31

R^2 = Allyl, Alkyl
up to 82%

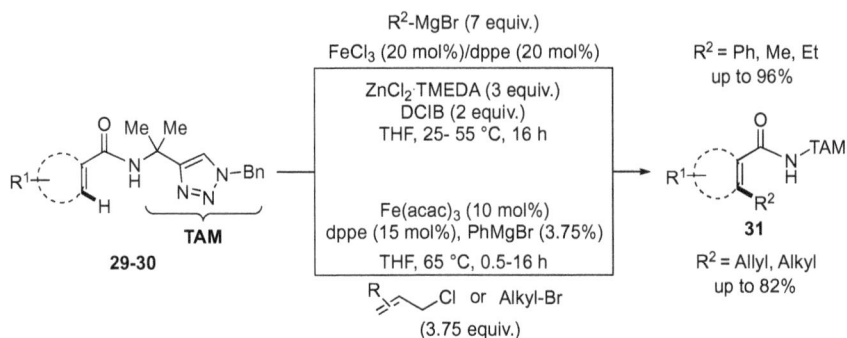

Scheme 6.11.   1,2,3-Triazole-based DGs.

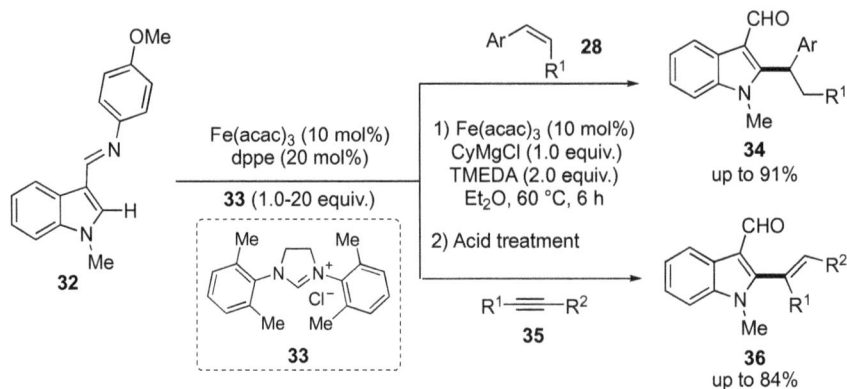

Fe(acac)₃ (10 mol%)
dppe (20 mol%)

33 (1.0-20 equiv.)

1) Fe(acac)₃ (10 mol%)
CyMgCl (1.0 equiv.)
TMEDA (2.0 equiv.)
Et₂O, 60 °C, 6 h

2) Acid treatment

$R^1 \equiv R^2$
35

32

33

34
up to 91%

36
up to 84%

Scheme 6.12.   Fe/NHC-catalyzed C–H alkylation and alkenylation of indoles.

Fe(acac)₃/dppe as catalyst and phenyl magnesium bromide as a base.

Besides the use of standard DGs, the appropriate choice of the ligand and/or type of substrate also allows selective C–H bond functionalization processes. As an example, in 2015, Yoshikai and coworkers reported the alkylation of indoles (**32**) with olefins (**28**) using a Fe/N-heterocyclic carbene (NHC) system through an imine-directed C–H activation, providing the products (**34**) in up to 91% yield (Scheme 6.12).[36] In this case, the active iron catalyst is generated *in situ* from Fe(acac)₃, an imidazolium salt (**33**) and a Grignard

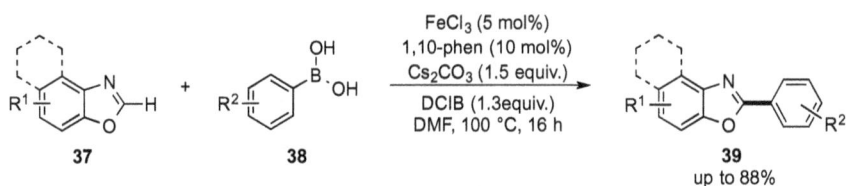

Scheme 6.13. FeCl$_3$-catalyzed C–H arylation of benzoxazoles with boronic acids.

reagent. Interestingly, the *cis*-β-methylstyrene reacted well, while the *trans*-isomer only poorly, suggesting the importance of an effective precoordination of the Fe-catalyst to the olefin. Under the optimal reaction conditions, the reaction of various diaryl, dialkyl, and silyl-substituted alkynes (**35**) proceeded smoothly with a good stereo- and regioselectivity toward the alkenylated products (**36**). More recently, the same group reported the ortho-arylation of pivalophe-none N-H imines using the Fe(acac)$_3$/dppe cataklytic system in combination with ZnCl$_2$•TMEDA/DCB (2,3-dichlorobutane)/PhMgBr as reagents, giving the corresponding arylated products in up to 98% yield.[37]

Additionally, Malakar and coworkers described the catalytic C–H arylation of benzoxazoles (**37**) with aryl boronic acids (**38**) (Scheme 6.13).[38] They employed the FeCl$_3$/1,10-phenantroline (1,10-phen) catalytic system in combination with DCIB as oxidant and CsCO$_3$ as base in dimethylformamide (DMF), obtaining the cor-responding coupling products (**39**) from 55% to 88% yield.

6.2.1.2 *Fe-catalyzed C–H coupling with organoaluminum reagents*

Nakamura and coworkers developed an iron-catalyzed *ortho*-methyl-ation of simple aryl carboxylic acids (**40**) with trimethylaluminum, employing the tridentate phosphine ligand (Me$_2$N-TP) without the need of a DG (Scheme 6.14). A wide range of *ortho*-methylated car-boxylic acids (**41**) were obtained in 36%–95% yield, where the steric hindrance of the substituents affected dramatically to the progress of the reaction. The same Fe/Me$_2$N-TP catalyst was also efficiently employed to methylate the *ortho*-C–H bond in aromatic ketones,

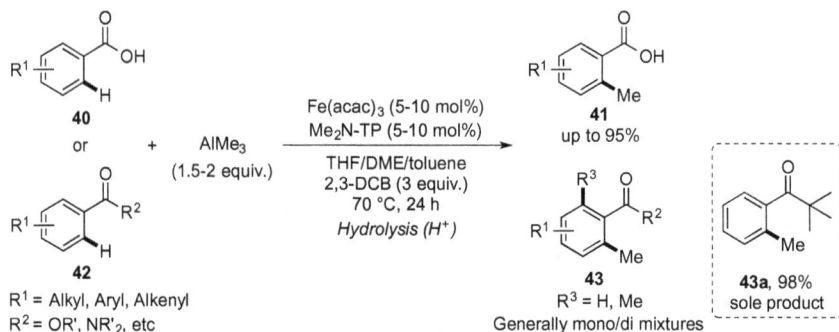

Scheme 6.14.   Methylation of carboxylic acid derivatives and ketones with AlMe₃.

esters, amides, and *N*-acetyl indoles (**42**) (Scheme 6.14).[39] As remarkable example, the reaction of *tert*-butyl phenyl ketone led exclusively to the monomethylated product (**43a**).

Lately, the group of Yoshikai also described the use of organo-aluminum reagents in the Fe(acac)₃/dppbz (dppbz = 1,2-bis (diphenylphosphino)benzene) catalyzed the methylation of N–H piv-alophenone through imine-directed C–H activation. Thus, the use of Me₃Al/DCB led to the corresponding methylated adducts in up to 90% yield.[37]

6.2.1.3 *Fe-catalyzed annulation reactions*

The synthesis of cyclic products can be also carried out by C–H func-tionalization using iron complexes as catalysts. Upon coordination of alkenes and alkynes to the low-valent iron species, the *in situ* formed ferracycle intermediate (**A**) *via* C–H activation can further react to give alkenylation or cyclization products (Scheme 6.15).[40] As a key example, Nakamura reported the synthesis of phenan-threnes (**45**) through a Fe-catalyzed [4 + 2]-benzannulation of 2-biaryl Grignard reagents (**44**) with alkynes (**35**) (Scheme 6.15).[41] The reaction was carried out with the Fe(acac)₃/dtbpy catalytic sys-tem and DCIB as oxidant, giving the 9,10-(di)substituted phenan-threnes in a range of 39%–96% yield.

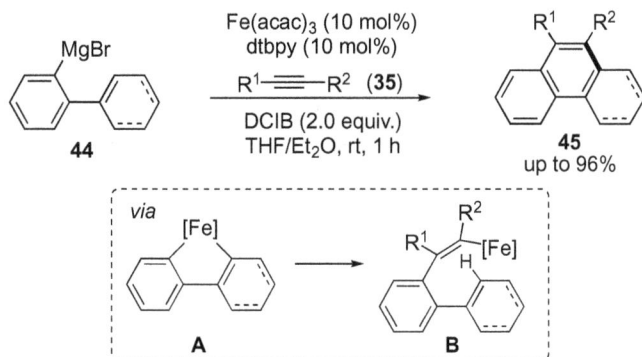

Scheme 6.15.   Fe-catalyzed [4 + 2]-benzannulation of alkynes with aryl Grignards.

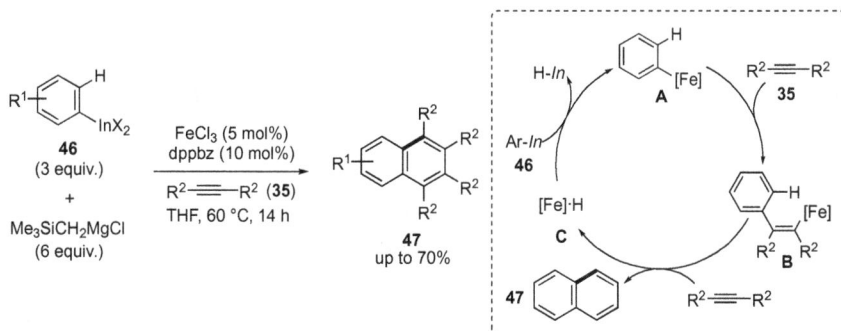

Scheme 6.16.   [2 + 2 + 2]-Annulation reaction with aryl-indium reagents.

The authors suggested the formation of a five-membered ring ferracycle intermediate (**A**), which undergoes alkyne insertion (**B**) and reductive elimination to form the cyclic product (**45**).

In 2012, Adak and Yoshikai developed a FeCl$_3$/dppbz catalyzed [2 + 2 + 2]-annulation reaction between aryl-indium reagents (**46**) and disubstituted alkynes (**35**) to form the corresponding naphthalene derivatives (**47**) (Scheme 6.16).[42] In this reaction, electron-rich ArInX$_2$ reacted faster and unsymmetrical aryl/alkyl alkynes led to high regioselectivities toward the 1,4-dialkyl-2,3-diphenyl products. As plausible mechanism, the authors proposed a first transmetalation

Scheme 6.17.   Fe-catalyzed the annulation of aryl/alkenyl 8-quinolylamides.

between the organoindium and the iron salt to form Fe-complex (**A**), followed by the insertion of the alkyne and generation of a ferracycle intermediate (**B**) upon carbometalation and C–H activation. Lastly, the reductive elimination in (**B**) provides the naphthalene adduct.

In 2016, Nakamura and coworkers made use again of the privileged 8-quinolylamide DG in aryl and alkenyl substrates (**20–21**) for the Fe-catalyzed annulation with alkynes (**35**) to build piperidinones or isoquinolinones (**48**) under oxidative conditions (Scheme 6.17).[43] For this transformation, the authors propose the formation of the metallacycle (**A**) as intermediate, in which a new C–C and a C–N bonds are formed in the overall process. Later on, the same authors described the synthesis of 1-indenones following the same methodology, but in the absence of an oxidant.[44] In this case, two new C–C bonds are formed by attack of the organoiron to the carbonyl group of the amide.

In 2018, Ackerman and coworkers reported the synthesis of substituted isoquinolinones (**51**) *via* an Fe-catalyzed C–H annulation by functionalization of monoprotected benzamides (**49**) bearing a triazolylmethylhexyl (TAH) amide as triazole-based DG with allenyl reagents (**50**) (Scheme 6.18).[45] In this transformation, the catalytic cycle starts with a C–H metalation step forming the intermediate (**A**), which upon a subsequent allene migratory insertion gives the iron complex (**B**). Then, an oxidation-induced reductive elimination allows the formation of the iron-allyl complex (**C**), which undergoes a 1, 4-iron migration by intramolecular C–H activation to generate the stabilized allylic-benzylic iron intermediate (**D**). To close the catalytic cycle, the iron-allyl complex (**D**) undergoes a

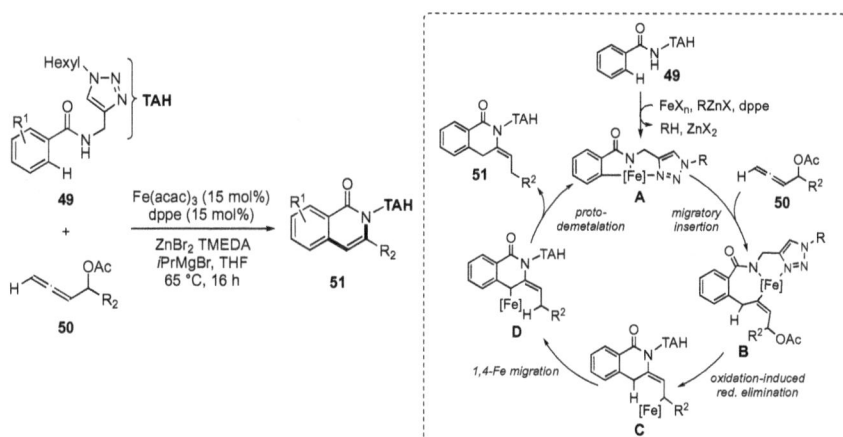

Scheme 6.18. Fe-catalyzed C–H annulation of TAH-benzamides with allenes.

proto-demetalation to generate the corresponding *exo*-methylene-3,4-dihydroisoquinolinone derivative (**51**). Later on, the same group carried out the annulation transformation employing propargyl acetates under the same reaction conditions (FeCl$_2$/dppe/ZnBr$_2$•TMEDA/*i*PrMgBr) at room temperature, giving the corresponding isoquinolone derivatives in up to 85% yield.[46]

Alternatively, the group of Panda described the synthesis of benzozepines (**53**) by intramolecular annulation of *o*-(aryloxy)diarylacetylenes (**52**) *via* a FeCl$_3$-catalyzed hydroarylation reaction in dichloroethane (DCE) (Scheme 6.19).[47] The authors proposed a mechanism in which the Fe-complex acts as both p-acid and Lewis acid catalyst. Thus, a first carbometalation step to the corresponding vinyl carbocationic intermediate (**A**) takes place, followed by an intramolecular Friedel–Crafts type reaction to give the 7-endo-dig annulation intermediate (**B**). Finally, protonation of (**C**) by the *in situ* generated HCl leads to the desired product (Scheme 6.19, right).

6.2.1.4 *Fe-catalyzed reactions with diazo-compounds*

In 2016, Díaz-Requejo, Costas, and Pérez developed an interesting approach for the functionalization of nonactivated arenes such as

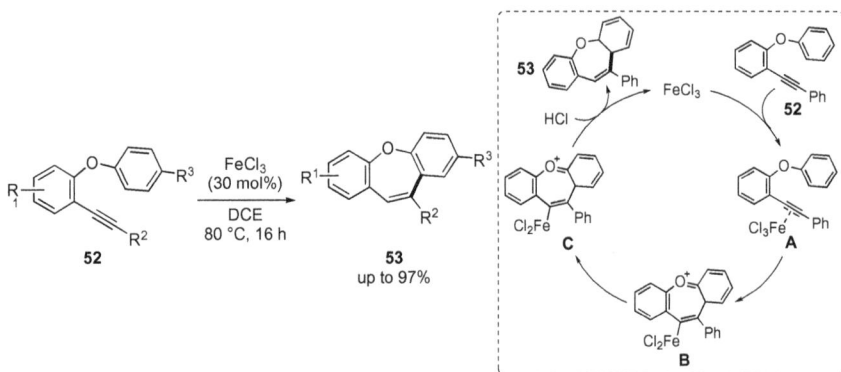

Scheme 6.19. Benzozepines synthesis by a Fe-catalyzed hydroarylation reaction.

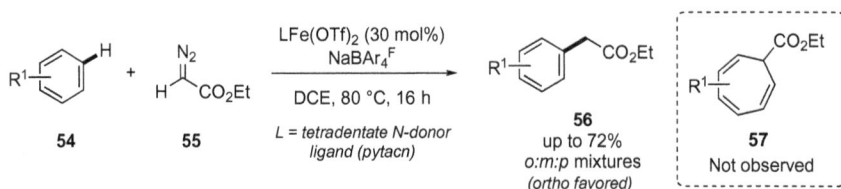

Scheme 6.20. Fe-catalyzed aromatic C–H functionalization *via* carbene insertion.

benzene (**54**), R^1 = H) with diazoacetates (**55**) using an iron salt as catalyst and $NaBAr_4^F$ (BAr_4^F = tetrakis(bis-3,5-trifluoromethyl-phenyl)borate) as triflate scavenger (Scheme 6.20).[48] Under these conditions, the C–H carbene insertion was achieved chemoselectively, hence forming the desired adduct (**56**) vs. the addition/ring expansion products (**57**) (*via* the Buchner reaction). Two years later, experimental and theoretical forces were joint to elucidate the mechanism of this transformation.[49] It was then proposed that the catalyst precursor $LFe(OTf)_2$ generates an electrophilic Fe-carbene complex that reacts with the arene in an electrophilic-type aromatic substitution.

6.2.2 *Csp³–H bond activation*

Iron-catalyzed Csp^3–H activation remains a challenging task. The reason may be attributed to the weak interaction of an iron catalyst with Csp³–H bonds, the poor stability of alkyliron species and the high tendency to undergo β-hydride elimination. As a consequence, to date there are only a few examples of iron-catalyzed Csp³–H activation.[50]

Similarly to the case of the Fe-catalyzed arylation of Csp²–H bonds, the α-Csp³–H bond arylation of ethers was found "unexpectedly" when THF was used as solvent in the cross-coupling reaction of 4-iodotoluene (**58**) with phenylzinc (Scheme 6.21, left).[51] Moreover, a similar transformation was achieved during the preparation of phenyl magnesium bromide from the corresponding bromobenzene (**60**) in the presence of a catalytic amount of Fe_2O_3 (Scheme 6.21, right).[52]

Based on these observations, a protocol for the iron-catalyzed α-arylation of aliphatic amines (**61**) through a 1,5-hydrogen transfer was developed by Nakamura and coworkers (Scheme 6.22).[51] In this case, the protection of the amine with the *o*-iodo or bromobenzyl

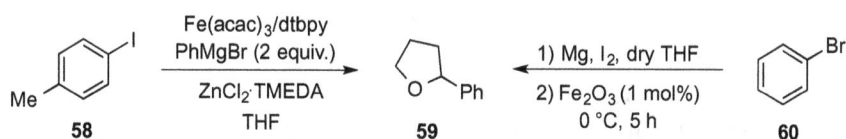

Scheme 6.21. Fe-catalyzed the arylation of THF with phenyl Grignard.

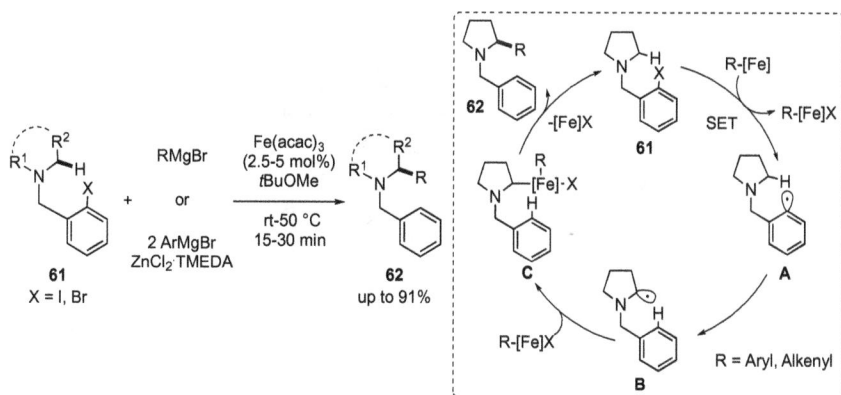

Scheme 6.22. Fe-catalyzed α-arylation of amines through 1,5-hydrogen transfer.

group was necessary for directing the reaction and avoiding the formation of side products. Regarding the mechanism of the transformation, the authors postulated the generation of an aryl radical (**A**) by reaction of the organoiron species with the halogen group through a single-electron transfer (SET). The intramolecular aryl radical is then capable to abstract the α-hydrogen of the amine *via* a 1,5-hydrogen transfer to generate an α-aminoalkyl radical (**B**). The recombination of (**B**) with the formed Fe-complex gives the intermediate (**C**), which leads to the corresponding α-arylamine product (**62**) by reductive elimination. Numerous aliphatic cyclic and acyclic amines could successfully be used. Moreover, two examples of the related alkenylation reaction could also be achieved using alkenyl Grignard reagents.[51]

Embracing a similar approach, Nakamura developed a $Fe(acac)_3$-catalyzed remote C–H bond arylation of 2-iodoalkylarenes (**63**) through a 1,5-hydrogen transfer step (**A** → **B**) to provide the arylated products (**64**) in up to 90% yield (Scheme 6.23).[53] An *in situ* formed NHC from the corresponding salt was here employed as ligand in combination with diphenyl zinc, formed *in situ* from PhMgBr and $ZnBr_2 \cdot$TMEDA.

In the field of directed Csp^3–H activation, Nakamura and coworkers reported the Fe(III)/dppbz-catalyzed arylation of various 2,2-disubstituted propionamides (**65**) bearing the 8-aminoquinolyl

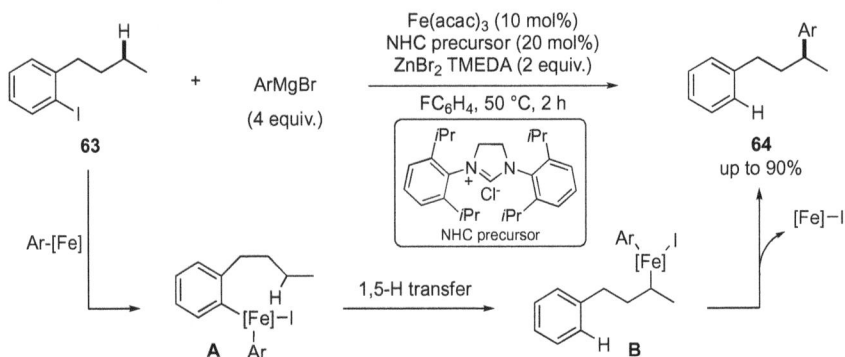

Scheme 6.23.   Arylation of aliphatic C–H bond *via* 1,5-hydrogen transfer.

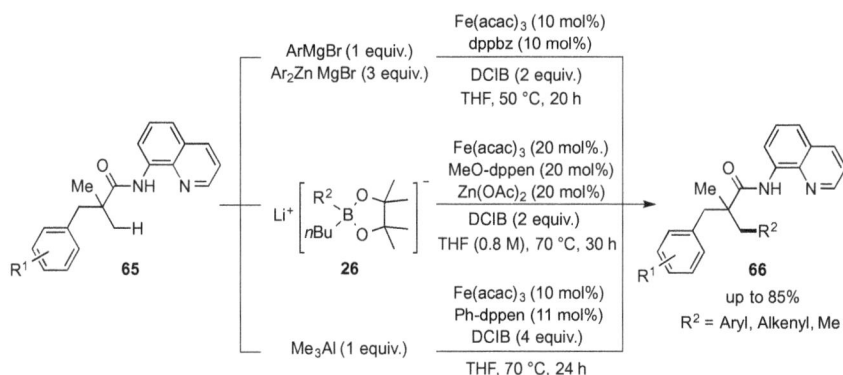

Scheme 6.24.   Fe-catalyzed β-functionalization of carboxamides.

DG with aryl zinc reagents and DCIB as oxidant (Scheme 6.24).[54] Under these conditions, the corresponding arylated products (66) were obtained from low to excellent yields. Alternatively, the same type of aliphatic carboxamides (65) could be β–arylated and alkenylated with borate reagents by a cocatalytic system involving $Fe(acac)_3$ (10 mol%) and $Zn(OAc)_2$ (20 mol%) (Scheme 6.24).[55]

It is important to note that the organoborate compounds are better donors compared to organozinc or Grignard reagents, providing hence better results. Moreover, the same group employed $AlMe_3$ as methylating agent for the monofunctionalization of the 2,2-disubstituted propionamides (65) using the $Fe(acac)_3$/Ph-dppen catalytic system (Scheme 6.24).[56]

Ackermann and coworkers extended this chemistry to the β–functionalization of aliphatic amides (67) bearing TAM as a 1,2,3-triazole-based bidentate DG (Scheme 6.25).[33,34] As in the previous case, the activation of the primary C–H bonds is overwhelmingly preferred, and a further functionalization on the secondary, benzylic C–H bond of the product did not occur.

Furthermore, the group of Nakamura developed a method for the arylation of Csp^3–H bonds of allylic substrates (69) and deactivated cyclic alkanes such as cyclohexane (70) (Scheme 6.26).[57] The reaction was cried out using $Fe(acac)_3$ as catalyst together with xantphos and mesityl iodide (MesI) as ligands in THF, obtaining the

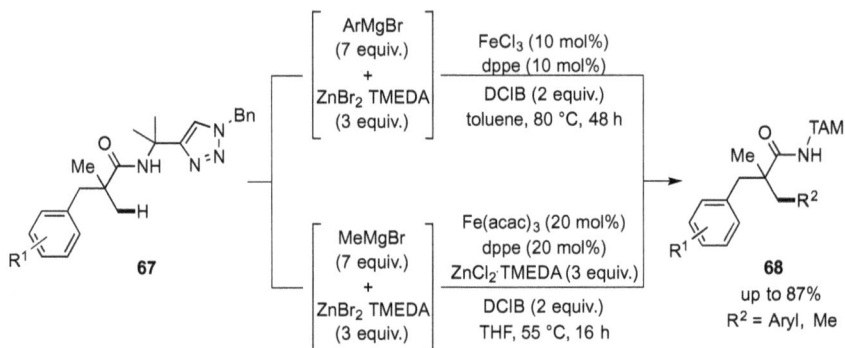

Scheme 6.25.   β-Functionalization of amides with 1,2,3-triazole DG.

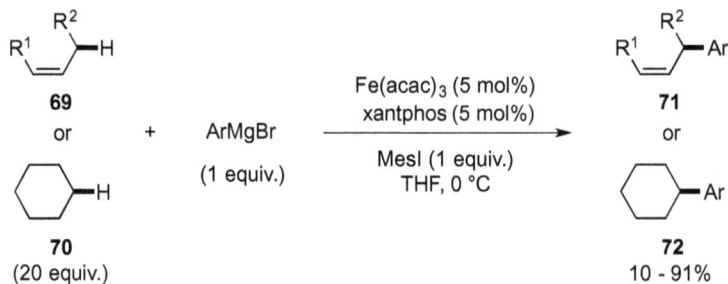

Scheme 6.26.   Fe-catalyzed arylation of allylic compounds and cyclohexane.

corresponding products (71)–(72) in a range from 10%–91% yield. These results already show that the Fe-catalyzed Csp³–H activation is more efficient when using a Zn cocatalytic system as presented before.

6.2.3 Oxidative C–H functionalization: C–C bond formation

The cross-dehydrogenative coupling (CDC) reaction has emerged as an attractive methodology for the formation of a new C–C bond from the direct coupling of two C–H bonds.[58] Although formally a H_2 molecule would be liberated, in most of the cases the hydrogen

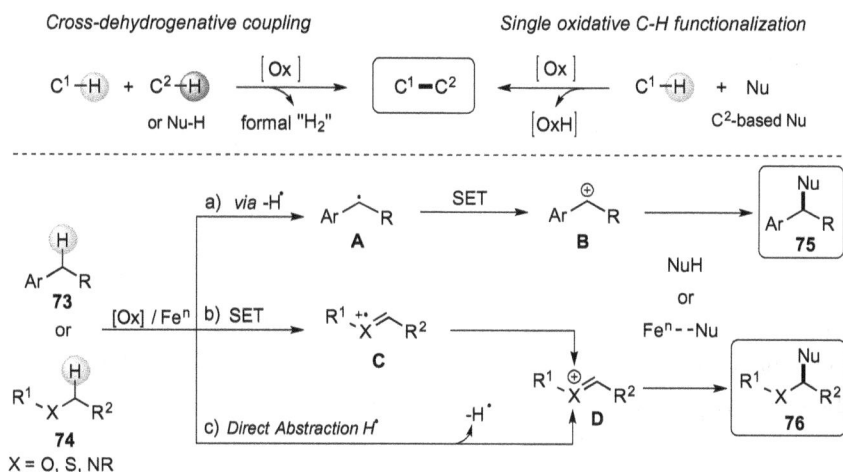

Scheme 6.27.   Oxidative approaches for C–C bond formation.

atoms cleaved during the reaction are trapped by the employed oxidant (Scheme 6.27, top-left).[58,59] An alternative approach in oxidative C–H functionalization is the coupling of a single C–H bond with a carbon-based nucleophile giving the coupling adduct (Scheme 6.27, top-right).

Depending on the nature of the substrate and the oxidant, the C–H functionalization can occur through different possible pathways (Scheme 6.27, bottom). Thus, in the case of the benzylic substrates (**73**), the formation of the corresponding benzyl radical (**A**) *via* hydrogen radical (H˙) abstraction by the oxidant takes commonly place (Scheme 6.27, path a). The generated carbon radical (**A**) is then further oxidized by the Fe-catalyst through a SET to give the cation (**B**).

Concerning the oxidative functionalization of Csp^3–H bonds adjacent to a heteroatom (**74**), two main pathways toward the formation of the resonance-stabilized carbenium adduct (**D**) can be proposed: through a i) Fe-catalyzed SET oxidation to give a radical cation intermediate (**C**), which is further oxidized into (**D**) (Scheme 6.27, path b), or a ii) direct hydride abstraction (Scheme 6.27, path c). Finally, the corresponding cationic adducts (**B**) or (**D**) reacts with the C-based nucleophile, activated or not by an ion-Lewis acid, to form the new C–C bond in (**75**) and (**76**), respectively.

In this section, the reported work on iron-catalyzed CDC reactions is presented and classified according to the different substrate classes.

6.2.3.1 *Oxidation of benzylic and allylic C–H bonds*

In 2007, Li, and Li reported the $FeCl_2$-catalyzed functionalization of benzylic substrates (**73**) with 1,3-dicarbonyl compounds (**77**) using di-*tert*-butyl peroxide (DTBP) as oxidant, providing the corresponding products (**78**) in moderate to excellent yields (up to 87%, Scheme 6.28, left-top).[60] In this reaction, the iron plays a double role as redox catalyst and Lewis acid. The authors postulated as plausible mechanism (Scheme 6.28, right), in which the iron catalyst (**A**) is responsible of the reductive homolytic cleavage of the peroxide into the *tert*-butoxyl radical, while a Fe(III) Lewis acid intermediate (**B**) activates the 1,3-dicarbonyl forming an enolate-type nucleophile (**C**). The *tert*-butoxyl radical abstracts then a hydrogen radical of the substrate forming the benzyl radical (**D**), which react with the Fe(III)-enolate to lead the product and regenerate the Fe(II) catalyst (**A**).

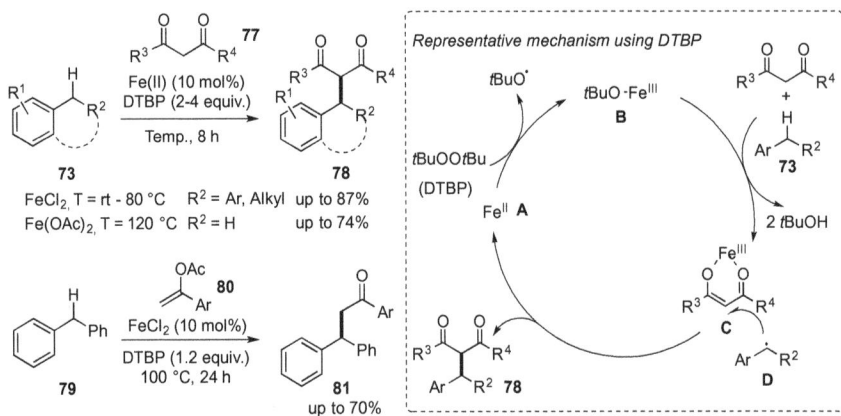

Scheme 6.28.　Fe-catalyzed CDC reaction of benzylic C–H bonds with peroxides.

Later on, Li extended the coupling reaction to toluene derivatives (**73**), R^2 = H) using $Fe(OAc)_2$ as catalyst.[61] Several toluenes were efficiently coupled with 1,3-dicarbonyl compounds obtaining the products in up to 74% isolated yield. Similarly, Gan and Shi described the $FeCl_2$-catalyzed functionalization of benzylic C–H bonds of diphenylmethane (**79**) with vinyl acetates (**80**) using DTBP as oxidant at 100 °C (Scheme 6.28, left-bottom).[62]

In 2009, Shi and coworkers carried out a cross-dehydrogenative arylation (CDA) of benzylic compounds with electron-rich arenes catalyzed by $FeCl_2$ using 2,3-dichloro-5,6-dicyano-1,4-benzoquinone (DDQ) as oxidant (Scheme 6.29, top-left).[63] Thus, the authors successfully achieved the coupling with a wide range of arenes and diarylmethanes (**73**) leading to (**82**) in 61%–96% yield. As mechanism, they postulated that the role of the iron was to assist the two consecutive SET oxidation steps to form fisrt the benzyl radical (**A**) and then the corresponding carbocation (**B**), which reacts then with the arene through a Friedel–Crafts type reaction *via* (**C**). Embracing similar reaction conditions ($FeCl_2$/DDQ), in 2014, Song and coworkers developed a methodology for the coupling of toluene derivatives (**83**) with 1,3-dicarbonyl compounds (**77**) (Scheme 6.29,

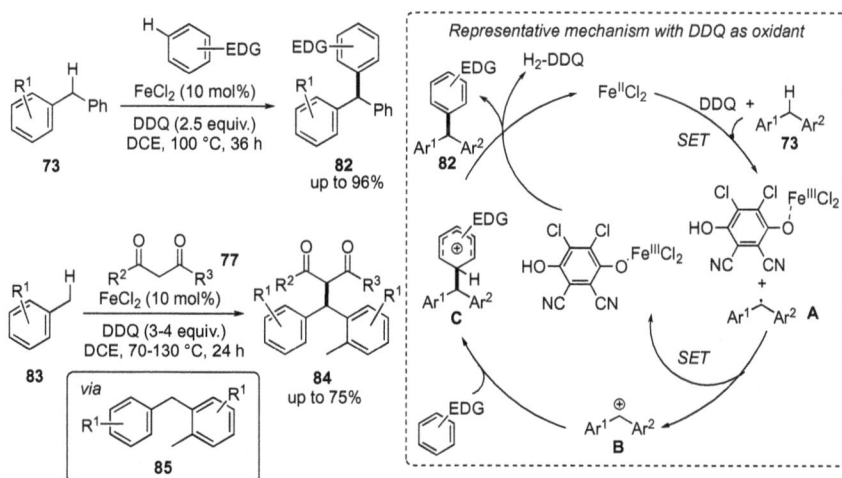

Scheme 6.29. $FeCl_2$-catalyzed CDC reaction of benzylic C–H bonds using DDQ.

Scheme 6.30.   Fe-catalyzed benzoannulation reaction *via* allylic oxidation.

bottom-left).[64] The authors suggested that the process is initiated by homocoupling of two aryl methane molecules to produce the corresponding diaryl methylene intermediates (85), followed by an analogous CDC reaction with the Lewis acid-activated nucleophile to give the final products (84).

A similar functionalization of benzylic compounds has been recently reported by the group of Arnold. In this case, an enzymatic Fe-porphyrin-catalyzed enantioselective benzylic C–H alkylation with diazoacetate is described, obtaining the alkylated products in up to 99:1 enantiomeric ratio (e.r.).[65]

Regarding the direct activation of allylic C–H bonds, scarce examples are reported there in the literature. Hence, in 2012, the group of Deng developed an $FeCl_3$-catalyzed CDC/benzoannulation tandem reaction of 1,2-aryl-propenes (86) with styrenes (87) using DDQ as oxidant for the synthesis of biologically polysubstituted naphthalene derivatives (88) in up to 80% yield (Scheme 6.30).[66]

6.2.3.2 *Csp³–H bonds alpha to oxygen*

Another type of Csp³–H bonds that can be successfully applied in CDC reactions are the one in alpha to a heteroatom such as oxygen or nitrogen (see Section 6.2.3.3). These a-C–H bonds can be readily oxidized to form the active oxonium or iminium ion intermediates, which can react with a variety of nucleophiles.[58,67]

Focusing on Csp³–H bonds alpha to oxygen, in 2008, Li and coworkers reported the direct a-Csp³–H functionalization of ethers and thioethers (74)[68] with dicarbonyl compounds such as ethyl

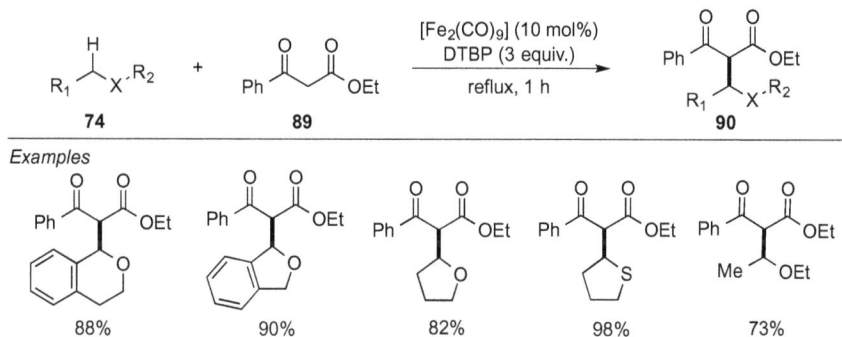

Scheme 6.31. CDC of linear and cyclic (thio)ethers catalyzed by [Fe$_2$(CO)$_9$].

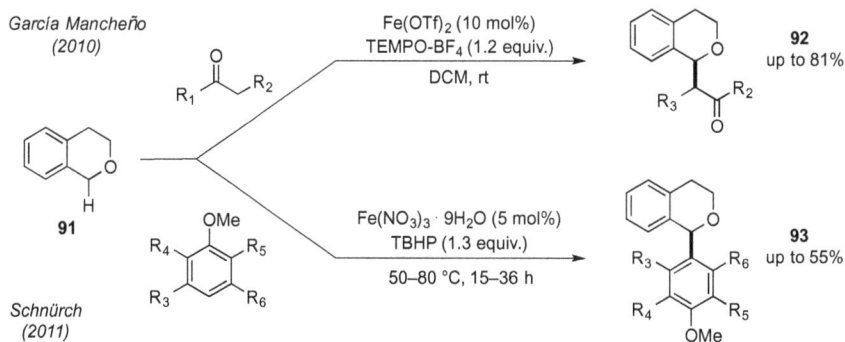

Scheme 6.32. Functionalization of isocromanes.

benzoyl acetate (**89**) catalyzed by Fe$_2$(CO)$_9$ using DTBP as oxidant (Scheme 6.31). With respect to the substrate scope, both cyclic and acyclic ether derivatives give the corresponding products (**90**) from moderate to excellent yields.

In 2010, the group of García Mancheño introduced an oxoammonium salt (TEMPO-BF$_4$) as mild, efficient hydride-abstractor for the Fe(OTf)$_2$-catalyzed oxidative a-C(sp^3)–H functionalization of isochromanes (**91**) with enolizable carbonyl compounds (Scheme 6.32, top).[69] In this case, the reactive oxonium intermediate was formed directly without going through a radical species, leading to the coupling products (**92**) in moderate to good yields. One year later,

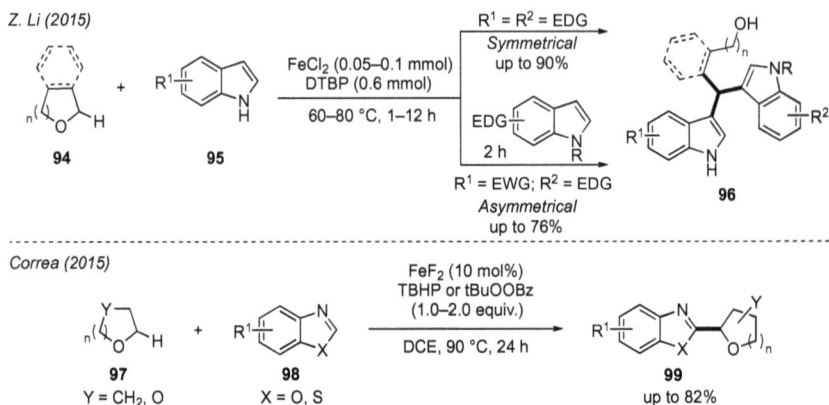

Scheme 6.33.   Iron-catalyzed α-C–H bond (hetero)arylation reactions of ethers.

Schnürch and coworkers presented the arylation of isochromanes (91) with electron-rich arenes using $Fe(NO_3)_3 \cdot 9H_2O$ as catalyst and *tert*-butyl hydroperoxide (TBHP) as oxidant (Scheme 6.32, bottom).[70]

This chemistry was also extended to the oxidative a-Csp^3–H functionalization of ethers with electron-rich arenes and heteroarenes (Scheme 6.33). In this regard, Li reported the iron-catalyzed oxidative coupling of ethers (94) with indoles (95) following a Friedel–Crafts reaction.[71] Interestingly, besides symmetric products, asymmetric 1,1-bis(indolyl)methane derivatives (96, $R^1 \neq R^2$) could be selectively obtained when electron-deficient and electron-rich indoles are used sequentially (Scheme 6.33, top).

More recently, the group of Correa developed a new approach for the direct α-heteroarylation of cyclic and acyclic ethers (97) with azoles (98) employing FeF_2 as catalyst and TBHP as oxidant (Scheme 6.33, bottom).[72] The C2-substitution product (99) was obtained as the major isomer in up to 82% yield. The authors proposed that the FeF_2 catalyst assists the reductive homolytic cleavage of the oxidant.

6.2.3.3 Csp³–H bonds alpha to nitrogen

The typical type of substrates employed on the Fe-catalyzed oxidative functionalization of Csp^3–H bonds adjacent to nitrogen

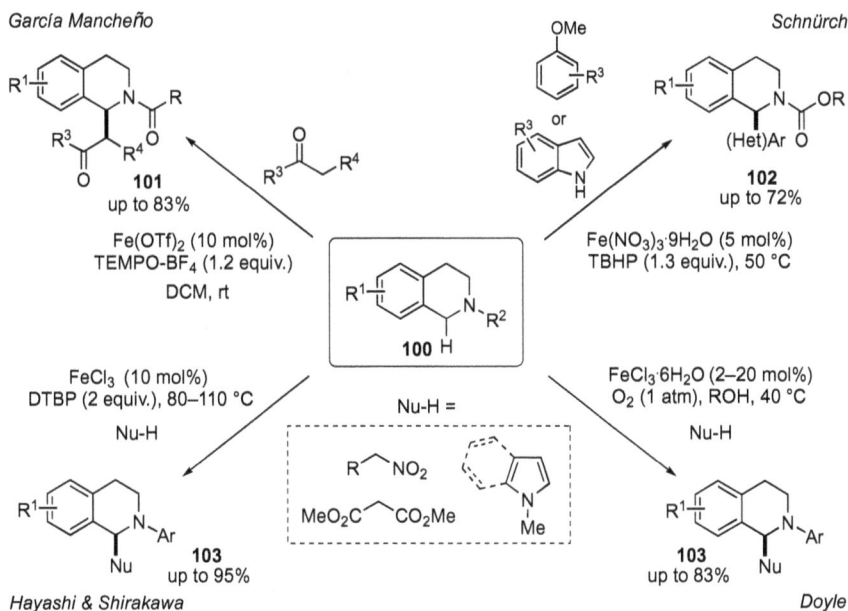

Scheme 6.34. Fe-catalyzed the CDC functionalization of tetrahydroquinolines.

comprise i) tetrahydroquinolines (THIQs), ii) N,N-dimethyl- or N-methylaniline, iii) aryl glycine derivatives, and iv) formamides or *N*-methylformamides.

Starting with THIQs (**100**) as substrates, García Mancheño[69] and Schnürch[70] presented separately the iron-catalyzed C–H functionalization with enolizable carbonyls and electron-rich arenes to provide substituted THIQs (**101**) and (**102**), respectively (Scheme 6.34, top). A similar work was presented by Hayashi, Shirakawa, and cow-orkers, using the FeCl$_3$/DTBP catalyst/oxidant system with nitroal-kanes, activated methylenes and indoles as coupling partners (Scheme 6.34, bottom-left).[73] Moreover, the group of Doyle pre-sented in 2013 the FeCl$_3$·6H$_2$O-catalyzed aerobic C–H oxidation for the Mannich reaction of THIQs with the corresponding nitroal-kanes, *N*-substituted indoles, and 2,3-dihydro-1*H*-pyrrole obtaining the corresponding α-substituted adducts (**103**) in good to excellent yields (Scheme 6.34, bottom-right).[74]

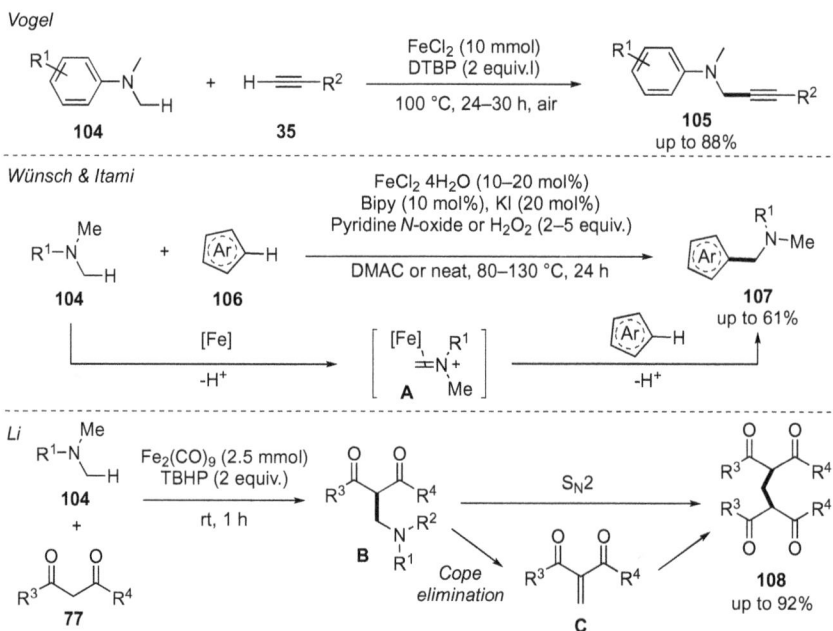

Scheme 6.35.   Fe-catalyzed the oxidative coupling of N-methyl anilines.

Continuing with mono- and di-methylanilines, Vogel and coworkers reported in 2009, the direct alkynylation of N-methyl tertiary amines (104) with terminal alkynes (36) to form the corresponding propargyl amines (105) (Scheme 6.35, top).[75] This transformation was carried out under solvent-free conditions in the presence of FeCl$_2$ as catalyst and DTBP as oxidant with excellent results. Moreover, the C–H alkylation of N-methylanilines has been recently studied by the group of Cai, employing the FeCl$_2$/3,4,7,8-tetramethyl-1,10-phenanthroline catalitic system and TBHP as oxidant, leading to the alkylated anilines derivatives in 43%–75% yield.[76] In 2009, Wünsch and Itami developed an iron-catalyzed oxidative coupling of N-methyl amines (104) with heteroarenes (106) using the FeCl$_2$·4H$_2$O/pyridine N-oxide system (Scheme 6.35, middle).[77] The oxidative coupling reaction was applied to the intermolecular coupling of thiophenes, furans, and indoles, obtaining predominantly the corresponding arylated adduct (107) at the N-methyl C–H bonds over other reactive bonds. Although the mechanism of this transformation is not well known, the author proposed the formation of an iron-activated iminium species (**A**), which can be trapped by nucleophilic heteroarenes giving the corresponding benzylic amines. The groups of Hayashi and Shirakawa,[73] and Doyle,[74] also succeeded to extend their methodology for THIQs to *N*-methyl tertiary amines and various electron-rich (hetero)arenes and silyloxy-furanes, respectively.

Additionally, Li and coworkers reported the use of *N,N*-dimethylanilines (**104**) as a one-carbon source by reaction with 2 equiv. of a 1,3-dicarbonyl compound (**77**) under $Fe_2(CO)_9$ catalysis (Scheme 6.35, bottom).[78] Under these conditions methylene-bridged dicarbonyl products (**108**) are formed in yields up to 92%. The authors proposed the formation of the intermediate (**B**) by CDC coupling, followed by a SN2 substitution by the second 1,3-dicarbonyl compound, in which the *N*-methylaniline unit acts as leaving group. The other reasonable pathway may be initiated by a Cope elimination from intermediate (**B**) to create the corresponding Michael acceptor (**C**), which give the final adduct by subsequent Michael addition.

A few examples of the C–H oxidation/functionalization of glycine derivatives (**109**) have been reported (Scheme 6.36). In 2011, the group of García Mancheño reported the one-pot synthesis of substituted quinolines (**110**) from *N*-aryl glycines (**109**) by a $FeCl_3$-catalyzed dehydrogenative Povarov/oxidation tandem reaction with mono- and 1,2-disubstituted aryl or alkyl olefins using TEMPO oxoammonium salt as mild and nontoxic oxidant (Scheme 6.36, left).[79] The same group further described the $Fe(OTf)_2$-catalyzed synthesis of dihydroquinazolines (**111**) by homocoupling.[80] Alternatively, the group of Hu presented in 2012, a $FeCl_3$-catalyzed oxidative coupling of *N*-arylglycines (**109**) with alkynes and cyclic ketones employing DTBP and DDQ as oxidants, respectively (Scheme 6.36, right).[81] The application of these methods provides the corresponding substituted quinolines (**112**) (up to 88% yield) and α-functionalized *N*-arylglycines (**113**) (up to 83% yield).

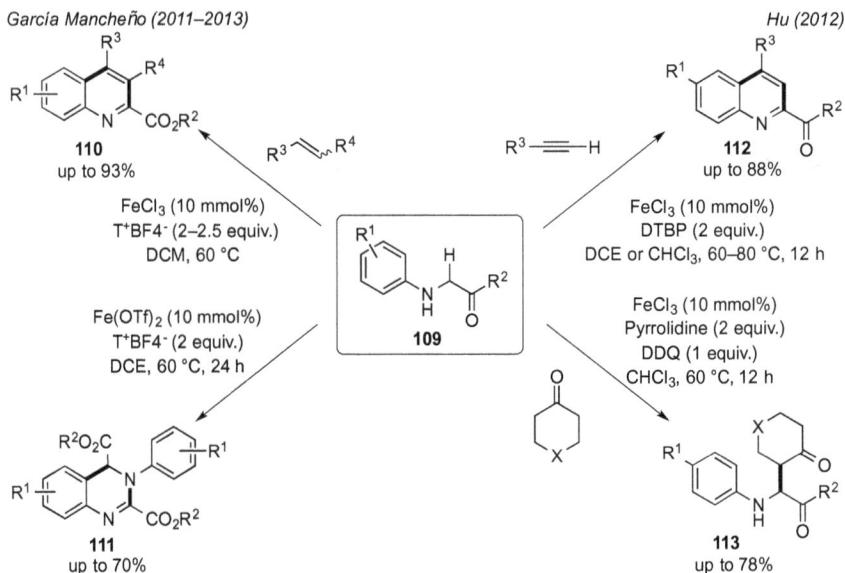

Scheme 6.36. Fe-catalyzed the oxidative C–H coupling of N-arylglycines.

Furthermore, N,N-dimethyl formamide (**114**, R^1 = H) and acetamide (**115**, R^1 = Me) have recently been successfully employed in CDC reactions as methylene sources. In this context, the groups of Xu and Li reported independently the $FeCl_3 \cdot 6H_2O$-catalyzed methylenation of benzylic Csp^3–H bonds (**116**)[82] and ketones (**117**)[83] with N,N-dimethylacetamide (**115**) and potassium persulfate ($K_2S_2O_8$) as oxidant, respectively (Scheme 6.37). These reactions involve the transfer of one methyl moiety of the N,N-dimethylacetamide (**115**), followed by elimination to generate the vinyl adducts (**119**) and (**120**) in up to 92% yield.

Similarly, the group of Wang reported in 2013 the iron-catalyzed vinylation of 2-methylquinolines (**118**) with and N,N-dimethyl formamide (**114**) in the presence of TBHP as oxidant (**121**) up to 79%).[84]

6.2.3.4 Csp³–H bonds alpha to a nitrile group

In 2016, Kim, Wu, and coworkers presented a method for the synthesis of α-cyanomethyl-β-dicarbonyl compounds (**124**) from

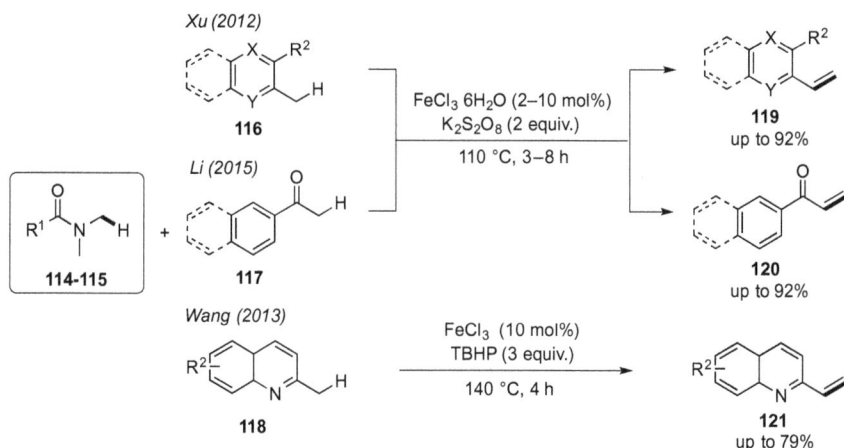

Scheme 6.37. Fe-catalyzed Csp^3–H bond methylenation with N,N-dimethyl amides.

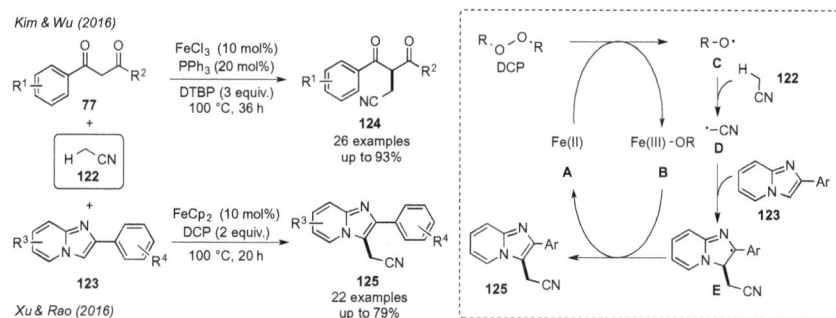

Scheme 6.38. Fe-catalyzed CDC with acetonitrile.

acetonitrile (**122**) employing FeCl$_3$ as catalyst, PPh$_3$ as ligand, and DTBP as oxidant (Scheme 6.38).[85] The application of this methodology well-tolerates a wide range of aryl-substituted 1,3-dicarbonyl compounds with different stereoelectronic properties, as well as hererocyclic substituents such as thiophene or furane giving up the products (**124**) in up to 93% yield. One year later, Xu, Rao, and coworkers reported the dehydrogenative Csp^2–Csp^3 coupling of acetonitrile (**122**) with 2-arylimidazo[1,2-a]pyridines (**123**). The reaction is catalyzed by FeCp$_2$ using DCP as oxidant, in which electron-rich imidazoles reacted significantly better than electron-poor deriva-tives (61%–79% vs. 24%–49% yield).[86]

As plausible mechanism, the authors proposed that the Fe(II) catalyst (**A**) reacts with DCP, generating the Fe(III) complex (**B**) and an alkoxy radical (**C**). This radical abstracts then a hydrogen atom from the acetonitrile giving the cyanomethyl radical (**D**), which is well-suited to add to 2-arylimidazo[1,2-*a*]pyridines (**X**) to form the radical (**E**). The final product (**125**) is afforded after the oxidation with (**B**) and subsequent deprotonation, regenerating the iron(II) catalyst and closing the catalytic cycle (Scheme 6.38, right).

6.2.3.5 *Functionalization of unactivated alkylic C–H bonds*

The only example reported to date with simple alkanes was pre-sented in 2007 by Li and Zhang (Scheme 6.39).[87] Hence, the alkylation of 1,3-dicarbonyl compounds (**77**) with inactive cycloal-kanes (**126**) was achieved under catalytic condition using the $FeCl_2 \cdot 4H_2O$ salt in the presence of DTBP as oxidant at 100 °C. This transformation follows the same general mechanism already shown with organic peroxides, in which the iron(II) salt initially catalyzes the hemolytic cleave of the oxidant to give an alkoxy radical and the Fe(III) complex that acts as Lewis acid to activate the dicarbonyl nucleophile (**A**). Under those conditions, the alkyl radical could be formed and trapped with the Fe-enolate to form the alkylated β-keto ester (**127**) and regenerate Fe(II) species.

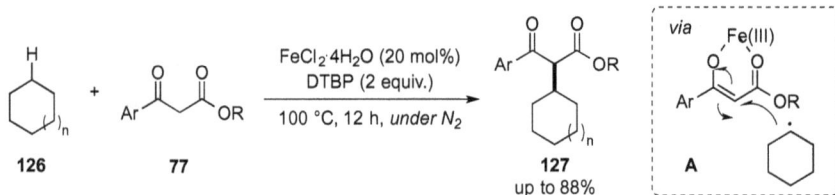

Scheme 6.39.   Fe-catalyzed oxidative C–H functionalization of cycloalkanes.

6.2.3.6 *Other reactions with aromatic and alkynylic substrates*

The group of Li carried out in 2009 the synthesis of substituted benzo-furans (**129**) through an $FeCl_3 \cdot 6H_2O$-catalyzed oxidative coupling and subsequent annulation reaction using phenols (**128**) and β-keto esters (**77**) using DTBP as oxidant (Scheme 6.40).[88] The postulated catalytic cycle starts with the formation of the intermediate (**A**), in which an iron salt [Fe^n] activates both the carbonyl and the phenolic group and facilitates the *ortho*-C–C bond formation. A reductive elimination giving intermediate (**B**), and subsequent tautomerization and condensation, provide the corresponding benzofuran derivatives in up to 75% yield.

Later on, Hajra and coworkers developed the $FeCl_3$-catalyzed dicarbonylation of imidazoheterocycles (**123**) or (**131**) with oxoal-dehydes (**130**) under ambient air, obtaining the coupling products (**132–133**) in high yields (up to 87% yield) (Scheme 6.41).[89] The

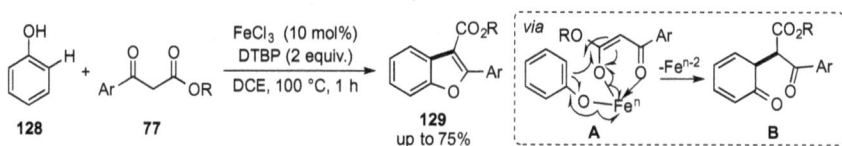

Scheme 6.40.　Fe-catalyzed the synthesis of substituted benzofurans *via* CDC.

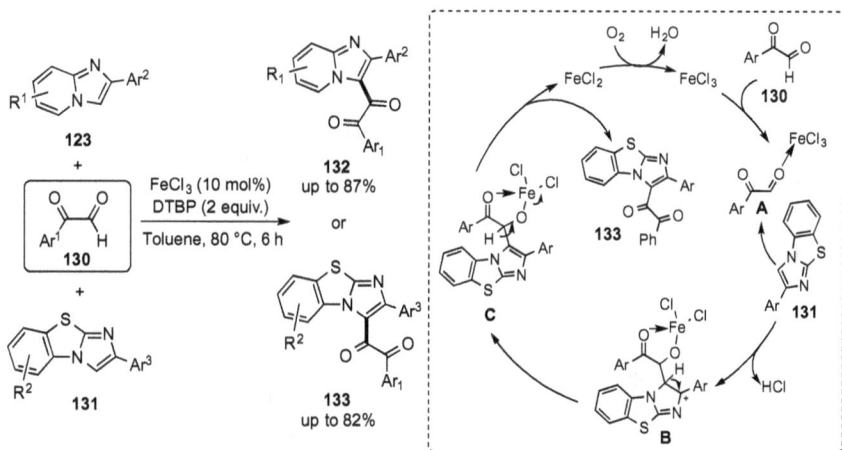

Scheme 6.41.　Fe-catalyzed dicarbonylation of imidazoheterocycles.

authors proposed a nonradical pathway, in which FeCl$_3$ catalyzed the addition of the arylglyoxal (**130**) to benzo[*d*]imidazo[2,1-*b*]thiazoles (**131**) at the C-3 postion to produce the Fe(III)-chelated intermediate (**A**. Fe(III) is converted into the oxo-Fe(IV) complex (**B**) *via* aerobic oxidation. After aromatization to (**C**), the product (**133**) is obtained through reductive elimination and Fe(II) is reoxidized to Fe(III) in the presence of oxygen to complete the catalytic cycle.

In 2017, Yu and coworkers reported the C–H alkylation of *S,S*-functionalized internal olefins (**134**) mediated by a catalytic amount of FeCl$_3$ using DTBP as the oxidant and 1,4-Diazabicyclo[2.2.2] octane (DABCO)·6 H$_2$O as additive (Scheme 6.42).[90] The alkylthio functionality was essential for the internal olefinic C–H bond to undergo alkylation with ethers (**135**) and benzylic C–H bonds (**136**) such as toluene derivatives. Moreover, the obtained tetrasubstituted olefins (**137–138**) could be employed for the formation of highly substituted pyrazoles and isoxazoles. The same group also presented the Fe(OAc)$_2$-catalyzed regioselective radical oxidative annulation of *S,S*-functionalized internal olefins (**134**) with β-ketoesters (**77**) for

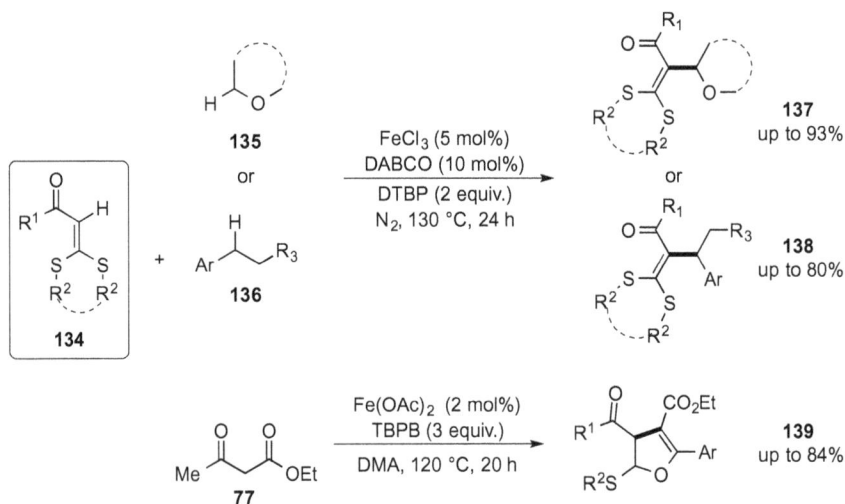

Scheme 6.42. Fe-catalyzed the C–H functionalization of *S,S*-functionalized olefins.

the synthesis of tetrasubstituted furans (**139**) (Scheme 6.42).[91] Regarding the mechanism of these reactions, the authors suggested a radical pathway for the Csp^2–H/Csp^3–H cross-coupling.

Concerning the role of iron catalyst, it sometimes acts as chelating agent, favoring the C–H functionalization step, instead of being directly involved in the C–H cleavage through insertion. In this regard, various FeCl$_3$-catalyzed chelation-induced C5-functionalization reactions of 8-aminoquinolines were developed (Scheme 6.43).

Accordingly, the halogenation of 8-amido quinolines (**20**) at the C5-position catalyzed by FeCl$_3$ using PhI(OAc)$_2$ as oxidant was reported by Sheng and Zhang.[92] Substituted arenes with electron withdrawing or electron donating groups, as well as non- and cyclic alkanes, behave well in this reaction providing the corresponding halogenated products (**140**) in up to 83% yield (Scheme 6.43, top). Furthermore, Zeng, Fe, and Sheng reported similar allylation[93] (Scheme 6.43, middle) and benzylation reactions employing *N*-benzylic sulfonamide[94] or benzylic acetates[95] (Scheme 6.43,

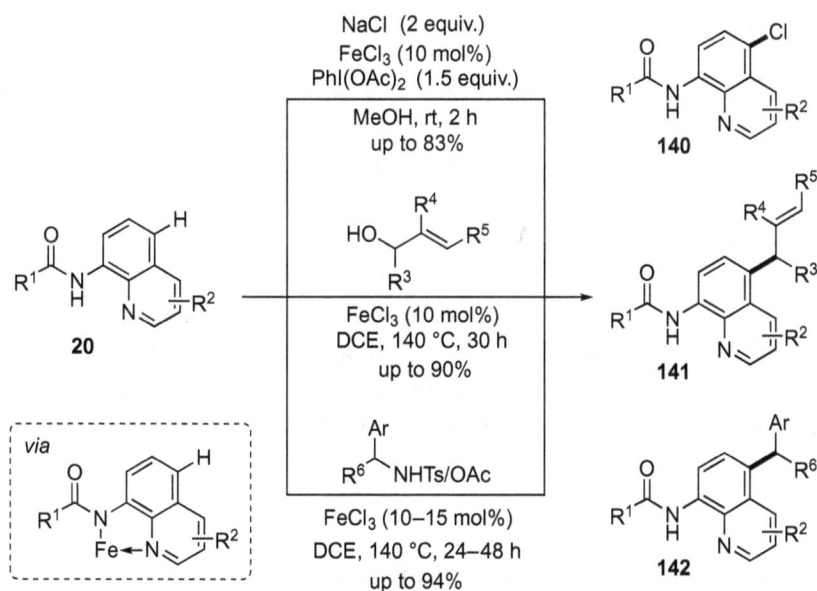

Scheme 6.43. Fe-catalyzed C5-functionalization of 8-amido quinolones.

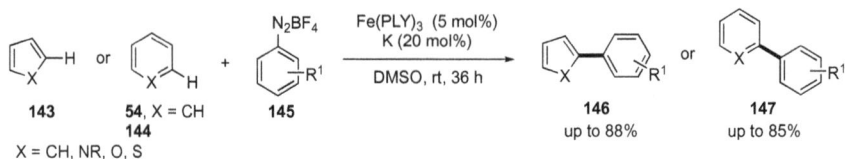

Scheme 6.44.   C–H arylation of (hetero)arenes with diazonium salts.

bottom), leading to the corresponding products (**141–142**) in up to 90% and up to 94%, respectively.

In 2018, Mandal presented the synthesis of a Fe(III)-based catalyst bearing a phenalenyl (PLY) ligand and catalytic potassium metal as external reducing agent to carry out the C–H arylation of (hetero) arenes compounds (**143–144**) with a variety of aryl diazonium salts (**145**) (Scheme 6.44).[96] The catalyst, which can be involved in reversible multielectron redox processes due to the PLY ligands, could be used in one-pot consecutive catalytic runs (until 10 times) without notably loss of the catalytic activity.

6.2.4 *C–N bond formation*

C–N bond formation through C–H activation/functionalization plays an important relevance in medicinal and pharmaceutical industry in the synthesis of bioactive building blocks.[97] However, the amination *via* iron-catalyzed C–H bond cleavage has been hardly studied so far until the past few years. This type of transformation can be carried out by three different catalytic approaches employing: i) a C–H bond activation, ii) the formation of a metal–nitrenoid species that reacts with the substrate in a concerted or stepwise manner,[98] or iii) a C–H oxidation to the corresponding cationic or radical intermediate species (Scheme 6.45).[99]

6.2.4.1 *Via C–H bond activation*

An important contribution in this field was made by Nakamura and coworkers, who reported, in 2014, the Fe-catalyzed *ortho*-amination of aromatic carboxamides (**20**) bearing the 8-quinonyl DG with *N*-chloro morpholine (**148**) and anilines (**149**) (Scheme 6.46,

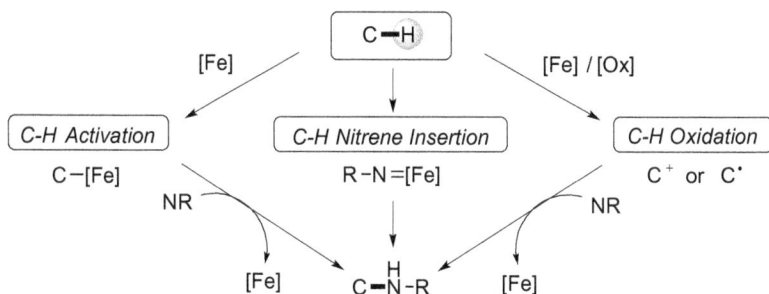

Scheme 6.45.   Activation modes for the C–N formation from C–H bonds.

Ortho amination of aromatic carboxamides

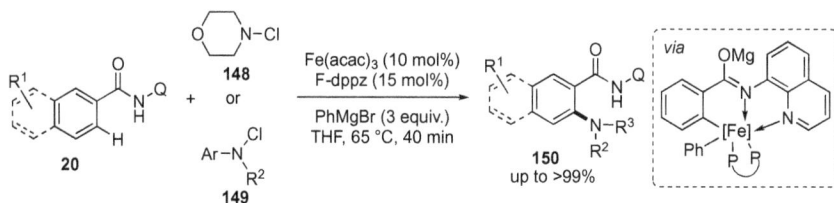

Intramolecular C–H amination

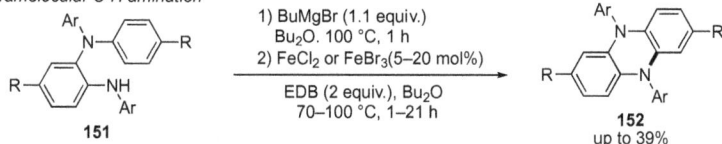

Scheme 6.46.   Fe-catalyzed aromatic C–H bond amination reactions.

top).[100] In 2015, the same group reported the intramolecular C–H amination of benzenediamines (**151**) for the synthesis of dihydrophenazines (**152**), however, harsher conditions (70–100 °C) were required to obtain only moderate results (Scheme 6.46, bottom).[101]

Other interesting example imploying a Fe(II)-catalyzed C–H activation/amination process was reported by Xu in 2017 (Scheme 6.47).[102] This reaction proceeds by a sulfur directed Csp^3–H bond amination through an initial Csp^2–S cross-coupling reaction between iodo- or bromoanilines (**153**) and benzylthiols (**154**) obtaining the corresponding adducts (**155**) up to 95% yield. The authors proposed that the iron complex (**A**), formed *in situ* from Fe(OAc)$_2$, undergoes oxidative addition into the C-halogen bond forming the Fe(III)

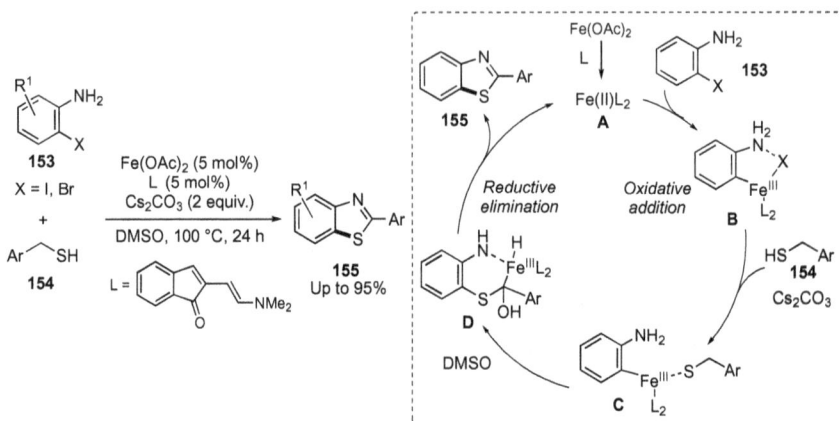

Scheme 6.47.   Csp^3–H bond amination *via* initial Csp^2-S cross-coupling reaction.

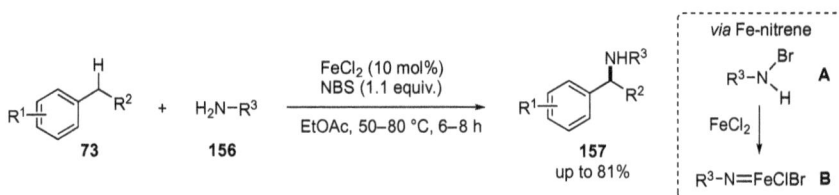

Scheme 6.48.   Fe-catalyzed amidation of benzylic Csp^3–H bonds with NBS.

intermediate (**B**). Next, complexation to benzylthiol (**C**) and subsequent intermolecular oxidative addition leads to chelate (**D**), which evolves to the coupling product upon final reductive elimination.

6.2.4.2 *Via metal-nitrenoid C–H insertion*

In 2008, Fu and coworkers reported a FeCl$_2$-catalyzed amidation of benzylic Csp^3–H bonds (**73**) with primary amines (**156**) in presence of *N*-bromosuccinimide (NBS) as oxidant.[98f] Under the optimized reaction conditions, up to 81% on the benzylic amides (**157**) were achieved. The proposed mechanism of the reaction involves the initial bromination of the amine to (**A**) and subsequent formation of the active iron-nitrene complex (**B**), which inserts in the benzylic C–H bond (Scheme 6.48).

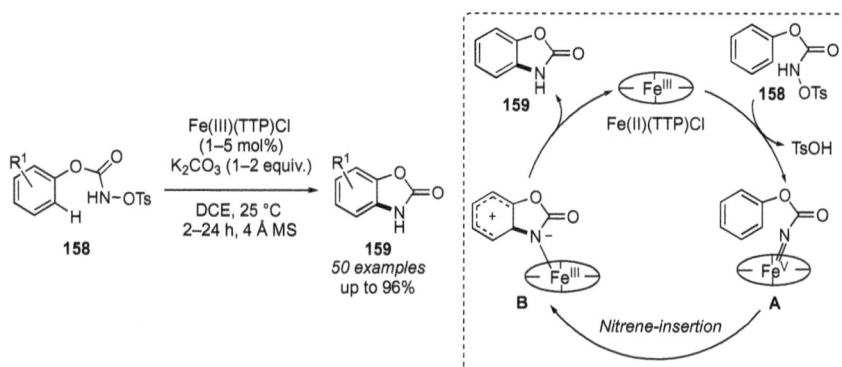

Scheme 6.49. Fe-catalyzed aryl Csp^2–H amidation reaction.

More recently, the group of Singh developed a straightforward synthesis of benzoxazolinones (**159**) *via* an intramolecular Csp^2–H amidation of arylcarbamates (**158**) catalyzed by a porphyrine-Fe(III) complex [Fe(III)(TTP)Cl, Scheme 6.49].[103] The *N*-tosyloxy group in the substrates were used as both nitrene precursor and internal oxidant (N–O bond). As a plausible mechanism, the authors postulated the formation of a Fe(V)–nitrenoid (**A**) as key intermediate, which undergoes the nitrene insertion onto the *ortho*-aromatic C–H bond (**B**).

In 2017, the intramolecular amination of aryl azides (**160**) to form indolines (**161**) catalyzed by the nucleophilic iron complex TBA[Fe] was reported by the group of Plietker (Scheme 6.50, top).[104] Similar transformations implying a Fe-nitrene intermediate (**A**) were also reported by Vlugt and Bruin,[105] by Betley[106] and coworkers, and by Che.[107] In these cases, the synthesis of substituted pyrrolidines (**163**) through an intramolecular C–H amination with aliphatic azides (**162**) and (**164**) employing an iron(II) complex (Fe-cat1 and Fe-cat2, respectively) and a redox active ligand was described (Scheme 6.50, middle). Moreover, in the case of Batley, a high selective diastereoselective method in favor of the *syn* products was developed. Additionally, the group of Che also reported the Fe(III)-porphyrin [Fe(TF$_4$DMAP)Cl] catalyzed intermolecular C–H amination of benzylic and alkylic C–H bonds (**165**) with aryl azides

Intramolecular C-H amination

Plietker (2017)

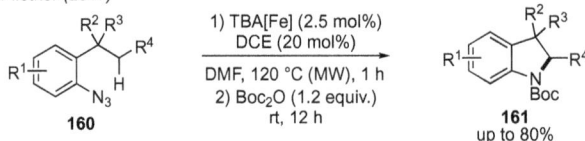

Vlugt & Bruin (2017)

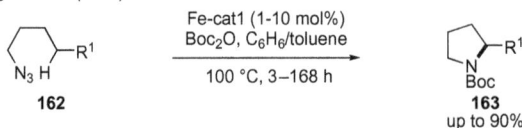

Betley/Che (2017/2019)

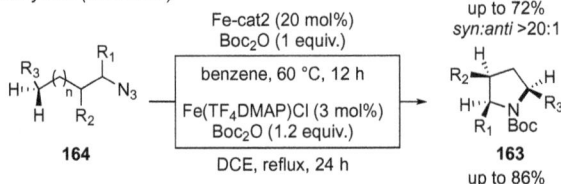

Intermolecular C-H amination

Che (2019)

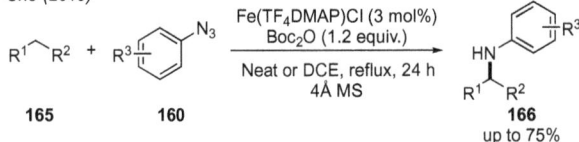

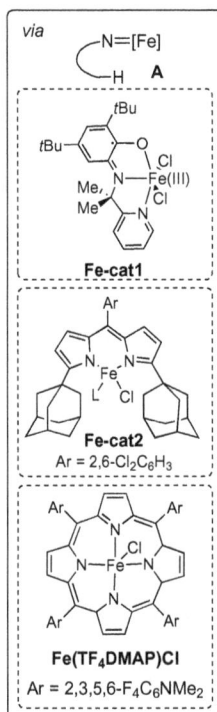

Scheme 6.50.   Fe-catalyzed intramolecular amination of aryl and alkyl azides.

(160) in DCE or under neat conditions, obtaining the amination adducts in up to 75% yield (Scheme 6.50, bottom).[107]

6.2.4.3 *Via C–H bond oxidation*

In 2010, Li and coworkers reported an $FeCl_3 \cdot 6H_2O$-catalyzed *N*-alkylation of azoles by oxidation of an α-Csp^3–H bond adjacent to an oxygen atom using TBHP as oxidant under neutral conditions (Scheme 6.51, top).[108] A wide variety of azoles (167) and ethers (135) afford the corresponding oxidative coupling products (168) in good to excellent yields (up to 97%).

Li (2010) / Wang & He (2015)

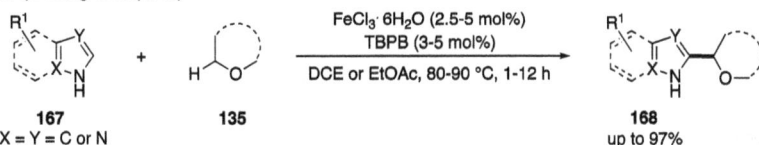

Chen (2011-2012)

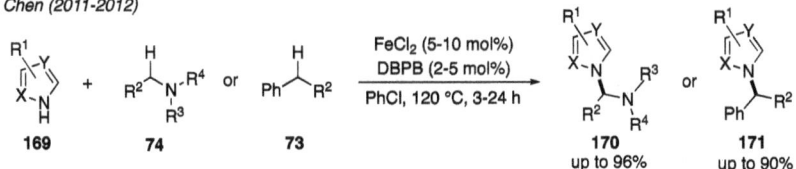

Reddy (2011)

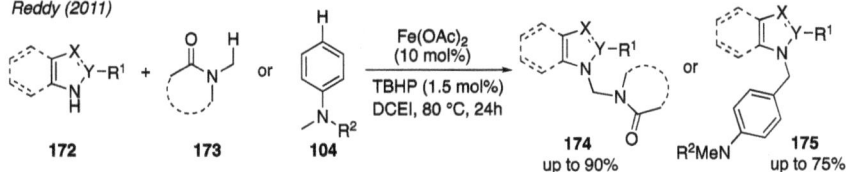

Scheme 6.51.　Fe-catalyzed *N*-alkylation of azoles.

Additionally, Wang, He, and coworkers presented in 2015 a similar transformation using aryl tetrazoles as the coupling partners.[109] Another related work was reported by the group of Chen, in which the *N*-alkylation of azoles (**169**) *via* oxidative cleavage of the α-Csp³–H bond of amides and sulfonamides (**74**),[110] or benzylic hydrocarbons (**73**)[111] catalyzed by FeCl₂ using DTBP as oxidant (Scheme 6.51, middle). Furthermore, the group of Reddy applied a similar methodology using *N*-alkylamides (**173**) and *N*-methyl anilines (**104**) (Scheme 6.51, bottom).[112]

On the other hand, the group of Cheng and Zhu studied the FeCl₂/bipyridine-catalyzed direct amination of *N*-substituted pyrrolidin-2-ones (**176**) with anilines (**177**) using TBHP as oxidant, obtaining the corresponding α-aminated adduct (**178**) with high regioselectivity and from moderate to high yields (up to 89%) (Scheme 6.52, top-left).[113] In 2014, Bao and coworkers reported a similar transformation, in this case catalyzed by FeCl₃ in the absence

Chen & Zu (2012)

Chen & Zu (2012)   Yang (2013)

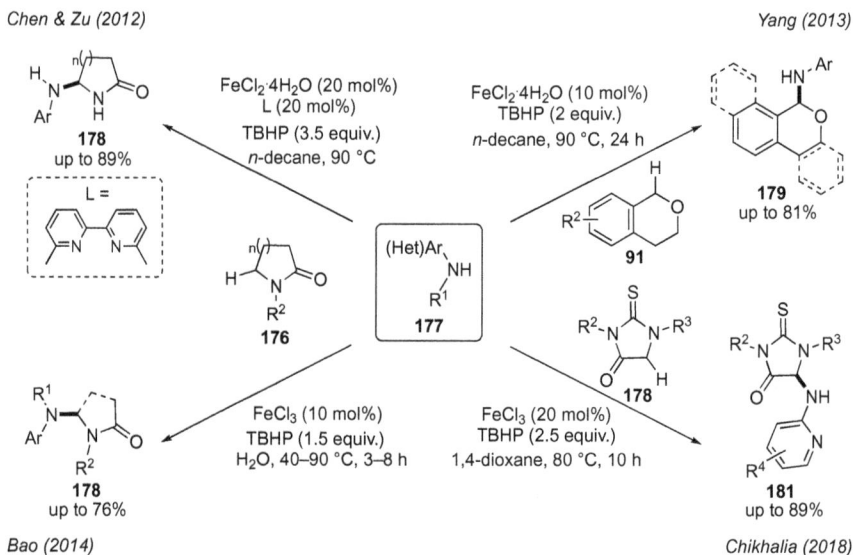

Scheme 6.52.   Fe-catalyzed α-amination of amides and ethers with (hetero)aryl amines.

of a ligand (Scheme 6.52, bottom-left).[114] Anilines (**177**) were also used by Yang and coworkers for the iron-catalyzed direct amination of isochromans (**91**) (Scheme 6.52, top-right).[115] A variety of cyclic hemiaminals (**178**) were obtained in good to excellent yields under mild conditions. More recently, Chikhalia developed an $FeCl_3$-catalyzed hetero-CDC of thiohydantoins (**180**) with 2-aminopyridines using TBHP as oxidant.[116] This method tolerates the presence of a wide range of functional groups and provides the corresponding *N*-substituted amines (**181**) in moderate to good yields (Scheme 6.52, bottom-right).

As alternative, Bolm and coworkers reported in 2014 the use of sulfoximines (**180**) as nitrogen source for Fe-catalyzed C–H amination reactions (Scheme 6.53).[117] Following a similar hetero-CDC methodology, the synthesis of *N*-alkylated sulfoximines (**183**) with branched substituents was achieved (up to 88% yield) by the reaction between NH-sulfoximines (**182**) and diarylmethanes (**73**).

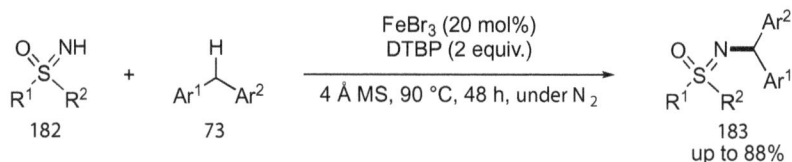

Scheme 6.53. Fe-catalyzed N-alkylation of sulfoximines.

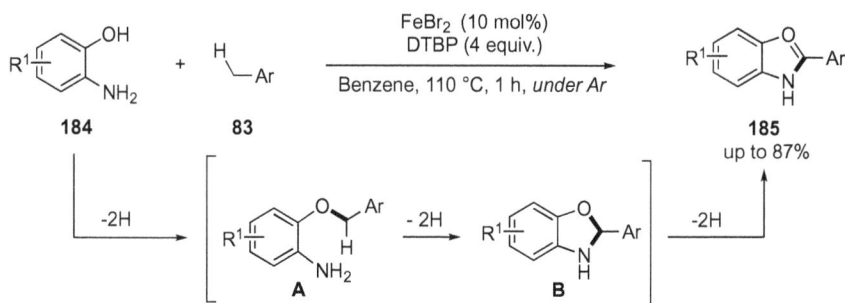

Scheme 6.54. Tandem oxidative cyclization of 2-aminophenols with arylmethanes.

The same year, the group of Gu and Li further developed the iron-catalyzed tandem oxidative cyclization with 2-aminophenols (**184**) and simple arylmethanes (**83**) (Scheme 6.54).[118] The reaction was carried out with FeCl$_2$ as catalyst and DTBP as oxidant, affording a wide variety of substituted benzoxazoles (**185**). In this transformation, the oxidative Csp^3–H/O–H coupling of 2-aminophenols (**184**) with toluene occurred first (**A**), followed by an oxidative Csp^3–H/N–H cyclization (**B**) and aromatization to form the final product.

In 2017, Jiao and coworkers described a mild FeBr$_2$-catalyzed C–H amination of substituted (hetero)arenes (**54**, R^1 = H, **184**) using a hydroxylamine triflate salt (**187**) as redox amination agent and AgNTf$_2$ as additive, giving the corresponding amination adducts (**188**) in up to 90% yield (Scheme 6.55).[119] As possible mechanism, the authors postulate a SET from FeBr$_2$ to the NH$_2$ reagent, which undergoes a N-O bond cleavage forming a N-radical cation (**A**). The

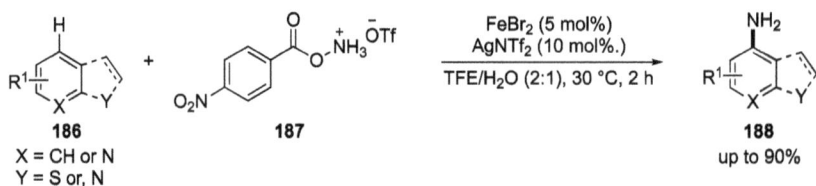

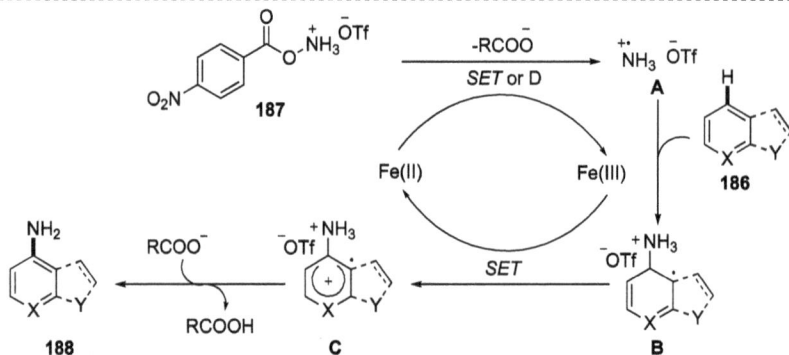

Scheme 6.55.   Fe-catalyzed C–H amination of (hetero)arenes.

radical addition of (**A**) to the arene forms the radical intermediate (**B**), which is oxidized to (**C**) by Fe(III) through a second SET. Finally, the cationic intermediate (**C**) rearomatized by deprotonation assisted by the generated benzoate ($ArylCO_2^-$), giving the corresponding aminated product.

At this point, it is worthy to mention that only the examples in which the Fe-catalyst is *directly* involved in the C–H bond cleavage or oxidation step have been considered. However, the Fe-catalyst might also lead to formal C–H functionalization products by *facilitating* the initial formation of a N-centered radical, which subsequently promotes a 1,5-hydrogen atom transfer by hydrogen hydride abstraction. Then, the formed C-centered radical further reacts, providing the corresponding C–N adduct. In this regard, the group of Zhu recently published the azidation of O-acyl oximes and N-acyloxy imidates employing $Fe(acac)_3$ as catalyst and $TMSN_3$ as azidation reagent. Consequently, the synthesis of the corresponding γ-azido ketones (in AcOH/*t*BuOH) and β-azido alcohols (EtOAc/NaOH/MeOH) was achieved in up to 94% and up to 84% yield, respectively.[120]

6.2.5 C–O bond formation

The selective C–H bond oxygenation of hydrocarbons represent a fundamental process in nature and is one of the most common and useful transformation in organic chemistry.[121] In one hand, C–H bond oxygenation processes are mediated in nature by both heme and nonheme metalloenzymes, normally based on Fe-porphyrins or Fe-amino acids coordination complexes. On the other hand, in synthetic chemistry, these transformations mainly relied on the well-known Gif[122] and Fenton[123] oxidations, where the iron-catalyzed C–H oxygenation involves the use of oxidants such as (air) oxygen or peroxides. The general mechanism implies two possible pathways: i) the formation of an organoiron intermediate through C–H activation by a concerted process, which takes place by abstraction of hydrogen and the formation of the new C–O bond; or ii) a radical C–H oxidation through a carbon centered radical species (Scheme 6.56).[124]

Considering the high importance of these oxygenation reactions, besides the classical Gif and Fenton oxidation, further studies in this field have recently been made aiming at developing more powerful, simple, and selective methods. In this regard, Bolm and coworkers presented in 2007 a mild oxidization reaction of benzylic C–H bonds (**73**) using aqueous TBHP as oxidant (Scheme 6.57).[125]

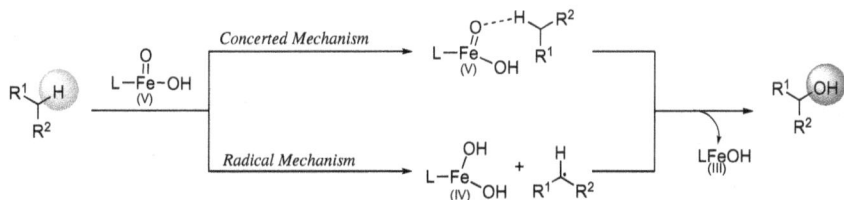

Scheme 6.56. Accepted mechanisms for the Fe-catalyzed C–H oxidation reaction.

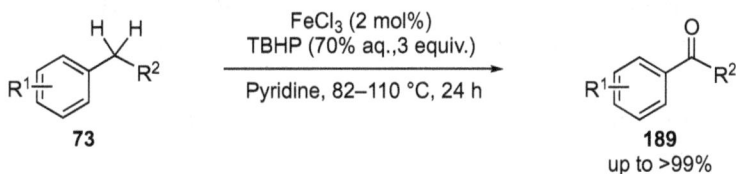

Scheme 6.57. Fe-catalyzed oxygenation of benzylic compounds into carbonyls.

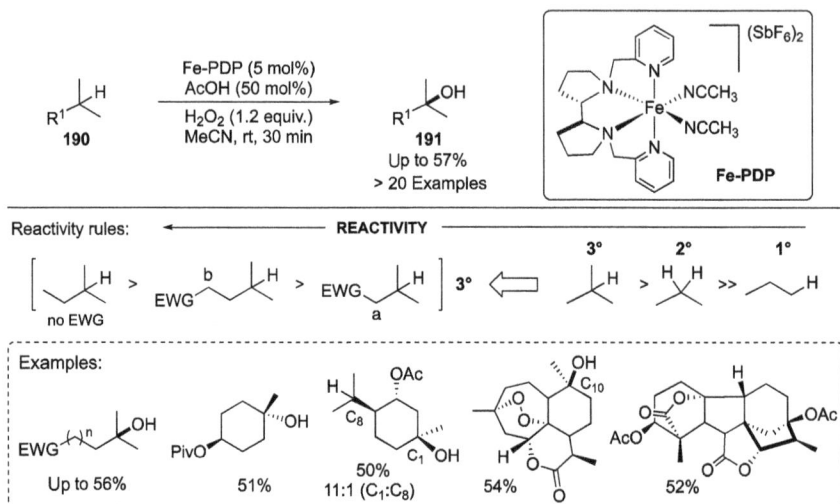

Scheme 6.58.   Fe-catalyzed oxidization of aliphatic C–H bonds.

Under these conditions, the targeted carbonyl derivatives (**189**) were obtained in moderate to high yields (30%–99%).

Roughly at the same time, White and coworkers carried out the Csp³–H oxidation of aliphatic compounds (**190**) using a nature-inspired Fe-PDP complex [PDP = 2-((*S*)-2-[(*S*)-1-(pyridin-2-ylmethyl)pyrrolidin-2-yl]pyrrolidin-1-ylmethyl)pyridine] and hydrogen peroxide as a mild, nontoxic oxidant (Scheme 6.58).[126a] A highly selective oxidation of nonactivated C–H bonds into the corresponding alcohols (**191**) was then obtained. The effect of the electronic and steric factors on the C–H selectivity was studied aiming at a determining the parameters for a predictable reactivity.

In general, it was found that electron-rich Csp³–H bonds are more prompt to be oxidized than electron-deficient positions. Thus, the electron-rich tertiary C–H sites are hydroxylated selectively, unless important steric hindrance around this position is present in the molecule. For the reaction on a secondary C–H site, an oxygenation to the corresponding ketone was mainly observed. Moreover, the late-stage oxidation of natural products, such as (+)-artemisinin, was also possible using this catalytic system.

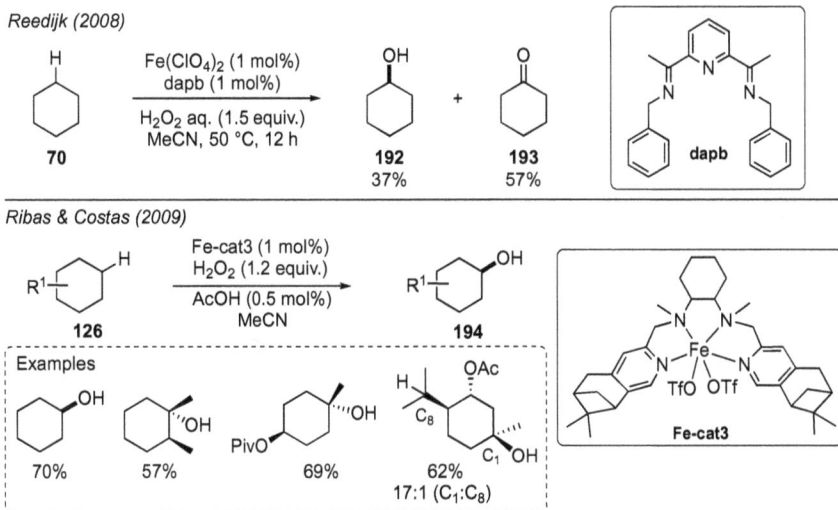

Scheme 6.59.   Fe-catalyzed selective C–H oxidation of cyclic hydrocarbons.

This chemistry was further exploited by Reedijk and coworkers for cyclohexene derivatives such as (**70**) using $Fe(ClO_4)_2$ as simple commercially available catalyst (Scheme 6.59, top).[127] However, this method suffered from low chemoselectivities (alcohol (**192**) and ketone (**193**) mixtures) and a narrow substrate scope. In 2009, Ribas and Costas presented another more general and effective nonheme mimetic Fe-complex [$Fe(CF_3SO_3)_2((S,S,R)$-mcpp), Fe-cat3] as catalyst (Scheme 6.59, bottom).[128a] Driven by the same intrinsic reactivity of aliphatic C–H bonds shown by White,[126] several derivatives (**126**) were hydroxylated (or oxygenated).[128] For example, in the study case of (-)-acetoxy-*p*-menthane, a regiospecific oxidation in the more accessible tertiary (C_1)-H bond was realized in a good yield using a low catalyst loading (1 mol%).

In 2017, the group of Gupta introduced a new peroxidase–Fe complex with a TAML framework (Fe–bTAML). This complex was able to carry out the selective C–H oxidation of a broad scope of unactivated tertiary and secondary alkanes (**190**) with *m*-chloroperoxybenzoic acid (CPBA) as oxidant, obtaining the desired alcohols (**191**) in up to 85% yield (Scheme 6.60).[129]

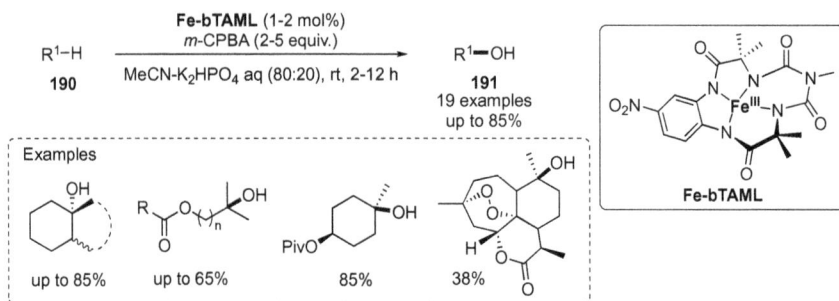

Scheme 6.60.   Fe-catalyzed selective C–H oxidation of unactivated alkanes.

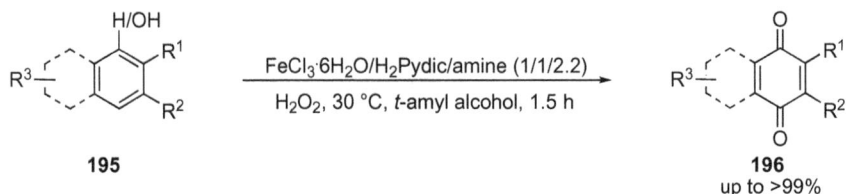

Scheme 6.61.   Fe-catalyzed selective oxidation of phenolic C–H bonds.

Although the selective C–H oxidation of alkanes has attracted great attention, the related oxidation of Csp^2–H bonds has also been studied. Thus, in 2010, Beller and coworkers reported on the selective oxidation of aromatic and phenolic C–H bonds (**195**) using $FeCl_3 \cdot 6H_2O$ as catalyst and hydrogen peroxide as oxidant (Scheme 6.61).[130] In this case, the corresponding benzoquinones (**196**) were obtained in excellent yields.

6.2.6 *C–S bond formation*

Concerning to the formation of C–S bonds by Fe-catalyzed C–H bond functionalization, only few examples have been reported to date in the literature.[131] Thus, besides the $FeBr_2$-catalyzed synthesis of thioesters (**199**) by radical coupling of thiols (**197**) with aldehydes (**198**) using TBHP as oxidant (Scheme 6.62, top),[131a] in 2012, Lei and coworkers reported the synthesis of 2-arylbenzomidazoles

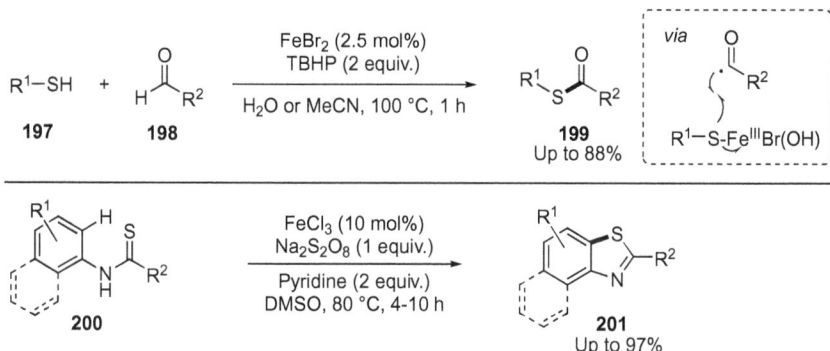

Scheme 6.62.   Fe-catalyzed C–H activation/cyclization of *N*-aryl benzothioamides.

(**199**) by intramolecular C–H activation of *N*-aryl benzothioamide (**200**) under mild conditions (Scheme 6.62, bottom).[131b]

In this case, FeCl$_3$ was used as catalyst in combination with Na$_2$S$_2$O$_8$ as stoichiometric oxidant to achieve good to high conversions and up to 97% yield. As previously mentioned in Section 6.2.4.3, in some studies the Fe-catalyst is not directly involved in the C–H activation step, but participates in the formation of a radical in a remote position, which is the responsible of the C–H functionalization by hydrogen radical abstraction. For example, the group of Cook recently carried out a Fe(OTf)$_2$-catalyzed C–H thioesterification of aliphatic silylperoxides with (non)substituted aromatic disulfides upon oxygen-centered radical formation followed by a 1,5-H transfer for the introduction of the sulfur atom. Under these conditions, the thioether adducts were achieved in 16%–72% yield.[132]

6.2.7 *C–B and C–Si bond formation*

The borylation and silylation of C–H bonds are highly appealing since they provide, in a straightforward manner, important intermediates in organic synthesis.[133] Despite its synthetic interest, only few examples in which iron species have been used as catalysts in C–H borylation and silylation reactions have been reported.

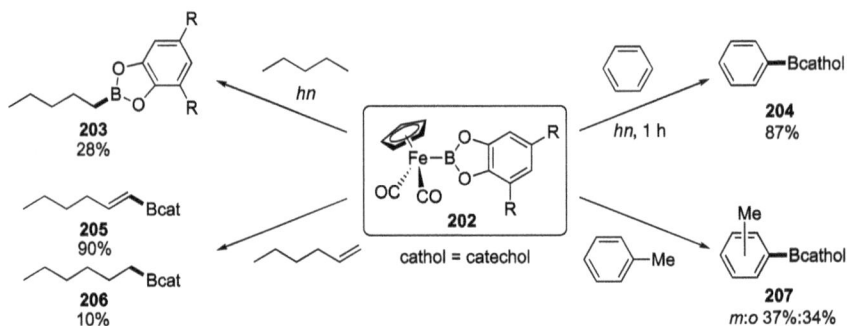

Scheme 6.63. Fe-catalyzed C–H borylation reactions.

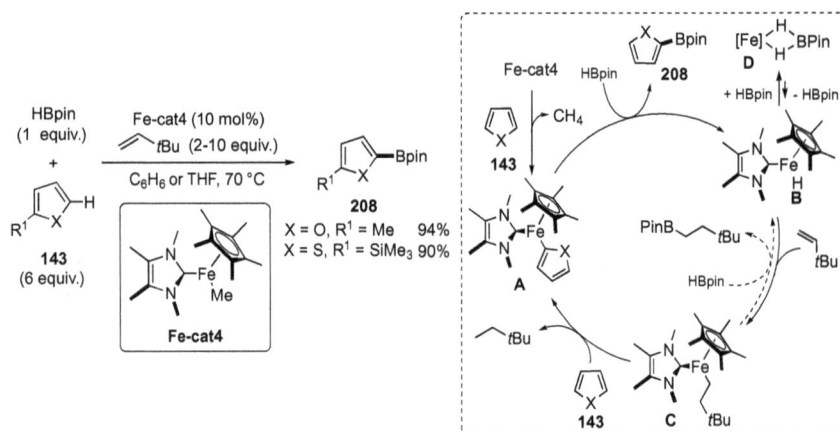

Scheme 6.64. Fe-catalyzed borylation of heteroarenes.

Regarding the C–B bond formation, Hartwig and coworkers already reported in the 1990s the borylation of alkanes, alkenes, and arenes promoted by stoichiometric amounts of an iron-boryl species (**202**) based on cyclopentadienyl iron-carbonyl complex (Scheme 6.63).[134]

However, this research area kept latent until 2010, when Tatsumi and coworkers presented an efficient catalytic approach for the borylation of heteroarenes (Scheme 6.64).[135] Thus, a half-sandwich NHC iron complex (Fe-cat4) was used to carry out the borylation of furans and thiophenes (**143**) with pinacolborane (HBPin) in the presence of an alkene as hydrogen acceptor.

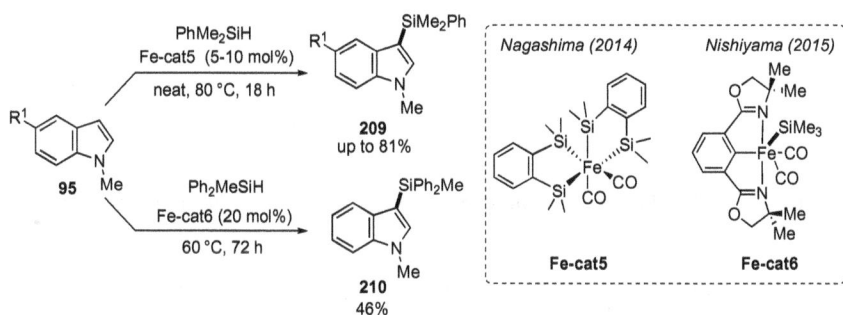

Scheme 6.65.   C–H bond silylation of *N*-methylindoles with phenylmethylsilanes.

The authors also studied the mechanism of the reaction, revealing that the catalyst Fe-cat4 inserts in the C–H bond of the heteroarene forming the corresponding aryl iron complex (**A**).

This species reacts with HBpin to generate the corresponding borylation product and the iron hydride (**B**), which can reversible be borylated giving the unproductive side complex (**D**). (**B**) can also add to the sacrificial olefin to form an alkyliron intermediate (**C**) that further reacts with the heteroarene to regenerate the active aryl iron complex (**A**).

Regarding the field of C–H bond silylation reactions,[136] the related Fe-catalyzed transformation was neglected until 2014, when Nagashima and coworkers developed the C_3-silylation of *N*-methylindoles (**95**) with phenyldimethylsilane catalyzed by an iron dicarbonyl complex containing a di-silylferracycle moiety (Fe-cat5) (Scheme 6.65, top).[137] Right after, in 2015, the group of Nishiyama reported a similar silylation reaction of indoles (**95**) catalyzed by an NCN-pincer-type iron-dicarbonyl silyl complex (Fe-cat6) (Scheme 6.65, bottom).[138] Although the latter reaction could be carried out at milder temperatures (60 vs. 80 °C), it required prolonged reaction times (72 vs. 18 h).

6.3   Heterogeneous Fe-Catalyzed C–H Functionalization

Heterogeneous catalysis has generated a considerable interest by the scientific community as the key to reduce the environmental impact that chemical processes generate. In this regard, although

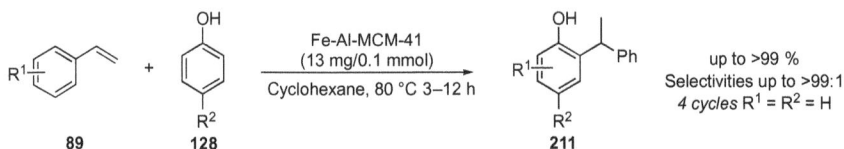

Scheme 6.66.   Fe-catalyzed heterogeneous C–H hydroarylation of styrenes.

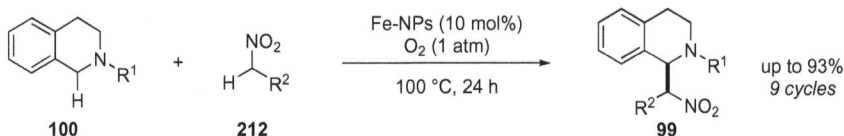

Scheme 6.67. Heterogeneous oxidative coupling between THIQs and nitroalkanes.

C–H activation/functionalization using Fe-complexes represents a high appealing approach, this field has only scarcely been explored during the last years.[115,139] In this regard, in 2010 Koner and coworkers described the synthesis of iron-containing mesoporous aluminosilicate Fe-Al_MCM-41 as heterogeneous catalyst for the hydroarylation of styrenes (**89**) with phenols (**128**) through C–H functionlization (Scheme 6.66).[139b] The corresponding coupling adducts (**211**) were obtained in good to high yields (65%–99%) and *ortho*-selectivities (95:5 to >99:1). Additionally, the catalyst could be recovered and reused in a next catalytic cycle up to four runs without apparent loss of the catalytic activity and selectivity.

The same year, Song, Li, and coworkers presented the oxidative coupling between Csp^3–H bond adjacent to nitrogen and Csp^3–H bond to nitroalkenes (**212**), catalyzed by iron nanoparticles (Fe_3O_4 NPs, <50 nm particle size) magnetically recovered using oxygent as mild oxidant (Scheme 6.67).[140] The studies for the recovery and reuse of the catalyst were carried out with the model reaction between nitromethane (**212**, R^2 = H) and *N*-phenyl-THIQ (**100**, R^1 = Ph) under the standard reaction conditions. The catalyst was reused in nine cycles without decrease in the catalytic activity, after being

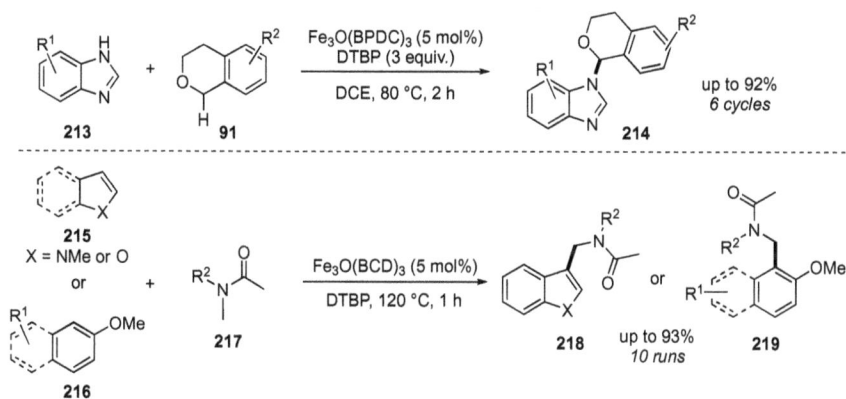

Scheme 6.68.   Fe-catalyzed heterogeneous coupling of isochromanes and amides.

adsorbed on the magnetic stirring bar and subsequently washed between each cycle. Moreover, no metal leaching of the iron nano-particles was detected by Plasma Atomic Emission Spectroscopy (ICP-AES).

In 2016, the group of Phan described the synthesis of an iron biphenyl-4,4'-dicarboxylate–organic framework [$Fe_3O(BPDC)_3$] as heterogeneous catalyst for the generation of azole derivatives (**214**) *via* direct C–N coupling of azoles (**213**) with ethers such as isochromanes (**91**) through an oxidative C–H reaction with DTBP as mild oxidant (Scheme 6.68, top).[141] A high catalytic efficiency in a wide range of substrates with different electronic properties was achieved. Moreover, the heterogeneous catalyst could be reused up to six times maintaining the same levels of catalytic efficiency. This Fe-MOF (Fe-metal organic framework) could also be employed for the direct C–C coupling of electron-rich aromatic compounds (**215**) or (**216**) with alkylamides (**217**) (Scheme 6.68, bottom).[139a]

The same group reported the use of heterogeneous iron-based MOF [$Fe_3(BTC)(NDC)_2\,6.65H_2O$] (BTC = 1,3,5-benzenetricarboxy-late; NDC = 2,6-napthalenedicarboxylate) for the functionalization of coumarins (**220**) with *N,N*-dimethylanilines (**104**) *via* a CDC reaction using TBHP as oxidant (Scheme 6.69).[142] A leaching test confirmed the heterogeneous nature of the catalyst. Moreover, the

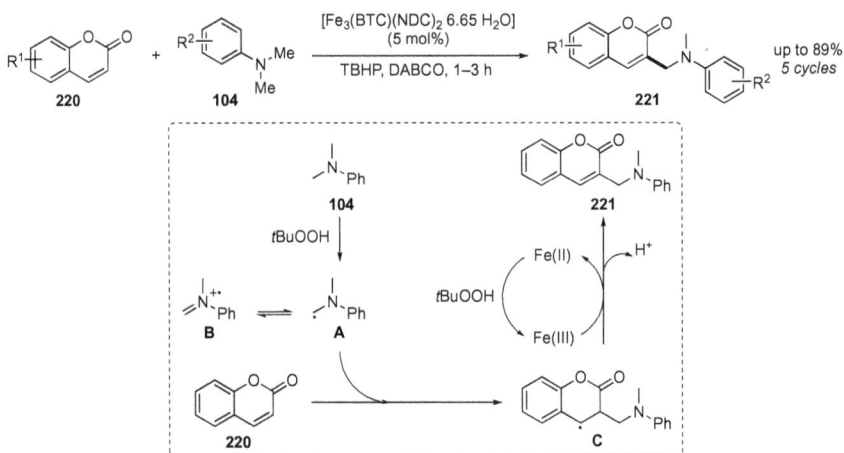

Scheme 6.69.   Iron-organic-frameworks in C–H couplings with coumarins.

catalyst could be recycled without a notable decrease in the catalytic activity until the fifth run. With respect to the mechanism of the reaction, the authors postulated a first hydrogen radical abstraction from the *N, N*-dimethylaniline (**104**). The generated radical (**A**) reacts then with the coumarin (**220** to form a stable benzylic radical (**C**), which upon oxidization and proton release leads to the product (**221**) and the Fe(II) catalyst.

Roughly at the same time, the group of Satishkumar presented an heterogeneous Fe-catalyst supported on silica (8 wt% Fe/SBA-15) to carry out Csp^3–H arylation reactions (Scheme 6.70).[143] In particular, diphenyl methane (**79**) was reacted with substituted arenes (**54**) in the presence of DDQ, whereas simple benzene (**54**, R^1, R^2 = H) was coupled with aryl boronic acids (**38**) using DTBP as oxidant. In both reactions, the products (**222**) and (**223**) were achieved moderate to good isolated yields (55%–70% and 30%–83%, respectively), being possible to recover of the catalyst and reuse it up to five times without apparently decrease in the catalytic activity. The authors proposed a SET, initiated mechanism similar to the homogeneous oxidative C–H functionalizations (e.g., CDC) with DTBP for the reaction between aryl boronic acids and nonactivated arenes. The *tert*-butoxy radical (**B**) is then formed from DTBP

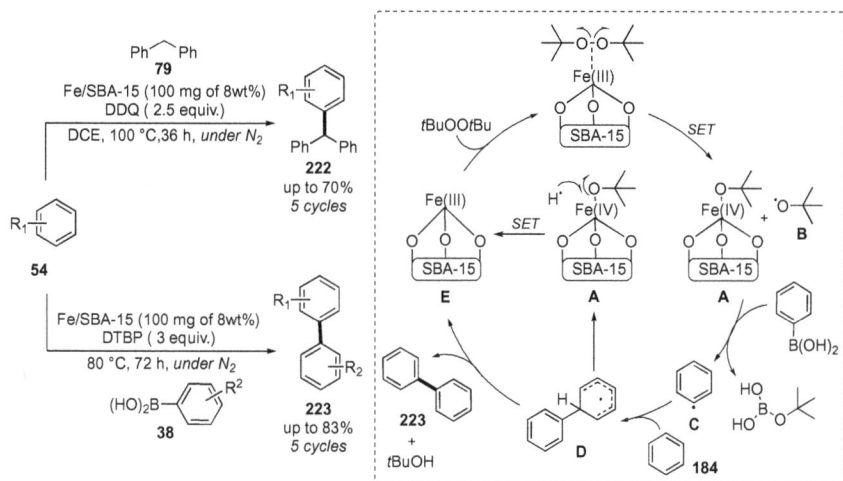

Scheme 6.70.   Fe-catalyzed heterogeneous C–H functionalization arenes.

by peroxide reductive cleavage with the Fe(III) catalyst, which is oxidized to Fe(IV) (**A**). The alkoxy radical intermediate (**B**) reacts with the aryl boronic acid, forming the aryl radical (**C**) that is further coupled with benzene to form the cyclohexadienyl radical (**D**). This radical is then oxidized by the Fe(IV), leading the coupled corresponding product (**223**) along with the regenerated supported Fe(III) catalyst. A similar catalytic cycle for the coupling with diphenyl methanes with DDQ as oxidant through a SET initiated mechanism was also proposed.

Lastly, in 2018, Islam and coworkers designed the synthesis of an iron naphthyl-azo complex catalyst supported on a polymer to carry out the oxidation of toluene derivatives (**83**) (Scheme 6.71).[144] Hydrogen peroxide (H_2O_2) was used as mild and environmentally friendly oxidant in the presence of a Ps-Fe-NAPA, an immobilized complex, giving the corresponding aldehydes (**224**) in high conversion (up to 96%).

The supported iron catalyst could be recovered and reused until the sixth cycle in the oxidation of toluene to benzaldehyde with no notable decrease in the catalytic activity. Additionally, after hot test

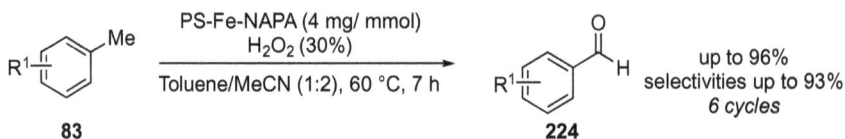

Scheme 6.71. Fe-catalyzed heterogeneous C–H oxidation of arylmethanes.

experiment no leaching was detected, reinforcing the hypothesize heterogeneous character of this catalytic system.

6.4 Conclusion and Perspectives

The catalytic direct functionalization reaction of C–H bonds is one of the most powerful approaches to introduce complexity in organic molecules. In this chapter, the work reported in the literature until Summer 2019 on the field of iron-catalyzed C–H activation/functionalization in organic synthesis is summarized. A wide range of organic transformations, together with their most accepted mechanisms with various catalytic systems including well-defined iron-complexes, have been presented.

Although a high reactivity and selectivity have been achieved for some systems and transformations, there are still several challenges that need to be addressed. Those include avoiding DGs in C–H activation reactions while using simple catalytic systems, going beyond the classical substrates for CDC reactions, developing efficient asymmetric reactions, expanding this chemistry to complex organic molecules and late-stage functionalization or scaling-up the processes for true synthetic applications and exploitation. Thus, it can be envisioned that future efforts will be set on the development of highly efficient and selective catalytic systems toward more versatile and practical Fe-catalyzed C–H functionalization processes.

References

1. T. Gensch, M. J. James, T. Dalton and F. Glorius, Increasing catalyst efficiency in C–H activation catalysis, *Angew. Chem. Int. Ed.* **57**, 2296–2306 (2018).

2. (a) E. M. Beccalli, G. Broggini, A. Fasana and M. Rigamonti, Palladium-catalyzed C–N bond formation via direct C–H bond functionalization. Recent developments in heterocyclic synthesis, *J. Organomet. Chem.* **696**, 277–295 (2010); (b) T. W. Lyons and M. S. Sanford, Palladium-catalyzed ligand-directed C–H functionalization reactions, *Chem. Rev.* **110**, 1147–1169 (2010); (c) C.-L. Sun, B.-J. Li and Z.-J. Shi, Pd-catalyzed oxidative coupling with organometallic reagents via C–H activation, *Chem. Commun.* **46**, 677–685 (2010); (d) K. M. Engle and J.-Q. Yu, Developing ligands for palladium(II)-catalyzed C–H functionalization: Intimate dialogue between ligand and substrate, *J. Org. Chem.* **78**, 8927–8955 (2013); (e) K. Ramachandiran, T. Sreelatha, N. V. Lakshmi, T. H. Babu, D. Muralidharan and P. T. Perumal, Palladium catalyzed C–H activation and its application to multi-bond forming reactions, *Curr. Org. Chem.* **17**, 2001–2024 (2013); (f) Y. Wu, J. Wang, F. Mao and F. Y. Kwong, Palladium-Catalyzed cross-dehydrogenative functionalization of C(sp²)-H bonds, *Chem. Asian J.* **9**, 26–47 (2014); (g) A. Dey, S. Agasti and D. Maiti, Palladium catalyzed meta-C–H functionalization reactions, *Org. Biomol. Chem.* **14**, 5440–5453 (2016); (h) V. P. Mehta and J.-A. Garcia-Lopez, σ-Alkyl-PdII species for remote C–H functionalization, *ChemCatChem* **9**, 1149–1156 (2017).

3. J. Choi and A. S. Goldman, Ir-catalyzed functionalization of C–H bonds, *Top. Organomet. Chem.* **34**, 139–168 (2011).

4. (a) D. A. Colby, A. S. Tsai, R. G. Bergman and J. A. Ellman, Rhodium catalyzed chelation-assisted C–H bond functionalization reactions, *Acc. Chem. Res.* **45**, 814–825 (2012); (b) G. Song and X. Li, Substrate activation strategies in rhodium(III)-catalyzed selective functionalization of arenes, *Acc. Chem. Res.* **48**, 1007–1020 (2015); (c) S. Motevalli, Y. Sokeirik and A. Ghanem, Rhodium-catalyzed enantioselective C–H functionalization in asymmetric synthesis, *Eur. J. Org. Chem.* 1459–1475 (2016); (d) X. Qi, Y. Li, R. Bai and Y. Lan, Mechanism of rhodium-catalyzed C–H functionalization: advances in theoretical investigation, *Acc. Chem. Res.* **50**, 2799–2808 (2017); (e) T. Yakura and H. Nambu, Recent topics in application of selective Rh(II)-catalyzed C–H functionalization toward natural product synthesis, *Tetrahedron Lett.* **59**, 188–202 (2018).

5. (a) P. B. Arockiam, C. Bruneau and P. H. Dixneuf, Ruthenium(II)-catalyzed C–H Bond activation and functionalization, *Chem. Rev.* **112**, 5879–5918 (2012); (b) C. Bruneau and P. H. Dixneuf, Ruthenium(II)-catalyzed functionalization of C–H bonds with alkenes: Alkenylation versus alkylation, *Top. Organomet. Chem.* **55**, 137–188 (2016); (c) S. Dana, M. R. Yadav and A. K. Sahoo, Ruthenium-catalyzed C–N and C–O bond-forming processes from C–H bond functionalization, *Top. Organomet. Chem.* **55**, 189–215 (2016); (d) P. Nareddy, F. Jordan and M. Szostak, Recent developments in ruthenium-catalyzed C–H arylation: Array of mechanistic manifolds, *ACS Catal.* **7**, 5721–5745 (2017).

6. J. A. Labinger, Platinum-catalyzed C–H functionalization, *Chem. Rev.* **117**, 8483–8496 (2017).

7. M. Hibino, T. Kimura, Y. Suga, T. Kudo and N. Mizuno, Oxygen rocking aqueous batteries utilizing reversible *topotactic* oxygen insertion/extraction in iron-based perovskite oxides $Ca_{1-x}La_xFeO_{3-\delta}$, *Sci. Rep.* **2**, 601 (2012).

8. (a) A. L. Feig and S. J. Lippard, Reactions of non-heme iron(II) centers with dioxygen in biology and chemistry, *Chem. Rev.* **94**, 759–805 (1994); (b) N. C. Andrews, Disorders of iron metabolism, *N. Engl. J. Med.* **341**, 1986–1995 (1999).

9. (a) S. Xiaoli, L. Jilai, H. Xuri and S. Chiachung, recent advances in iron-catalyzed C–H bond activation reactions, *Curr. Inorg. Chem.* **2**, 64–85 (2012); (b) G. Cera and L. Ackermann, Iron-catalyzed C–H functionalization processes, *Top Curr. Chem.* **374**, (2016); (c) M. Su, C. Li and J. Ma, Iron-catalyzed C–H activation, *J. Chin. Chem. Soc.* **63**, 828–840 (2016).

10. J. Müller and M. Bröring, Iron catalysis in organic chemistry: Reactions and applications, ed. Plietker, B., In Chapter 2, *Iron Catalysis in Biological and Biomimetic Reactions*, Wiley-VCH Verlag GmbH & Co. KGaA, Germany, 29–72 (2008).

11. P. Gandeepan, T. Müller, D. Zell, G. Cera, S. Warratz and L. Ackermann, 3d transition metals for C–H activation, *Chem. Rev.* **119**, 2191–2452 (2018).

12. N. V. Tzouras, I. K. Stamatopoulos, A. T. Papastavrou, A. A. Liori and G. C. Vougioukalakis, sustainable metal catalysis in C–H activation, *Coord. Chem. Rev.* **343**, 25–138 (2017).

13. (a) I. P. Beletskaya and A. V. Cheprakov, Metal complexes as catalysts for C–C cross-coupling reactions, *Compr. Coord. Chem. II*, **9**, 305–368 (2004); (b) T. Kohei and N. Miyaura, *Introduction to Cross-Coupling Reactions.* ed. N. Miyaura, In *Cross-Coupling Reactions: A Practical Guide*, Top. Curr. Chem. Springer, Berlin, Heidelberg (2002).

14. (a) C.-L. Sun, B.-J. Li and Z.-J. Shi, Direct C–H transformation via iron catalysis, *Chem. Rev.* **111**, 1293–1314 (2011); (b) R. Shang, L. Ilies and E. Nakamura, Iron-catalyzed C–H bond activation, *Chem. Rev.* **117**, 9086–9139 (2017).

15. G. Hata, H. Kondo and A. Miyake, Ethylenebis(diphenylphosphine) complexes of iron and cobalt. Hydrogen transfer between the ligand and iron atom, *J. Am. Chem. Soc.* **90**, 2278–2281 (1968).

16. (a) J. W. Rathke and E. L. Muetterties, Phosphine chemistry of iron(0) and -(II), *J. Am. Chem. Soc.* **97**, 3272–3273 (1975); (b) M. V. Baker and L. D. Field, Reaction of carbon-hydrogen bonds in alkanes with bis(diphosphine) complexes of iron, *J. Am. Chem. Soc.* **109**, 2825–2826 (1987); (c) D. H. R. Barton and D. Doller, The selective functionalization of saturated hydrocarbons: Gif chemistry, *Acc. Chem. Res.* **25**, 504–512 (1992).

17. S. Camadanli, R. Beck, U. Flörke and H.-F. Klein, C–H activation of imines by trimethylphosphine-supported iron complexes and their reactivities, *Organometallics* **28**, 2300–2310 (2009).

18. J. Norinder, A. Matsumoto, N. Yoshikai and E. Nakamura, Iron-catalyzed direct arylation through directed C–H bond activation, *J. Am. Chem. Soc.* **130**, 5858–5859 (2008).

19. L. Ilies, S. Asako and E. Nakamura, Iron-catalyzed stereospecific activation of olefinic C–H bonds with grignard reagent for synthesis of substituted olefins, *J. Am. Chem. Soc.* **133**, 7672–7675 (2011).

20. S. C. Bart, E. J. Hawrelak, E. Lobkovsky and P. J. Chirik, Low-valent α-diimine iron complexes for catalytic olefin hydrogenation, *Organometallics* **24**, 5518–5527 (2005).

21. J. J. Sirois, R. Davis and B. DeBoef, Iron-catalyzed arylation of heterocycles via directed C–H bond activation, *Org. Lett.* **16**, 868–871 (2014).

22. N. Yoshikai, S. Asako, T. Yamakawa, L. Ilies and E. Nakamura, Iron-catalyzed C–H bond activation for the ortho-arylation of aryl

pyridines and imines with grignard reagents, *Chem. Asian J.* **6**, 3059–3065 (2011).

23. (a) Y. Sun, H. Tang, K. Chen, L. Hu, J. Yao, S. Shaik and H. Chen, Two state reactivity in low-valent iron-mediated C–H activation and the implications for other first-row transition metals, *J. Am. Chem. Soc.* **138**, 3715–3730 (2016); (b) R. Shang, L. Ilies, S. Asako and E. Nakamura, Iron-catalyzed $C(sp^2)$–H bond functionalization with organoboron compounds, *J. Am. Chem. Soc.* **136**, 14349–14352 (2014).

24. G. Rouquet and N. Chatani, Catalytic functionalization of $C(sp^2)$–H and $C(sp^3)$–H bonds by using bidentate directing groups, *Angew. Chem. Int. Ed.* **52**, 11726–11743 (2013).

25. V. G. Zaitsev, D. Shabashov and O. Daugulis, Highly regioselective arylation of sp^3 C–H bonds catalyzed by palladium acetate, *J. Am. Chem. Soc.* **127**, 13154–13155 (2005).

26. (a) Y. Kobayashi and R. Mizojiri, Nickel-Catalyzed coupling reaction of lithium organoborates and aryl mesylates possessing an electron withdrawing group, *Tetrahedron Lett.* **37**, 8531–8534 (1996); (b) T. Hatakeyama, T. Hashimoto, Y. Kondo, Y. Fujiwara, H. Seike, H. Takaya, Y. Tamada, T. Ono and M. Nakamura, Iron-catalyzed suzuki–miyaura coupling of alkyl halides, *J. Am. Chem. Soc.* **132**, 10674–10676 (2010).

27. T. Doba, T. Matsubara, L. Ilies, R. Shang and E. Nakamura, Homocoupling-free iron-catalysed twofold C–H activation/cross-couplings of aromatics via transient connection of reactants, *Nat. Catal.* **2**, 400–406 (2019).

28. S. Asako, L. Ilies and E. Nakamura, Iron-catalyzed ortho-allylation of aromatic carboxamides with allyl ethers, *J. Am. Chem. Soc.* **135**, 17755–17757 (2013).

29. R. Jana, T. P. Pathak and M. S. Sigman, Advances in transition metal (Pd, Ni, Fe)-catalyzed cross-coupling reactions using alkyl-organo-metallics as reaction partners, *Chem. Rev.* **111**, 1417–1492 (2011).

30. E. R. Fruchey, B. M. Monks and S. P. Cook, A unified strategy for iron-catalyzed ortho-alkylation of carboxamides, *J. Am. Chem. Soc.* **136**, 13130–13133 (2014).

31. L. Ilies, S. Ichikawa, S. Asako, T. Matsubara and E. Nakamura, Iron-catalyzed directed alkylation of alkenes and arenes with alkylzinc halides, *Adv. Synth. Catal.* **357**, 2175–2179 (2015).

32. L. Ilies, Y. Zhou, H. Yang, T. Matsubara, R. Shang and E. Nakamura, Iron-catalyzed directed alkylation of carboxamides with olefins via a carbometalation pathway, *ACS Catal.* **8**, 11478–11482 (2018).

33. Q. Gu, H. H. Al Mamari, K. Graczyk, E. Diers and L. Ackermann, Iron-catalyzed $C(sp^2)$–H and $C(sp^3)$–H arylation by triazole assistance, *Angew. Chem. Int. Ed.* **53**, 3868–3871 (2014).

34. K. Graczyk, T. Haven and L. Ackermann, Iron-catalyzed $C(sp^2)$–H and $C(sp^3)$–H methylations of amides and anilides. *Chem. Eur. J.* **21**, 8812–8815 (2015).

35. G. Cera, T. Haven and L. Ackermann, Expedient Iron-catalyzed C–H allylation/alkylation by triazole assistance with ample scope, *Angew. Chem. Int. Ed.* **55**, 1484–1488 (2016).

36. M. Y. Wong, T. Yamakawa and N. Yoshikai, Iron-catalyzed directed C2-alkylation and alkenylation of indole with vinylarenes and alkynes, *Org. Lett.* **17**, 442–445 (2015).

37. W. Xu and N. Yoshikai, Iron-catalyzed ortho C–H arylation and methylation of pivalophenone N-H imines, *ChemSusChem.* **12**, 3049–3053 (2019).

38. N. Vodnala, R. Gujjarappa, A. K. Kabi, M. Kumar, U. Beifuss and C. C. Malakar, Facile protocols towards C2-arylated benzoxazoles using Fe(III)-catalyzed $C(sp^2$-H) functionalization and metal-free domino approach, *Synlett* **29**, 1469–1478 (2018).

39. R. Shang, L. Ilies and E. Nakamura, Iron-catalyzed ortho C–H methylation of aromatics bearing a simple carbonyl group with methylaluminum and tridentate phosphine ligand, *J. Am. Chem. Soc.* **138**, 10132–10135 (2016).

40. (a) B. D. Sherry and A. Furstner, The promise and challenge of iron-catalyzed cross coupling, *Acc. Chem. Res.* **41**, 1500–1511 (2008); (b) M. D. Greenhalgh, A. S. Jones and S. P. Thomas, Iron-catalysed hydrofunctionalisation of alkenes and alkynes, *ChemCatChem.* **7**, 190–222 (2015).

41. A. Matsumoto, L. Ilies and E. Nakamura, Phenanthrene synthesis by iron-catalyzed [4 + 2] benzannulation between alkyne and biaryl or 2-alkenylphenyl grignard reagent, *J. Am. Chem. Soc.* **133**, 6557–6559 (2011).

42. L. Adak and N. Yoshikai, Iron-catalyzed annulation reaction of arylindium reagents and alkynes to produce substituted naphthalenes, *Tetrahedron* **68**, 5167–5171 (2012).

43. T. Matsubara, L. Ilies and E. Nakamura, Oxidative C–H activation approach to pyridone and isoquinolone through an iron-catalyzed coupling of amides with alkynes, *Chem. Asian J.* **11**, 380–384 (2016).

44. L. Ilies, Y. Arslanoglu, T. Matsubara and E. Nakamura, Iron-catalyzed synthesis of indenones through cyclization of carboxamides with alkynes, *Asian J. Org. Chem.* **7**, 1327–1329 (2018).

45. J. Mo, T. Mueller, J. C. A. Oliveira and L. Ackermann, 1,4-iron migration for expedient allene annulations through iron-catalyzed C–H/N–H/C–O/C–H functionalizations, *Angew. Chem. Int. Ed.* **57**, 7719–7723 (2018).

46. J. Mo, T. Müller, J. C. A. Oliveira, S. Demeshko, F. Meyer and L. Ackermann, Iron-catalyzed C–H activation with propargyl acetates: Mechanistic insights into iron(II) by experiment, kinetics, mössbauer spectroscopy, and computation, *Angew. Chem. Int. Ed.* **58**, 12874–12878 (2019).

47. N. Panda, I. Mattan, S. Ojha and C. S.Purohit, Synthesis of medium-sized (6-7-6) ring compounds by iron-catalyzed C–H activation/annulation, *Org. Biomol. Chem.* **16**, 7861–7870 (2018).

48. A. Conde, G. Sabenya, M. Rodríguez, V. Postils, J. M. Luis, M. M. Díaz-Requejo, M. Costas and P. J. Pérez, Iron and manganese catalysts for the selective functionalization of arene $C(sp^2)$–H bonds by carbene insertion, *Angew. Chem., Int. Ed.* **55**, 6530–6534 (2016).

49. V. Postils, M. Rodriguez, G. Sabenya, A. Conde, M. M. Diaz-Requejo, P. J. Perez, M. Costas, M. Sola and J. M. Luis, Mechanism of the selective fe-catalyzed arene carbon-hydrogen bond functionalization, *ACS Catal.* **8**, 4313–4322 (2018).

50. (a) O. Baudoin, Transition metal-catalyzed arylation of unactivated $C(sp^3)$–H Bonds, *Chem. Soc. Rev.* **40** 4902–4911 (2011); (b) B. J. Li, Z. J. Shi, From $C(sp^2)$–H to $C(sp^3)$–H: Systematic studies on transition metal-catalyzed oxidative C–C formation, *Chem. Soc. Rev.* **41**, 5588–5598 (2012).

51. N. Yoshikai, A.Mieczkowski, A. Matsumoto, L. Ilies and E. Nakamura, Iron-catalyzed C–C bond formation at α-position of aliphatic amines via C–H bond activation through 1,5-hydrogen transfer, *J. Am. Chem. Soc.* **132**, 5568–5569 (2010).

52. P. P. Singh, S. Gudup, S. Ambala, U. Singh, S. Dadhwal, B. Singh, S. D. Sawant and R. A. Vishwakarma, Iron oxide mediated direct C–H arylation/alkylation at α-position of cyclic aliphatic ethers, *Chem. Commun.* **47**, 5852–5854 (2011).

53. B. Zhou, H. Sato, L. Ilies and E. Nakamura, Iron-catalyzed remote arylation of aliphatic C–H bond via 1,5-hydrogen shift, *ACS Catal.* **8**, 8–11 (2018).

54. R. Shang, L. Ilies, A. Matsumoto and E. Nakamura, β-arylation of carboxamides via iron-catalyzed C(sp³)–H bond activation, *J. Am. Chem. Soc.* **135**, 6030–6032 (2013).

55. L. Ilies, Y. Itabashi, R. Shang and E. Nakamura, Iron/Zinc-Co-catalyzed directed arylation and alkenylation of C(sp³)–H bonds with organoborates, *ACS Catal.* **7**, 89–92 (2017).

56. R. Shang, L. Ilies and E. Nakamura, Iron-catalyzed directed C(sp²)–H and C(sp³)–H functionalization with trimethylaluminum, *J. Am. Chem. Soc.* **137**, 7660–7663 (2015).

57. M. Sekine, L. Ilies and E. Nakamura, Iron-catalyzed allylic arylation of olefins via C(sp³)–H activation under mild conditions, *Org. Lett.* **15**, 714–717 (2013).

58. (a) C. J. Li, Cross-dehydrogenative coupling (CDC): Exploring C–C bond formations beyond functional group transformations, *Acc. Chem. Res.* **42**, 335–344 (2009); (b) S. A. Girard, T. Knauber and C.-J. Li, The cross-dehydrogenative coupling of C$_{sp3}$–H bonds: A versatile strategy for C–C bond formations, *Angew. Chem. Int. Ed.* **53**, 74–100 (2014).

59. L. Lv and Z. Li, Fe-catalyzed cross-dehydrogenative coupling reactions, *Top. Curr. Chem.* **374**, 1–39 (2016).

60. Z. Li, L. Cao and C. J. Li, FeCl$_2$-catalyzed selective C–C bond formation by oxidative activation of a benzylic C–H bond, *Angew. Chem., Int. Ed.* **46**, 6505–6507 (2007).

61. S. Pan, J. Liu, Y. Li and Z. Li, Iron-catalyzed benzylation of 1,3-dicarbonyl compounds by simple toluene derivatives, *Chin. Sci, Bull.* **57**, 2382–2386 (2012).

62. C. X. Song, G. X. Cai, T. R. Farrell, Z. P. Jiang, H. Li, L. B. Gan and Z. J. Shi, Direct functionalization of benzylic C–Hs with vinyl acetates via Fe-catalysis, *Chem. Commun.* **40**, 6002–6004 (2009).

63. Y. Z. Li, B. J. Li, X. Y. Lu, S. Lin and Z. J. Shi, Cross dehydrogenative arylation (CDA) of a benzylic C–H bond with arenes by iron catalysis, *Angew. Chem. Int. Ed.* **48**, 3817–3820 (2009).

64. K. Yang and Q. Song, Fe-catalyzed double cross-dehydrogenative coupling of 1,3-dicarbonyl compounds and arylmethanes, *Org. Lett.* **17**, 548–551 (2015).

65. R. K. Zhang, K. Chen, X. Huang, L. Wohlschlager, H. Renata and F. H. Arnold, Enzymatic assembly of carbon–carbon bonds via iron-catalysed sp^3 C–H functionalization, *Nature* **565**, 67–72 (2019).

66. H. Liu, L. Cao, J. Sun, J. S. Fossey and W.-P. Deng, Iron-catalysed tandem cross-dehydrogenative coupling (CDC) of terminal allylic C(sp^3) to C(sp^2) of styrene and benzoannulation in the synthesis of polysubstituted naphthalenes, *Chem. Commun.* **48**, 2674–2676 (2012).

67. (a) S. I. Murahashi, N. Komiya and H. Terai, Ruthenium-catalyzed oxidative cyanation of tertiary amines with hydrogen peroxide and sodium cyanide, *Angew. Chem. Int. Ed.* **44**, 6931–6933 (2005); (b) X. Guo, Z. Li, C. J. Li, Cross-dehydrogenative-coupling (CDC) reaction, *Prog. Chem.* (in Chinese) **22**, 1434–1441 (2010).

68. Z. Li, R. Yu and H. Li, Iron-catalyzed C–C bond formation by direct functionalization of C–H bonds adjacent to heteroatoms, *Angew. Chem. Int. Ed.* **47**, 7497–7500 (2008).

69. H. Richter and O. García Mancheño, Dehydrogenative functionalization of C(sp^3)–H bonds adjacent to a heteroatom mediated by oxoammonium salts, *Eur. J. Org. Chem.* 4460–4467 (2010).

70. M. Ghobrial, M. Schnürch and M. D. Mihovilovic, Direct functionalization of (un)protected tetrahydroisoquinoline and isochroman under iron and copper catalysis: two metals, two mechanisms, *J. Org. Chem.* **76**, 8781–8793 (2011).

71. X. Guo, S. Pan, J. Liu and Z. Li, One-pot synthesis of symmetric and unsymmetric 1,1-bis-indolylmethanes via tandem iron-catalyzed C–H bond oxidation and C–O bond cleavage, *J. Org. Chem.* **74**, 8848–8851 (2009).

72. A. Correa, B. Fiser and E. Gómez-Bengoa, Iron-catalyzed direct α-arylation of ethers with azoles, *Chem. Commun.* **51**, 13365–13368 (2015).

73. E. Shirakawa, T. Yoneda, K. Moriya, K. Ota, N. Uchiyama, R. Nishikawa and T. Hayashi, Iron-catalyzed oxidative coupling of

alkylamines with arenes, nitroalkanes, and 1,3-dicarbonyl compounds, *Chem. Lett.* **40**, 1041–1043 (2011).

74. M. O. Ratnikov, X. Xu and M. P. Doyle, Simple and sustainable iron-catalyzed aerobic C–H functionalization of N,N-dialkylanilines, *J. Am. Chem. Soc.* **135**, 9475–9479 (2013).

75. C. M. R. Volla and P. Vogel, Chemoselective C–H bond activation: Ligand and solvent free iron-catalyzed oxidative C–C cross-coupling of tertiary amines with terminal alkynes. reaction scope and mechanism, *Org. Lett.* **11**, 1701–1704 (2009).

76. Z.-L. Li, K.-K. Sun, P.-Yu. Wu and C. Cai, Iron-catalyzed regioselective α-C–H alkylation of N-methylanilines: Cross-dehydrogenative coupling between unactivated $C(sp^3)$–H and $C(sp^3)$–H bonds via a radical process, *J. Org. Chem.* **84**, 6830–6839 (2019).

77. M. Ohta, M. P. Quick, J. Yamaguchi, B. Wünsch and K. Itami, Fe-catalyzed oxidative coupling of heteroarenes and methylamines, *Chem. Asian J.* **4**, 1416–1419 (2009).

78. H. Li, Z. He, X. Guo, W. Li, X. Zhao and Z. Li, Iron-catalyzed selective oxidation of N-Methyl amines: highly efficient synthesis of methylene-bridged bis-1,3-dicarbonyl compounds, *Org. Lett.* **11**, 4176–4179 (2009).

79. H. Richter and O. García Mancheño, TEMPO oxoammonium salt-mediated dehydrogenative povarov/oxidation tandem reaction of N-alkyl anilines, *Org. Lett.* **13**, 6066–6069 (2011).

80. R. Rohlmann, T. Stopka, H. Richter and O. García Mancheño, Iron-catalyzed oxidative tandem reactions with TEMPO oxoammonium salts: Syn-thesis of dihydroquinazolines and quinolines, *J. Org. Chem.* **78**, 6050–6064 (2013).

81. P. Liu, Z. Wang, J. Lin and X. Hu, An efficient route to quinolines and other compounds by iron-catalysed cross-dehydrogenative coupling reactions of glycine derivatives, *Eur. J. Org. Chem.* 1583–1589 (2012).

82. S.-J. Lou, D.-Q. Xu, D.-F. Shen, Y.-F. Wang, Y.-K. Liu and Z.-Y. Xu, Highly efficient vinylaromatics generation via iron-catalyzed sp^3 C–H bond functionalization CDC reaction: A novel approach to preparing substituted benzo[α]phenazines, *Chem. Commun.* **48**, 11993–11995 (2012).

83. Y.-M. Li, S.-J. Lou, Q.-H. Zhou, L.-W. Zhu, L.-F. Zhu and L. Li, Iron-catalyzed α-methylenation of ketones with N,N-dimethylacetamide: An approach for α,β-unsaturated carbonyl compounds, *Eur. J. Org. Chem.* 3044–3047 (2015).

84. Y. Li, F. Guo, Z. Zha and Z. Wang, Iron-catalyzed synthesis of 2-vinylquinolines via sp³ C–H functionalization and subsequent C–N cleavage, *Chem. Asian J.* **8**, 534–537 (2013).

85. C. Wang, Y. Li, M. Gong, Q. Wu, J. Zhang, J. K. Kim, M. Huang and Y. Wu, Method for direct synthesis of α-cyanomethyl-β-dicarbonyl compounds with acetonitrile and 1,3-dicarbonyls, *Org. Lett.* **18**, 4151–4153 (2016).

86. H. Su, L. Wang, H. Rao and H. Xu, Iron-catalyzed dehydrogenative sp³-sp² coupling via direct oxidative C–H activation of acetonitrile, *Org. Lett.* **19**, 2226–2229 (2017).

87. Y. Zhang and C. J. Li, Highly efficient direct alkylation of activated methylene by cycloalkanes, *Eur. J. Org. Chem.* 4654–4657 (2007).

88. X. Guo, R. Yu, H. Li and Z. Li, Iron-catalyzed tandem oxidative coupling and annulation: An efficient approach to construct polysubstituted benzofurans, *J. Am. Chem. Soc.* **131**, 17387–17393 (2009).

89. S. Samanta, S. Mondal, S. Santra, G. Kibriya and A. Hajra, FeCl$_3$-catalyzed cross-dehydrogenative coupling between imidazoheterocycles and oxoaldehydes, *J. Org. Chem.* **81**, 10088–10093 (2016).

90. Q. Wang, J. Lou, P. Wu, K. Wu and Z. Yu, Iron-mediated oxidative C–H alkylation of S,S-functionalized internal olefins via C(sp^2)–H/C(sp^3)–H cross-coupling, *Adv. Synth. Catal.* **359**, 2981–2998 (2017).

91. J. Lou, Q. Wang, K. Wu, P. Wu and Z. Yu, Iron-catalyzed oxidative C–H functionalization of internal olefins for the synthesis of tetrasubstituted furans, *Org. Lett.* **19**, 3287–3290 (2017).

92. Y. Guan, K. Wang, J. Shen, J. Xu, C. Shen and P. Zhang, Iron-catalyzed C5 halogenation of 8-amidoquinolines using sodium halides at room temperature, *Catal. Lett.* **147**, 1574–1580 (2017).

93. X. Cong and X. Zeng, Iron-catalyzed, chelation-induced remote C–H allylation of quinolines via 8-amido assistance, *Org. Lett.* **16**, 3716–3719 (2014).

94. M. Cui, J.-H. Liu, X.-Y. Lu, X. Lu, Z.-Q. Zhang, B. Xiao and Y. Fu, Iron-mediated remote C–H bond benzylation of 8-aminoquinoline amides, *Tetrahedron Lett.* **58**, 1912–1916 (2017).

95. T.-J. Niu, J.-D. Xu, B.-Z. Ren, J.-H. Liu and G.-Q. Hu, Iron-catalyzed regioselective C5-H benzylation of 8-aminoquinolines with benzylic acetates, *ChemistrySelect* **4**, 4682–4685 (2019).

96. S. Chakraborty, J. Ahmed, B. K. Shaw, A. Jose and S. K. Mandal, An iron-based long-lived catalyst for direct C–H arylation of arenes and heteroarenes, *Chem. Eur. J.* **24**, 17651–17655 (2018).

97. (a) V. S. Thirunavukkarasu, S. I. Kozhushkov and L. Ackermann, C–H nitrogenation and oxygenation by ruthenium catalysis, *Chem. Commun.* **50**, 29–39 (2014); (b) J. Jiao, K. Murakami and K. Itami, Catalytic methods for aromatic C–H amination: An ideal strategy for nitrogen-based functional molecules, *ACS Catal.* **6**, 610–633 (2016).

98. (a) S. M. Paradine and M. C. White, Iron-catalyzed intramolecular allylic C–H amination, *J. Am. Chem. Soc.* **134**, 2036–2039 (2012); (b) L. Zhang and L. Deng, C–H bond amination by iron-imido/nitrene species, *Chin. Sci, Bull.* **57**, 2352–2360 (2012); (c) E. T. Hennessy and T. A. Betley, Complex N-heterocycle synthesis via iron-catalyzed, direct C–H bond amination, *Science* **340**, 591–595 (2013); (d) Y. Liu, X. Guan, E. L.-M. Wong, P. Liu, J.-S. Huang and C.-M. Che, Nonheme iron-mediated amination of C(sp^3)–H bonds. Quinquepyridine-supported iron-imide/nitrene intermediates by experimental studies and DFT calculations, *J. Am. Chem. Soc.* **135**, 7194–7204 (2013); (e) D. A. Iovan and T. A. Betley, Characterization of iron-imido species relevant for N-group transfer chemistry, *J. Am. Chem. Soc.* **138**, 1983–1993 (2016); (f) Z. Wang, Y. Zhang, H. Fu, Y. Jiang and Y. Zhao, Efficient intermolecular iron-catalyzed amidation of C–H bonds in the presence of N-bromosuccinimide, *Org. Lett.* **10**, 1863–1866 (2008); (g) T. Hatakeyama, R. Imayoshi, Y. Yoshimoto, S. K. Ghorai, M. Jin, H. Takaya, K. Norisuye, Y. Sohrin and M. Nakamura, Iron-catalyzed aromatic amination for nonsymmetrical triarylamine synthesis, *J. Am. Chem. Soc.* **134**, 20262–20265 (2012).

99. T. Zhang and W. Bao, Synthesis of 1*H*-indazoles and 1*H*-pyrazoles via FeBr$_3$/O$_2$ mediated intramolecular C–H amination, *J. Org. Chem.* **78**, 1317–1322 (2013).

100. T. Matsubara, S. Asako, L. Ilies and E. Nakamura, Synthesis of anthra-nilic acid derivatives through iron-catalyzed ortho amination of aromatic carboxamides with N-chloroamines, *J. Am. Chem. Soc.* **136**, 646–649 (2014).

101. Y. Aoki, R. Imayoshi, T. Hatakeyama, H. Takaya and M. Nakamura, Synthesis of 2,7-disubstituted 5,10-diaryl-5,10-dihydrophenazines via iron-catalyzed intramolecular ring-closing C–H amination, *Heterocycles* **90**, 893–900 (2015).

102. F.-F. Duan, S.-Q. Song and R.-S. Xu, Iron(II)-catalyzed sulfur directed $C(sp^3)$–H bond amination/C–S cross coupling reaction, *Chem. Commun.* **53**, 2737–2739 (2017).

103. A. V. G. Prasanthi, S. Begum, H. K. Srivastava, S. K. Tiwari and R. Singh, Iron-catalyzed arene C–H amidation using functionalized hydroxyl amines at room temperature, *ACS Catal.* **8**, 8369–8375 (2018).

104. I. T. Alt, C. Guttroff and B. Plietker, Iron-catalyzed intramolecular aminations of $C(sp^3)$–H bonds in alkylaryl azides, *Angew. Chem. Int. Ed.* **56**, 10582–10586 (2017).

105. B. Bagh, D. L. J. Broere, V. Sinha, P. F. Kuijpers, N. P. van Leest, B. de Bruin, S. Demeshko, M. A. Siegler and J. I. van der Vlugt, Catalytic synthesis of N-heterocycles via direct $C(sp^3)$–H amination using an air-stable iron(III) species with a redox-active ligand, *J. Am. Chem. Soc.* **139**, 5117–5124 (2017).

106. D. A. Iovan, M. J. T. Wilding, Y. Baek, E. T. Hennessy and T. A. Betley, diastereoselective C–H bond amination for disubstituted pyrrolidines, *Angew. Chem. Int. Ed.* **56**, 15599–15602 (2017).

107. Y.-D. Du, Z.-J. Xu, C.-Y. Zhou and C.-M. Che, An effective $[Fe^{III}(TF_4DMAP)Cl]$ catalyst for C–H bond amination with aryl and alkyl azides, *Org. Lett.* **21**, 895–899 (2019).

108. S. Pan, J. Liu, H. Li, Z. Wang, X. Guo and Z. Li, Iron-catalyzed N-alkylation of azoles via oxidation of C–H bond adjacent to an oxygen atom, *Org. Lett.* **12**, 1932–1935 (2010).

109. K.-Q. Zhu, L. Wang, Q. Chen and M.-y. He, Iron-catalyzed oxidative dehydrogenative coupling of ethers with aryl tetrazoles, *Tetrahedron Lett.* **56**, 4943–4946 (2015).

110. Q. Xia and W. Chen, Iron-catalyzed N-alkylation of azoles via cleavage of an sp^3 C–H bond adjacent to a nitrogen atom, *J. Org. Chem.* **77**, 9366–9373 (2012).

111. Q. Xia, W. Chen and H. Qiu, Direct C–N Coupling of imidazoles and benzylic compounds via iron-catalyzed oxidative activation of C–H bonds, *J. Org. Chem.* **76**, 7577–7582 (2011).

112. G. Saidulu, R. A. Kumar and K. R. Reddy, Iron-catalyzed C–N bond formation via oxidative C_{sp^3}–H bond functionalization adjacent to nitrogen in amides and anilines: Synthesis of *N*-alkyl and *N*-benzyl azoles, *Tetrahedron Lett.* **56**, 4200–4203 (2015).

113. X. Mao, Y. Wu, X. Jiang, X. Liu, Y. Cheng and C. Zhu, A highly regioselective sp³ C–H amination of tertiary amides based on Fe(II) complex catalysts, *RSC Adv.* **2**, 6733–6735 (2012).

114. M. Sun, T. Zhang and W. Bao, $FeCl_3$ catalyzed sp³ C–H amination: Synthesis of aminals with arylamines and amides, *Tetrahedron Lett.* **55**, 893–896 (2014).

115. D. Chen, F. Pan, J. Gao and J. Yang, Iron-catalyzed direct C(sp³)–H amination reactions of isochroman derivatives with primary arylamines under mild conditions, *Synlett* **24**, 2085–2088 (2013).

116. S. S. Mudaliar, A. P. Patel, J. J. Patel and K. H. Chikhalia, Iron-catalyzed cross-dehydrogenative C–N coupling of thiohydantoins with various amines, *Tetrahedron Lett.* **59**, 734–738 (2018).

117. Y. Cheng, W. Dong, L. Wang, K. Parthasarathy and C. Bolm, Iron-catalyzed hetero-cross-dehydrogenative coupling reactions of sulfoximines with diarylmethanes: A new route to N-alkylated sulfoximines, *Org. Lett.* **16**, 2000–2002 (2014).

118. L. Gu, W. Wang, Y. Xiong, X. Huang and G. Li, Synthesis of 2-arylbenzoxazoles through oxidation of C–H bonds adjacent to oxygen atoms, *Eur. J. Org. Chem.* 319–322 (2014).

119. J. Liu, K. Wu, T. Shen, Y. Liang, M. Zou, Y. Zhu, X. Li, X. Li and N. Jiao, Fe-catalyzed amination of (hetero)arenes with a redox-active aminating reagent under mild conditions, *Chem. Eur. J.* **23**, 563–567 (2017).

120. R. O. Torres-Ochoa, A. Leclair, Q. Wang and J. Zhu, Iron-catalysed remote C(sp³)–H azidation of *O*-Acyl oximes and *N*-acyloxy imidates enabled by 1,5-hydrogen atom transfer of iminyl and imidate radicals: Synthesis of γ-azido ketones and β-azido alcohols, *Chem. Eur. J.* **25**, 9477–9484 (2019).

121. (a) M. C. White and J. Zhao, Aliphatic C–H oxidations for late-stage functionalization, *J. Am. Chem. Soc.* **140**, 13988–14009 (2018); (b) A. C. Lindhorst, S. Haslinger, F. E. Kühn, Molecular iron complexes as catalysts for selective C–H bond oxygenation reactions, *Chem. Commun.* **51**, 17193–17212 (2015).

122. (a) C. Knight and M. J. Perkins, Concerning the mechanism of 'Gif' oxidations of cycloalkanes, *J. Chem. Soc. Chem. Commun.* 925–927 (1991); (b) D. H. R. Barto and D. Doller, The selective functionalization of saturated hydrocarbons: Gif chemistry, *Acc. Chem. Res.* **25**, 504–512 (1992); (c) D. H. R. Barton, On the mechanism of Gif reactions, *Chem. Soc. Rev.* **25**, 237–239 (1996); (d) M. J. Perkins, A radical reappraisal of Gif reactions, *Chem. Soc. Rev.* **25**, 229–236 (1996); (e) D. H. R. Barton, Gif chemistry: The present situation, *Tetrahedron* **54**, 5805–5817 (1998); (f) P. Stavropoulos, R. Çelenligil-Çetin and A. E. Tapper, The gif paradox, *Acc. Chem. Res.* **34**, 745–752 (2001).

123. (a) D. T. Sawyer, A. Sobkowiak and T. Matsushita, Metal [ML$_x$; M = Fe, Cu, Co, Mn]/hydroperoxide-induced activation of dioxygen for the oxygenation of hydrocarbons: oxygenated fenton chemistry, *Acc. Chem. Res.* **29**, 409–416 (1996); (b) P. A. MacFaul, D. D. M. Wayner and K. U. Ingold, A radical account of "oxygenated fenton chemistry," *Acc. Chem. Res.* **31**, 159–162 (1998); (c) C. Walling, intermediates in the reactions of fenton type reagents, *Acc. Chem. Res.* **31**, 155–157 (1998); (d) S. Goldstein and D. Meyerstein, Comments on the mechanism of the "fenton-like" reaction, *Acc. Chem. Res.* **32**, 547–55 (1999).

124. I. Prat, A. Company, V. Postils, X. Ribas, L. Que, J. M. Luis and M. Costas, The mechanism of stereospecific C–H oxidation by Fe(Pytacn) complexes: Bioinspired non-heme iron catalysts containing *cis*-labile exchangeable sites, *Chem. Eur. J.* **19**, 6724–6738 (2013).

125. M. Nakanishi and C. Bolm, Iron-catalyzed benzylic oxidation with aqueous *tert*-butyl hydroperoxide, *Adv. Synth. Catal.* **349**, 861–864 (2007).

126. (a) M. S. Chen and M. C. White, A predictably selective aliphatic C–H oxidation reaction for complex molecule synthesis, *Science* **318**, 783–787 (2007); (b) P. E. Gormisky and M. C. White, Catalyst-controlled aliphatic C–H oxidations with a predictive model for site-selectivity, *J. Am. Chem. Soc.* **135**, 14052–14055 (2013).

127. B. Retchera, J. S. Costa, J. Tang, R. Hage, P. Gameza and J. Reedijk, Unexpected high oxidation of cyclohexane by Fe salts and dihydrogen peroxide in acetonitrile, *J. Mol. Catal. A: Chem.* **286**, 1–5 (2008).

128. (a) L. Gómez, I. Garcia-Bosch, A. Company, J. Benet-Buchholz, A. Polo, X. Sala, X. Ribas and M. Costas, Stereospecific C–H oxidation

with H_2O_2 catalyzed by a chemically robust site-isolated iron catalyst, *Angew. Chem. Int. Ed.* **48**, 5720–5723 (2009); (b) L. Gómez, M. Canta, D. Font, I. Prat, X. Ribas and M. Costas, Regioselective oxidation of nonactivated alkyl C–H groups using highly structured non-heme iron catalysts, *J. Org. Chem.* **78**, 1421–1433 (2013).

129. S. Jana, M. Ghosh, M. Ambule and S. Sen Gupta, Iron complex catalyzed selective C–H bond oxidation with broad substrate scope, *Org. Lett.* **19**, 746–749 (2017).

130. K. Möller, G. Wienhöfer, K. Schröder, B. Join, K. Junge and M. Beller, Selective iron-catalyzed oxidation of phenols and arenes with hydrogen peroxide: Synthesis of vitamin E intermediates and vitamin K3, *Chem. Eur. J.* **16**, 10300–10303 (2010).

131. (a) Y.-T. Huang, S.-Y. Lu, C.-L. Yi and C.-F. Lee, Iron-catalyzed synthesis of thioesters from thiols and aldehydes in water, *J. Org. Chem.* **79**, 4561–4568 (2014); (b) H. Wang, L. Wang, J. Shang, X. Li, H. Wang, J. Gui and A. Lei, Fe-catalysed oxidative C–H functionalization/C–S bond formation, *Chem. Commun.* **48**, 76–78 (2012).

132. B. J. Groendyke, A. Modak and S. P. Cook, Fenton-inspired C–H functionalization: peroxide-directed C–H thioetherification, *J. Org. Chem.* **84**, 13073–13091 (2019).

133. (a) I. A. I. Mkhalid, J. H. Barnard, T. B. Marder, J. M. Murphy and J. F. Hartwig, C–H activation for the construction of C-B bonds, *Chem. Rev.* **110**, 890–931 (2010); (b) J. F. Hartwig, Borylation and silylation of C–H bonds: A platform for diverse C–H bond functionalizations, *Acc. Chem. Res.* **45**, 864–873 (2012).

134. (a) K. M. Waltz, X. He, C. Muhoro and J. F. Hartwig, Hydrocarbon functionalization by transition metal boryls, *J. Am. Chem. Soc.* **117**, 11357–11358 (1995); (b) K. M. Waltz and J. F. Hartwig, Selective functionalization of alkanes by transition-metal boryl complexes, *Science* **277**, 211–213 (1997); (c) K. M. Waltz, C. N. Muhoro and J. F. Hartwig, C–H activation and functionalization of unsaturated hydrocarbons by transition-metal boryl complexes, *Organometallics* **18**, 3383–3393 (1999); (d) C. E. Webster, Y. Fan, M. B. Hall, D. Kunz and J. F. Hartwig, Experimental and computational evidence for a boron-assisted, σ-bond metathesis pathway for alkane borylation, *J. Am. Chem. Soc.* **125**, 858–859 (2003).

135. T. Hatanaka, Y. Ohki and K. Tatsumi, C–H bond activation/boryla-tion of furans and thiophenes catalyzed by a half-sandwich iron N-heterocyclic carbene complex, *Chem. Asian J.* **5**, 1657–1666 (2010).

136. C. Cheng and J. F. Hartwig, Catalytic silylation of unactivated C–H bonds, *Chem. Rev.* **115**, 8946–8975 (2015).

137. Y. Sunada, H. Soejima and H. Nagashima, Disilaferracycle dicarbonyl complex containing weakly coordinated η^2-(H–Si) ligands: applica-tion to C–H functionalization of indoles and arenes, *Organometallics* **33**, 5936–5939 (2014).

138. J. Ito, S. Hosokawa, H. B. Khalid and H. Nishiyama, Preparation, characterization, and catalytic reactions of NCN pincer iron com-plexes containing stannyl, silyl, methyl, and phenyl ligands, *Organometallics* **34**, 1377–1383 (2015).

139. (a) S. H. Doan, K. D. Nguyen, P. T. Huynh, T. T. Nguyen and N. T. S. Phan, Direct C–C coupling of indoles with alkylamides via oxidative C–H functionalization using $Fe_3O(BDC)_3$ as a productive heteroge-neous catalyst, *J. Mol. Catal. A: Chem.* **423**, 433–440 (2016); (b) S. Haldar and S. Koner, Iron-containing mesoporous aluminosilicate: A highly active and reusable heterogeneous catalyst for hydroarylation of styrenes, *J. Org. Chem.* **75**, 6005–6008 (2010); (c) P. H. Pham, S. H. Doan, H. T. T. Tran, N. N. Nguyen, A. N. Q. Phan, H. V. Le, T. N. Tu and N. T. S. Phan, A new transformation of coumarins via direct C–H bond activation utilizing an iron-organic framework as a recyclable catalyst, *Catal. Sci. Technol.* **8**, 1267–1271 (2018).

140. T. Zeng, G. Song, A. Moores and C.-J. Li, Magnetically recoverable iron nanoparticle catalyzed cross-dehydrogenative coupling (CDC) between two Csp^3–H bonds using molecular oxygen, *Synlett*, 2002–2008 (2010).

141. K. D. Nguyen, S. H. Doan, A. N. V. Ngo, T. T. Nguyen and N. T. S. Phan, Direct C–N coupling of azoles with ethers via oxidative C–H activation under metal–organic framework catalysis, *J. Ind. Eng. Chem.* **44**, 136–145 (2016).

142. S. H. Doan, V. H. H. Nguyen, T. H. Nguyen, P. H. Pham, N. N. Nguyen, A. N. Q. Phan, T. N. Tu and N. T. S. Phan, Cross-dehydrogenative coupling of coumarins with Csp^3–H bonds using an

iron-organic framework as a productive heterogeneous catalyst, *RSC Adv.* **8**, 10736–10745 (2018).

143. C. Rajendran and G. Satishkumar, A sustainable heterogeneous iron catalyst Fe/SBA-15 towards direct arylation of unactivated C(sp^3)–H and C(sp^2)–H bonds with arenes, *ChemCatChem* **9**, 1284–1291 (2017).

144. P. Basu, T. K. Dey, A. Ghosh and S. M. Islam, Designing of a new heterogeneous polymer supported naphthyl-azo iron catalyst for the selective oxidation of substituted methyl benzenes, *J. Inorg. Organomet. Polym. Mater.* **28**, 1158–1170 (2018).

https://doi.org/10.1142/9781786349620_0007

Chapter 7

Iron-Catalyzed C–H Functionalization Reactions *via* Carbene and Nitrene Transfer Reactions

Claire Empel, Sripati Jana, and Rene M. Koenigs*

Institute of Organic Chemistry, RWTH Aachen University, Landoltweg 1, D-52074 Aachen, Germany
rene.koenigs@rwth-aachen.de

Iron complexes are currently emerging as versatile group of catalysts to conduct efficient C–H functionalization reactions and different strategies have been developed thus far. In this chapter, we discuss the recent advances made in the area of iron-catalyzed carbene and nitrene transfer reactions to conduct C–H functionalization reactions. We give an overview of privileged iron catalysts in carbene transfer reactions and a brief overview of the development of enzymatic carbene transfer reactions. We conclude with the discussion of C–H functionalization reaction *via* iron-catalyzed carbene transfer and discuss the advances in made in last five years on enzymatic transformations. Finally, we discuss iron-catalyzed nitrene transfer reactions using azides as nitrene precursor to enabling C–H amination reactions.

7.1 Iron and Its Privileged Role

The direct functionalization of unreactive C–H bonds was long considered one of the main challenges to synthetic organic chemists
and over the past decades this research area flourished leading to
the development of a plethora of synthetic methods for the functionalization reaction of C–H bonds, e.g., *via* the directing group-
assisted C–H activation, the innate reactivity of organic molecules,
or *via* direct C–H functionalization (Scheme 7.1).[1–4] The latter
describes a highly desirable process as it does not necessitate the
prefunctionalization of substrates or the removal of directing groups
in a following synthesis step that would impact on the step-economy
of the overall C–H functionalization reaction. In light of the demand
to streamline organic synthesis procedures, the direct C–H functionalization is highly attractive to reduce both economic and environmental impact of organic synthesis. Although highly desired, the
direct functionalization of C–H bonds represents a highly challenging transformation, as inert C–H bonds need to be activated and
addressed selectively. For this purpose, elegant methods have been
developed in last seven years, mainly relying on the use of expensive
and mostly toxic precious metal catalysts based on rhodium, irid-
ium, palladium, and others.[5–7] The removal of trace amounts of
these noble metal catalysts from drugs and agrochemicals

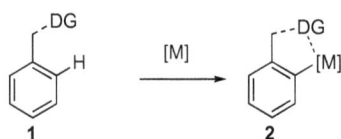

(A) Directing group assisted C–H activation   (B) Innate reactivity of organic molecules

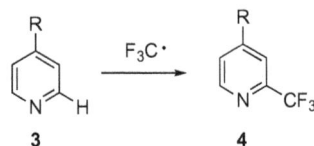

(C) Direct C–H functionalization of C–H bonds

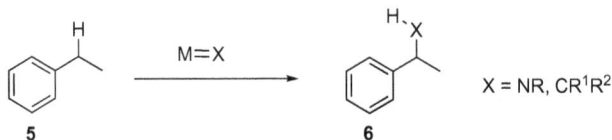

Scheme 7.1.   Strategies for the functionalization of C–H bonds.

represents another challenge associated with applications in an industrial environment.

Against this background, the development of the C–H functionalization reaction with third-row transition metal complexes has received significant attention to reducing the economic and environmental footprint of organic synthesis methodology.[8] Iron is the most abundant transition metal on Earth and the natural occurrence of iron-containing minerals is by far larger compared to other first-row transition metals such as cobalt, manganese, or copper.[9] The electron configuration of iron enables a broad number of possible oxidation and spin states making iron an attractive metal for the development of novel C–H functionalization reactions.[10] In nature, iron-containing enzymes are widespread and can be found in P450 enzymes, iron–sulfur clusters and play a pivotal role in oxygen transport in mammals or in the detoxification of xenobiotics.[11]

The high natural abundance of iron in the Earth's crust, cost-effectiveness, chemical reactivity, and low toxicity thus make iron catalysts a formidable choice to conduct C–H functionalization reactions.[11] In this chapter, we will discuss the advances made in this research area with a focus on C–H functionalization reactions with carbenes and nitrenes.

7.2 Synthetic Iron Complexes in Carbene Transfer Reactions

Carbene transfer reactions with diazoalkanes have developed to an important strategy to conduct C–H functionalization reactions with exquisite chemo- and enantioselectivity. Early reports date back to 1942 and 1959, when Meerwein and Doering reported on photochemical C–H functionalization with diazoalkanes. Yet missing control on the reactivity of the free carbene intermediates led to unselective reactions.[12,13]

These shortcomings led to the development of metal-catalyzed carbene transfer reactions, which are today most often conducted using noble metals, such as Ru(II), Ir(III), Au(I), Pd(II), or Cu(I).[14–17] A particularly important class of catalysts are Rh(II) paddlewheel

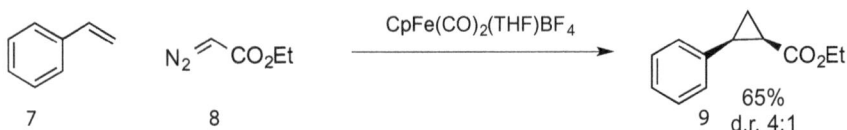

Scheme 7.2.   Strategies for the functionalization of C–H bonds.

complexes that are required to stabilize and control the reactivity of the electronically unsaturated carbene fragment to enable highly chemo- and enantioselective C–H functionalization reactions with diazoalkanes.[18]

The first carbene transfer reaction with iron complexes was reported by Hossain *et al.* using the cationic $CpFe(CO)_2(THF)BF_4$ complex in the cyclopropanation reaction of styrene (**7**) with ethyl diazoacetate (**8**).[19] Following this report, different groups reported on their efforts to conduct iron-catalyzed carbene transfer reactions (Scheme 7.2).[20,21]

Hemine is a cofactor in many enzymes and is a naturally occurring iron porphyrin complex. The latter were applied in iron-catalyzed carbene transfer reaction first by Woo *et al.*, who reported on *trans*-selective cyclopropanation reaction of styrene derivatives with ethyl diazoacetate (**8**) using an iron porphyrin complex in 1995.[22] In this report and in further reports by Che *et al.* and Aviv *et al.*, it was demonstrated that the porphyrin ligand can be fine-tuned to alter reactivity and selectivity of the iron porphyrin complex in carbene transfer reactions (Scheme 7.3).[23,24] The axial ligand of iron porphyrin complexes plays an important role in carbene transfer reactions and several reports describe the influence of additives such as triphenyl phosphine or arsine, dimethylamino pyridine, and *N*-methyl imidazole[23] and the first systematic study on their influence on cyclopropanation reactions was described by Carreira and Morandi.[25] Today, synthetic iron porphyrin complexes are one of the most important catalysts based on iron to conduct carbene transfer reactions.

Further developments of synthetic iron complexes in carbene transfer reactions focused on the application of either nitrogen- or

Iron porphyrin complexes   Iron corrole complexes   Hemine

10a, Ar = Ph
10b, Ar = C$_6$F$_5$

10c, Ar =

11, Ar = C$_6$F$_5$, X = OEt$_2$ **12**

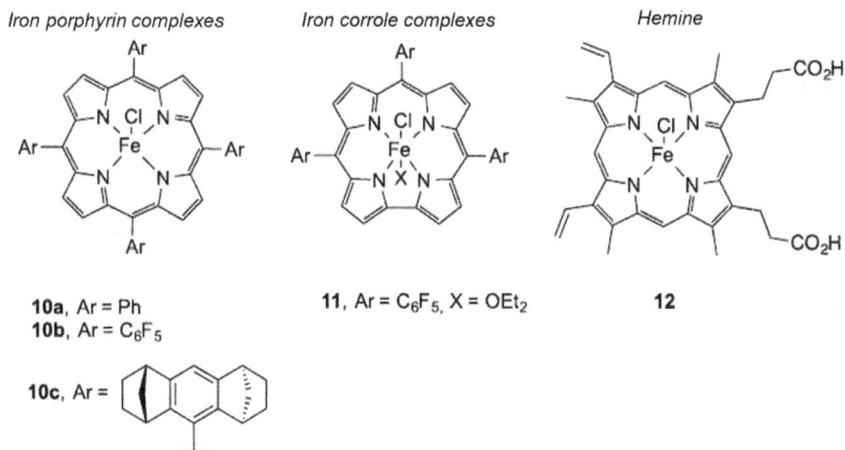

Scheme 7.3. Overview on iron porphyrin and corrole complexes used in carbene transfer reactions.

Privileged (chiral) ligands for complexes of iron salts

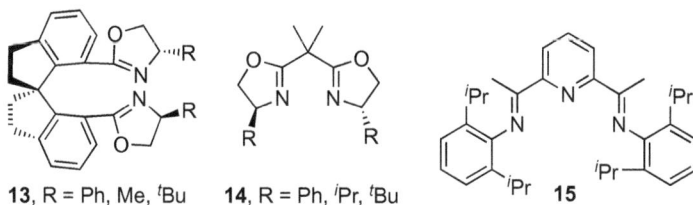

13, R = Ph, Me, ^{t}Bu   **14**, R = Ph, ^{i}Pr, ^{t}Bu   **15**

Scheme 7.4. Overview on iron porphyrin and corrole complexes used in carbene transfer reactions.

phosphorous-based ligands, including di- and tridentate ligands (Scheme 7.4), the application of which will be discussed at a later point in this chapter.[25–28]

    Iron porphyrin complexes are today important catalysts to conduct carbene transfer reactions, which proceed *via* a highly reactive iron carbene intermediate. Khade and Zhang studied the formation of this key intermediate by quantum chemical studies, which revealed an initial coordination of ethyl phenyldiazoacetate to the iron porphyrin complex (**16a**), followed by extrusion of nitrogen and formation of the iron carbene complex (**16d**) (Scheme 7.5). In

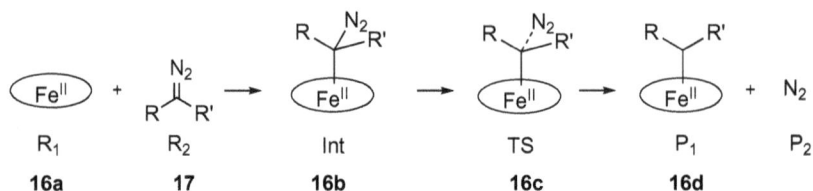

Scheme 7.5.   Mechanism of the formation of iron carbene complexes.

these studies, the authors also studied the influence of the axial ligand and showed that *N*-methyl imidazole reduces both the reaction energy ΔG and the activation energy ΔG^{TS}, thus resulting in stabilization of the iron carbene intermediate.[29]

In the next sections, the applications of iron catalysts in C–H functionalization reactions with diazoalkanes and azides will be discussed.

7.3 Iron-Catalyzed C–H Functionalization with Carbenes

The direct functionalization of the inert C–H bonds allows one of the most step economic transformations and thus shortens synthesis strategies. Transition metal-catalyzed C–H activation reaction has been used as an efficient tool for the atom and step-economical organic transformation. During the past decades, the C–H bond activation strategy has been successfully realized mostly with expensive 4d and 5d transition metals.[14-17] In contrast, naturally abundant first-row transition metal complexes have been thus far underutilized in C–H activation strategy. In the last four years, reports on cost-effective 3d transition metal such as copper, manganese, and iron are used in the catalytic activation of inert C–H bonds have increased noticeably.[8] A common strategy for the functionalization of C–H bonds is the metal-mediated insertion reactions of carbene fragments into the inert C–H bond. The application of organoiron species for C–H bond activation was first reported in 2008 by Nakamura and co-workers, and since then the number of iron-catalyzed C–H functionalization reactions rapidly increased (Scheme 7.6).[1]

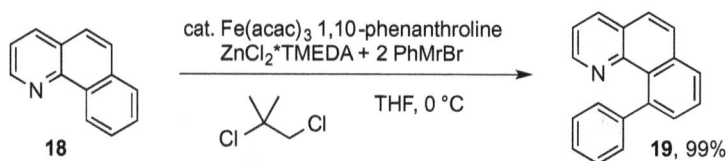

Scheme 7.6. Iron-catalyzed C–C bond forming reaction by C–H activation described by Nakamura *et al.* in 2008.

The high reactivity of the organoiron complexes often enables reaction at room temperature and offers the opportunity for low catalyst loadings as high turnover numbers (TON) are reported. Another positive aspect using iron in C–H bond functionalization reactions is often higher chemoselectivity compare to equivalent cross-coupling reactions. The reduced reactivity of the C–H bond compared to the C–X bond enables a more selective activation step and increases the chemoselectivity of the reaction.[30]

7.3.1 *Early stoichiometric examples on C–H functionalization with carbenes*

In 1996, Helquist *et al.* described the stoichiometric application of iron complexes in the functionalization of aliphatic C–H bonds *via* an intramolecular C–H insertion reaction to furnish cyclopentane derivatives (**21**). Using this approach, different *trans* substituted bi- and tricyclic systems can be synthesized from the preexisting six-, seven-, and eight- membered ring. Interestingly, only one diastereomer was obtained when the C–H insertion reaction took place at the benzylic and allylic position (Scheme 7.7).[31]

In further studies, the same group investigated the diastereoselectivity of the reaction. When using alkyl side chains the diastereoselectivity increased with sterically demanding substituents. Moreover, different substituents at the insertion side were introduced and even sterically equivalent substituents showed high diastereoselectivity. A possible explanation for the observed diastereoselectivity is the reaction pathway *via* a concerted C–H insertion including a chair-like cyclic transition state (**23**) as illustrated in Scheme 7.8.[32,33]

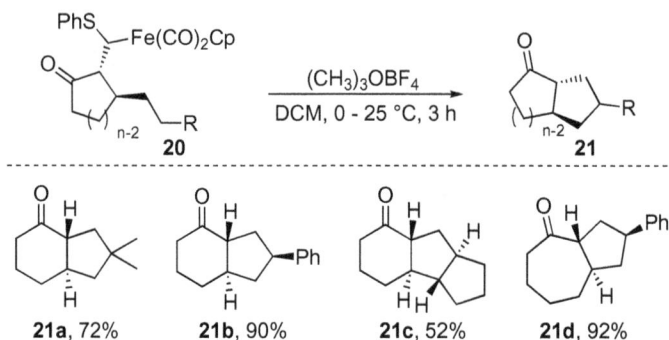

Scheme 7.7.　Intramolecular C–H insertion reactions for the synthesis of cyclopen-tane derivatives (**21**) by the Helquist group.

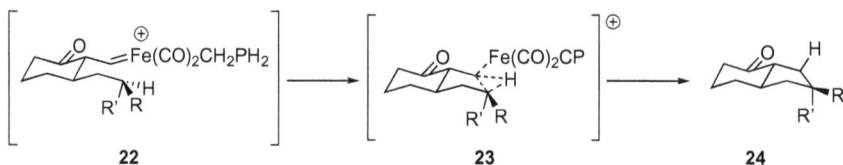

Scheme 7.8.　Investigation on the diastereoselectivity of the intramolecular C–H insertion reaction by Helquist *et al.*

7.3.2 *C–H functionalization reaction of aromatic C–H bonds*

The direct iron-catalyzed functionalization of benzene and alkyl benzenes *via* a carbene intermediate can provide three different products—insertion into a $C(sp^2)$-H bond, insertion into a $C(sp^3)$-H bond, and the addition reaction (Scheme 7.9).

Studies from Luis, Pérez, and coworkers using iron(II)-complexes bearing pytacn ligands and ethyl diazoacetate as carbene precursors showed high chemoselectivity and the insertion into the aromatic $C(sp^2)$-H bonds of benzene and alkyl benzene was detected as the main product. The formation of the cyclohep-tatriene could be observed in traces when using benzene (<1%) (Scheme 7.10, right).[34]

A scope of benzene derivatives (**30**) was investigated by the Pérez group; several monoalkyl substituted benzene derivatives (**30**)

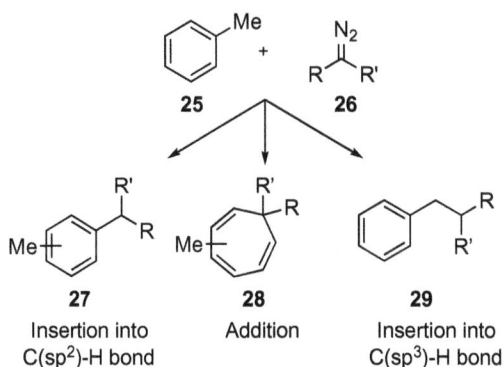

Scheme 7.9. Possible product in the iron-catalyzed functionalization of toluene (**25**) by carbene insertion.

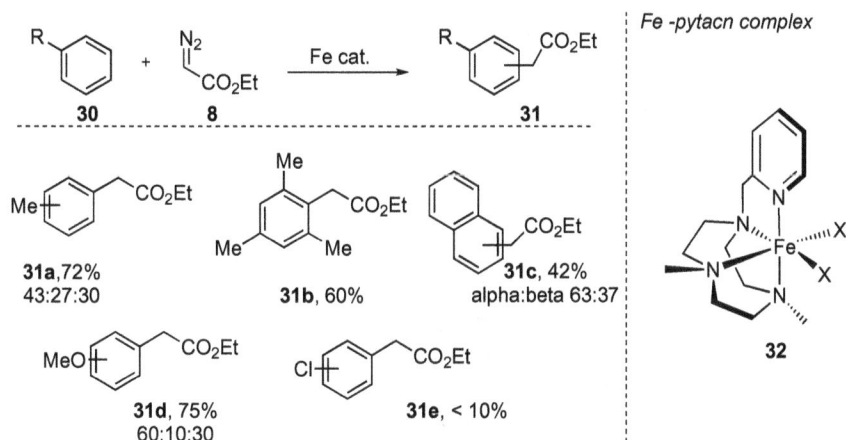

Scheme 7.10. Left: functionalization of arenes using ethyl diazoacetate; Right: iron-pytacn-complex.

formed exclusively the corresponding $C(sp^2)$-H bond insertion product in good yield. In all cases, a mixture of the three possible isomers was obtained. When using electron-withdrawing groups (e.g., Cl) the corresponding product was isolated in reduced yield (Scheme 7.10, left).[35]

Furthermore, the Pérez group investigated the mechanisms of the reaction; in their proposed mechanism, a dicationic species is

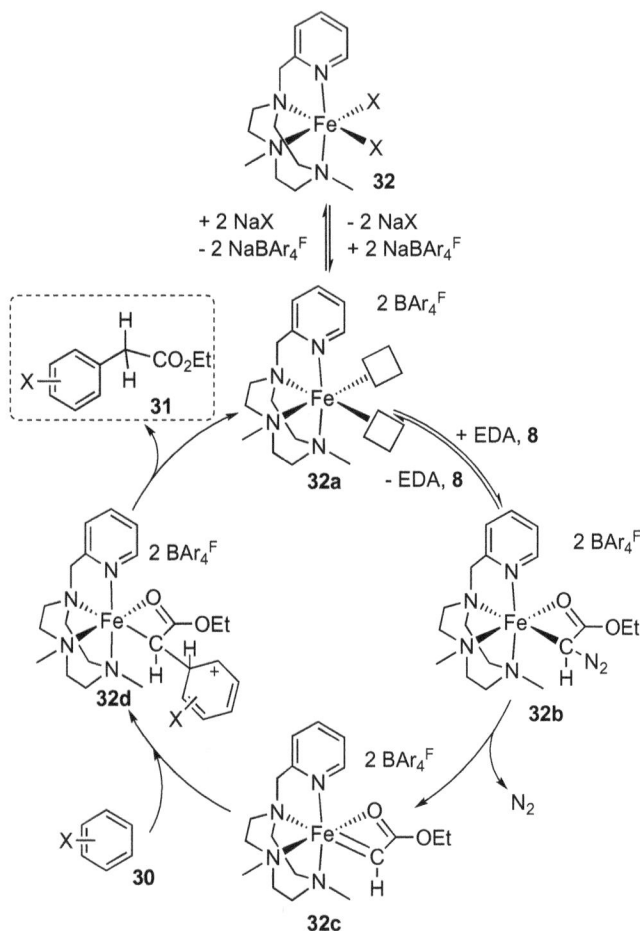

Scheme 7.11.   Proposed reaction pathway for the C–H functionalization reaction of arenes (**30**) using ethyl diazoacetate (**8**).

pointed out to be the reactive species (**32a**). The diazo compound (**8**) coordinates to the active species—several different coordination to the metal center are possible in this step and the coordination illustrated in Scheme 7.11 should only give an impression of a possible coordination (**32b**). In a next step, nitrogen gas is released and the metal carbene species (**32c**) is formed, that later is attacked by the arene (**30**). A Wheland-type intermediate (**32d**) is formed, which is proposed to undergo a 1,2-H shift to release the product (**31**) and the active catalytic species (**32a**) as described by Doyle, Padwa, and coworkers (Scheme 7.11).[35,36]

The Woo group investigated the C–H insertion reaction of benzene derivatives—their approach involves iron porphyrin carbenes, which are generated from different diazo compounds. Following a similar approach, $C(sp^2)$-H bond and $C(sp^3)$-H bond insertion reaction took place. Using an acceptor/acceptor diazo compound (**34**) the generated iron carbene complex smoothly underwent the $C(sp^2)$-H and $C(sp^3)$-H bond insertion reaction. Different benzene derivatives such as toluene, mesitylene, *p*-Cl-toluene (**33**), and anisole reacted under the present reaction conditions. The chemoselectivity between $C(sp^2)$-H bond and $C(sp^3)$-H bond insertion reaction using iron porphyrin complexes is relatively low and selectivity in the range of 2:1–10:1 were observed with the $C(sp^3)$-H bond insertion reaction product as the major product (Scheme 7.12).[37]

The iron-catalyzes C–H insertion reaction of donor/acceptor diazo compounds (**37**) with electron-rich benzene rings was investigated by Zhou *et al.* in 2015. A simple, nonporphyrin iron complex was generated *in situ* from $FeCl_3$, 1, 10-phenanthroline and $NaBAr_F$ enables the arylation of *N, N*-dimethylaniline (**38**) mild conditions. The applicability of this transformation was further investigated by studying various phenyl diazoesters and *N,N*-dimethylaniline derivatives—moreover, scale up experience were conducted (Scheme 7.13).[38]

The generation of iron carbene complexes using diazo compounds was also investigated by Deng and coworkers. The approach

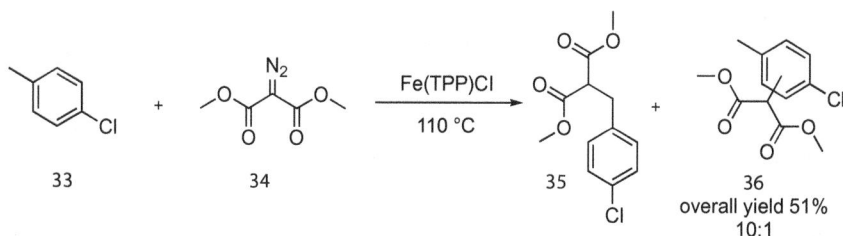

Scheme 7.12. Chemoselectivity between $C(sp^2)$-H bond and $C(sp^3)$-H bond insertion reaction investigated by Woo and co-workers.

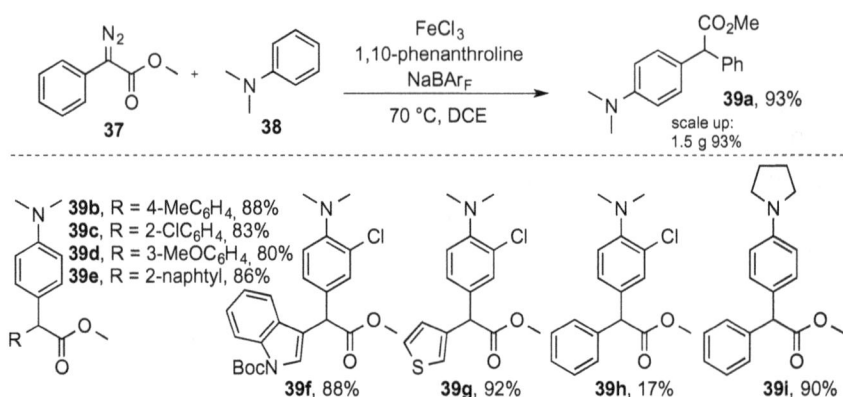

Scheme 7.13. Iron-catalyzed arylation of donor/acceptor diazo compound (37) with N,N-dimethylaniline (38) by the Zhou group.

involves an iron carbene complex generated from donor/acceptor diazo compound (37) and a bis(imino)pyridine iron complex. Following this strategy, the direct C–H functionalization of N,N-dimethylaniline (38) was carried out, leading to only one product (39a) in 71% yield.[28]

7.3.3 C–H functionalization reaction of heteroaromatic C–H bonds

N-heteroaromatic compounds like indoles or pyrroles are privileged moieties in drug molecules and natural products. The variety of synthetic strategies for the construction and the functionalization of these frameworks is a long-standing research field for organic chemists. Especially, the direct C–H bond functionalization of heteroaromatic would allow molecular diversity in late-stage functionalization reactions. This approach in combination with low-biotoxic iron catalysts and low catalyst loadings now enables application in medicinal chemistry, total synthesis, and agrochemistry.

The C–H functionalization of the indole moiety was extensively studying under precious metal-catalyzed reaction conditions—enabling the direct and selective C–H functionalization at various positions. First studies date back to the 1980s when Itahara, Ikeda, and

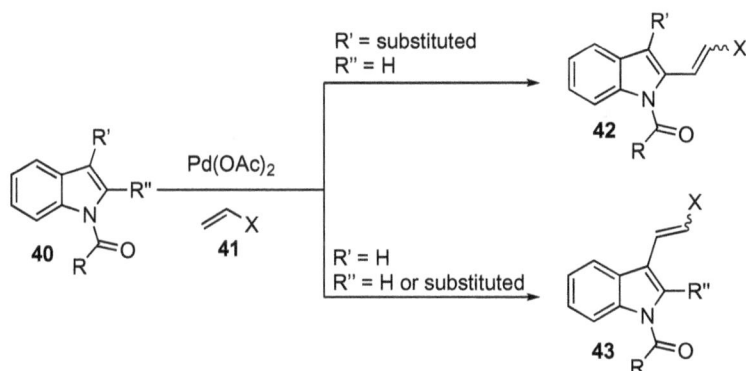

Scheme 7.14. Early investigations on C–H functionalization of indole heterocycles.

Sakakibara reported on the alkenylation of 1-acylindoles (**40**) using palladium acetate (Scheme 7.14).[39]

The implementation of functional groups in the C2- and C3-position is investigated in more detail compare to the other positions of the indole framework.[40,41]

In the last two years, iron carbene transfer reactions were studied resulting in valuable work from the Koenigs and the Fasan group for the direct C–H functionalization of indole derivatives with diazo compounds as carbene precursors.[42,43]

Early studies on iron porphyrin-catalyzed functionalization on indole (**44**) using ethyl diazoacetate (**8**) as carbene precursor and Fe(tetraphenylporphyrin, TPP) Cl as carbene transfer catalyst did not provide any reaction as described by Woo *et al.* in 2006. While for the close relative pyrrole (**45**), the C–H insertion product (**46**) was obtained in 37% yield (Scheme 7.15).[44]

This observation points at a special reactivity of the indole heterocycle. And only in 2019, Koenigs *et al.* reported on an iron-catalyzed C–H functionalization reaction of indoles in the C3-position using the same catalyst as described by the Woo group—Fe(TPP) Cl—and *in situ* generated diazo acetonitrile (**48**). Following this approach, protected and unprotected indoles and indazole were selectively functionalized in the C3-position, while no reaction was observed when blocking the C3-position (Scheme 7.16).

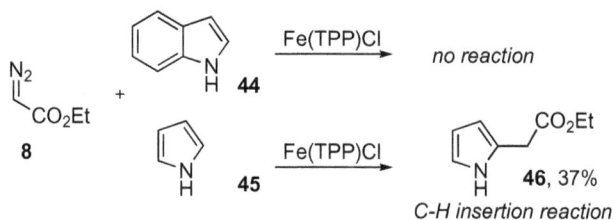

Scheme 7.15. Iron-catalyzed reaction of ethyl diazoacetate (**8**) with indole (**44**) and pyrrole (**45**) reported by Woo *et al.*

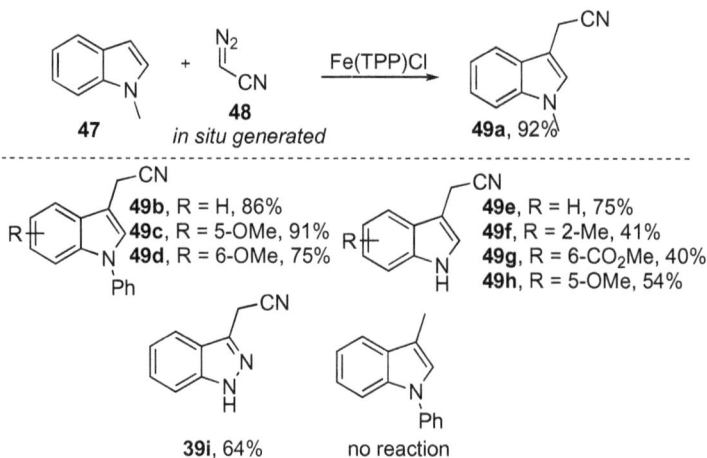

Scheme 7.16. Reaction of diazoacetonitrile (**48**) with indole heterocycles using Fe(TPP)Cl reported by Koenigs and coworkers.

The C–H functionalization of the indole heterocycle was also investigated using the heme-containing *Escherichia coli* protein YfeX. A library of YfeX variants and the wildtype protein were investigated. The TON of the YfeX wildtype (TON = 37) could be improved to 80 with YfeX I230A as the mutation of the wildtype. Moreover, the YfeX I230A enzyme was investigated using ethyl diazoacetate (**8**) instead of diazo acetonitrile (**48**), which leads to an increased TON of 236 (Scheme 7.17).[42]

A similar approach was performed by the Fasan group; different variants of myoglobin (Mb) were studied as biocatalyst in the

Koenigs and coworkers

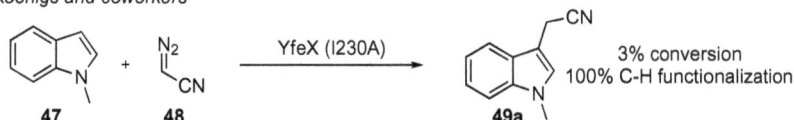

Fasan and coworkers

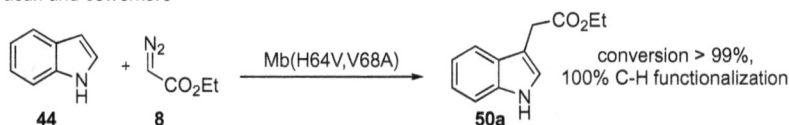

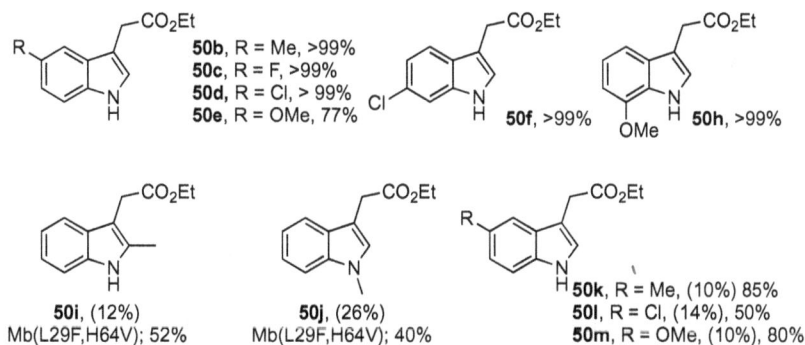

Scheme 7.17. Enzyme-catalyzed C3-alkylation reaction of indoles by the Koenigs and the Fasan group.

C3-functionalization of unprotected indoles using ethyl diazoacetate (**8**) as carbene precursor. In fist studies, the wildtype Mb and Mb(H64V) were tested and no C–H functionalization product was obtained. In a next step, the Mb wildtype was modified at the active side. The variant of Mb(H64V,V68A) in cells showed the highest conversion and a TON of 82 was observed detecting the C–H functionalization product (**50a**) only. The functional group tolerance of the enzyme-catalyzed transformation was investigated– different substitutions on the indole framework were well tolerated. When turning to *N*-protected indoles, the yield decreases to only 5%–26% using Mb(H64V,V68A) as catalyst, which indicated limitations of the substrate scope. The change to Mb(L29F,H64V) variant with a higher activity for the substrate improved the yield of the N-protected indoles to a moderate to good level (Scheme 7.17).[43]

In similar studies from Arnold *et al.* cytochrome P450 variants were investigated in the C–H functionalization of indole heterocycle. The approach of directed evolution was used to develop "carbene transferase" enzymes for the direct functionalization of inert C–H bonds. In a first step, cytochrome P450 variants for the C3-alkylation of the indole framework were investigated using a UV-Vis spectrophotometric-bases high-throughput screening, which opened up the opportunity for several thousand enzyme variants per generation. During the screening, site-saturation and random mutagenesis were investigated, which leads to a set of cytochrome P411 variants. The engineered cytochrome P411 variants showed enantioselective C–H functionalization. Moreover, the cytochrome P411 variants were tested in reactions with pyrrole (**53**)—a regioselective C–H functionalization in the C3-position or the C2-position could be conducted using different P411 variants (Scheme 7.18).[44]

The Zhou group reported on an enantioselective approach using $Fe(ClO_4)_2$ as iron source and chiral spirobisoxazoline ligands (**58**) for the C–H functionalization reaction of the C3-position of the indole heterocycle (**56**). Under relatively mild reaction conditions, various α-aryl-α-diazoester (**37**) underwent the C–H functionalization and

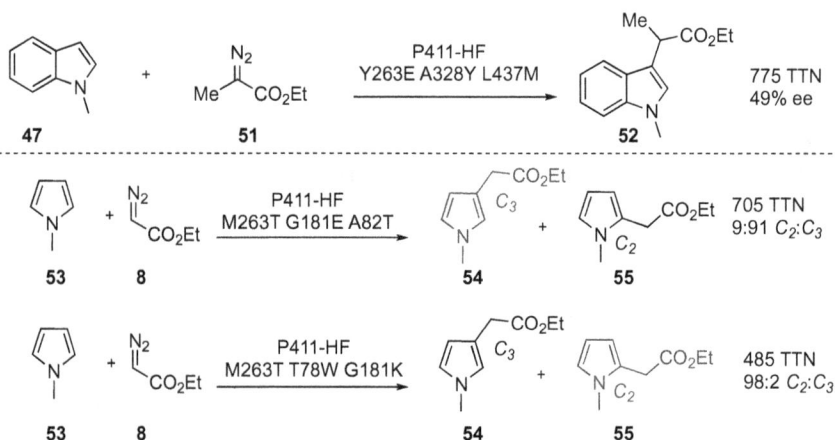

Scheme 7.18.   Scope of C3-alkyltion using engineered P411-HF enzymes; Regioselective alkylation of 1-methylpyrrole (**53**) reported by Arnold and coworkers.

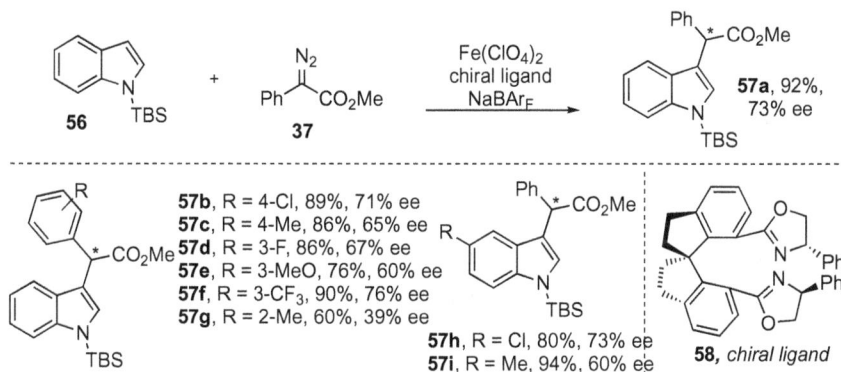

Scheme 7.19. Enantioselective iron-catalyzed C–H functionalization of indoles (**56**) with α-aryl-α-diazoesters (**37**) described by Zhou *et al.*

the corresponding reaction products were isolated in high yields and high enantiomer excess. The introduction of functional groups on the indole framework did not affect the reaction (Scheme 7.19).[27]

A variety of different mechanistic proposals based on experimental studies for the reaction mechanism of the iron-catalyzed functionalization of the indole framework are proposed. The utilization of deuterium labeling and competition experiments is a common way to determine possible reaction mechanisms. The Koenigs group used those strategies to investigate the C3-functionalization reaction of indole with diazo acetonitrile (**8**) in more detail. The reaction of 3-deutero-*N*-methylindole (***d*-47**) was studied with both catalysts—Fe(TPP) Cl and YfeX—the product (***d*-49**) was obtained with 55% and 35% deuterium label, respectively. Competition experiments of a 1:1 mixture of *N*-methylindole (**49**) and 3-deutero-*N*-methylindole (***d*-49**) resulted in a 5:1 mixture of the unlabeled and labeled product. The Zhou group reported on similar hydrogen -deuterium exchange (H/D) ratios for competition experiments in the reaction of *N*-benzylindole with donor/acceptor diazo compounds and proposed an addition/[1, 2]-proton transfer mechanism.[27] To further investigate the potential of a radical reaction pathway, the radical scavenger TEMPO (**59**) was added to the reaction mixture, resulting in complete inhibition of the product

Reaction with deutero indole

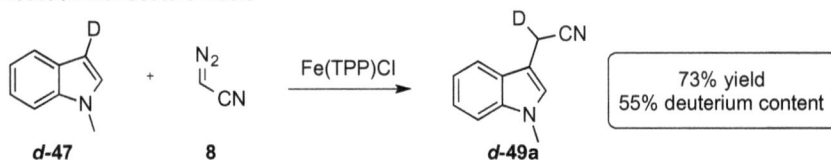

d-47   8   *d-49a*

73% yield
55% deuterium content

Competition experiments

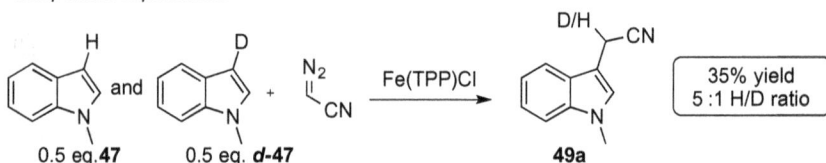

0.5 eq.**47**   0.5 eq. *d-47*   **49a**

35% yield
5 :1 H/D ratio

Reaction in the presence of radical scavenger

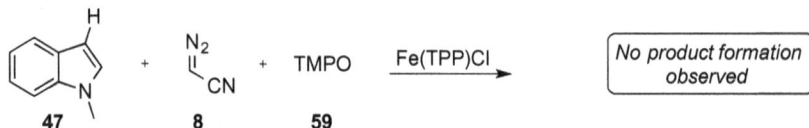

47   **8**   **59**

No product formation
observed

Scheme 7.20.   Experience for mechanistic investigations for the C3-functionalization of indole (**47**) by Koenigs and coworkers.

formation. Based on those experimental data the authors postulated a radical reaction pathway, yet unclear in which particular step the radical intermediate is involved (Scheme 7.20).[42]

The Fasan group investigated the mechanism of their studies for the direct functionalization of the unprotected indole (**44**) with ethyl diazoacetate (**8**). Their proposed mechanism involves the formation of a heme carbene intermediate from the catalytically active ferrous Mb and diazo. The carbene intermediate undergoes a nucleophilic attack by the indole substrate (**44**) to generate a zwitterionic intermediate (**60/61**). Production of the final product can be explained in two different ways either *via* a 1, 2-proton shift from the C3-atom to the ester α-carbon atom or the dissociation of the zwitterionic species and protonation by the solvent molecules. To justify the proper reaction pathways, the reaction between 1-methyl-3-deutero-indole (*d-44*) and ethyl diazoacetate (**8**) was investigated under enzymatic conditions. The product (**50a**) was obtained with complete loss of the deuterium label. Competition experiment of a 1:1 mixture of N-methylindole (**44**) and 3-deutero-*N*-methylindole (d-**44**) resulted in no enrichment of protonated substrate over time, which indicates a lack of kinetic isotope effect (KIE). This competition experiment resulted in no retention of deuterium label in the product (**50a**) and absence of KIE, which rule out a carbene C–H insertion process. Based on these experimental data, the occurrence of solvent-mediated protonation (D/H exchange) during the Mb(H64V,V68A)-catalyzed reaction is postulated (Scheme 7.21).[43]

The Zhou group reported on a proposed mechanism involving the generation of an iron carbene (**63**) in a first step, that reacts with the indole (**47**) to form a zwitterionic intermediate (**64/65**). In the next step, a proton migration step occurs—similar to the mechanism described by Fasan *et al.* The authors investigated competition studies and observed a KIE of 5.06. The significant first-order KIE value suggest the proton transfer as the rate-determining step (Scheme 7.22).[27]

The mechanistic studies in the C–H functionalization reaction of the indole heterocycle involves different reaction pathways—zwitterionic or radical-mediated. To uncover the exact reaction mechanism, further studies should be investigated. Moreover, the implementation of density-functional theory (DFT) calculations could be beneficial.

7.3.4 *C–H functionalization reaction of aliphatic C–H bonds*

The C–H functionalization reaction of aliphatic C–H bonds opens up new strategies for the synthesis of complex molecules. As the bond strains of the C–H bond increases from the $C(sp^2)$-H bonds or X–H bonds to the $C(sp^3)$-H bond, the functionalization becomes more challenging. Metal-mediated insertion reactions of carbene fragments are a common way for the functionalization of these bonds resulting in newly formed C–C bonds. A variety of precious metal-catalyzed $C(sp^3)$-H bond functionalization is known in literature—ranging from site-selective C–H functionalization to late-stage functionalization of complex molecules (Scheme 7.23).[5,45,46]

Deuterium labeling experiment

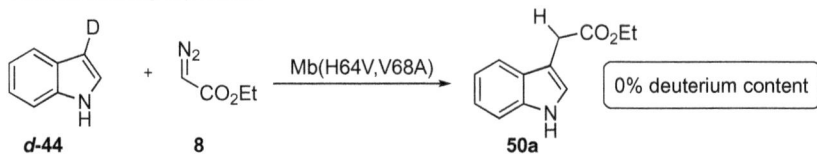

Competition experiment

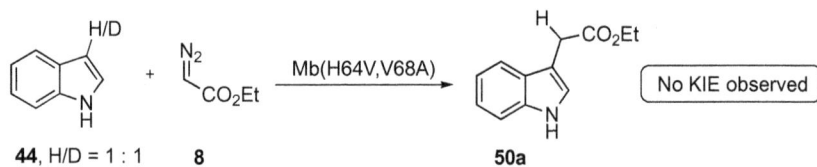

Mecanistic proposal

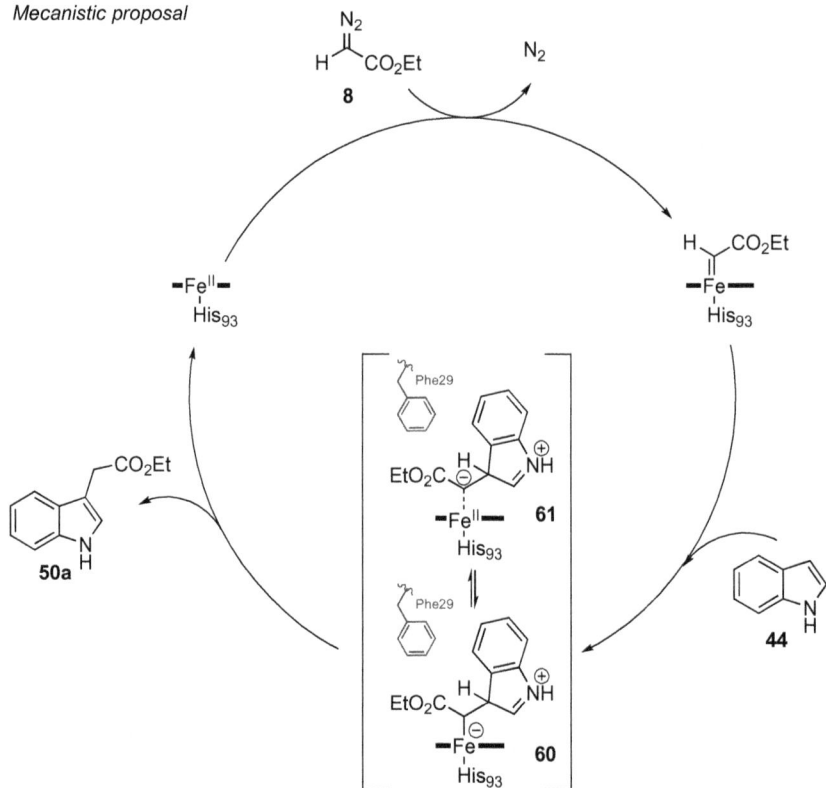

Scheme 7.21.   Mechanistic investigations and proposed catalytic cycle by the Fasan group.

Competition experiment

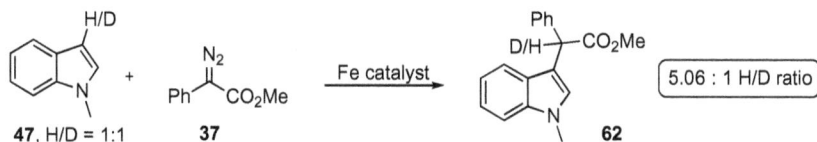

Mecanistic proposal

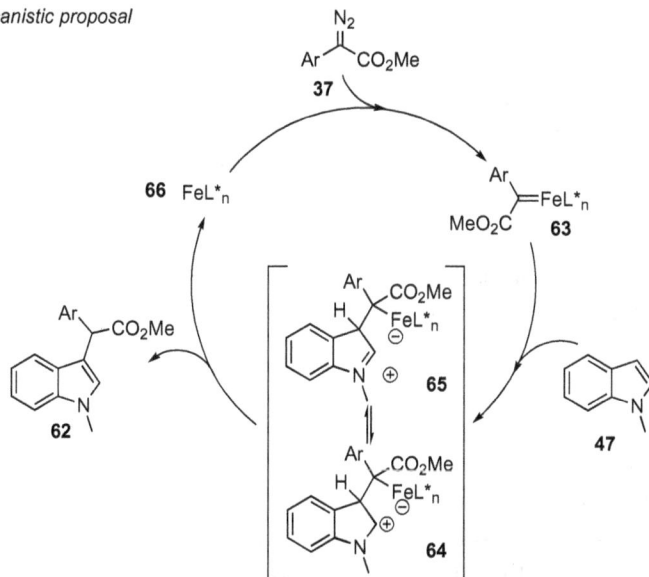

Scheme 7.22. Competition experiment and proposed mechanism for the C3-functionalization of *N*-methylindole (**47**) by Zhou *et al.*

The application of low-biotoxic, highly active iron catalysts blossomed up during the last five years and enables the efficient transformations under mild reaction conditions. Owning the potential to revolutionize organic chemistry, several groups reported on their efforts of the direct functionalization of inert, aliphatic C–H bonds, and the research field has expired rapidly.

The catalytic C–H insertion reaction using iron(III) porphyrins is described by Woo *et al.*—in their studies, they investigated cyclohexane (**72**) and tetrahydrofuran (**74**) as substrates. The authors utilized a donor/acceptor diazo compound (**37**) as a carbene precursor

Site-selective functionalization of methyl ethers

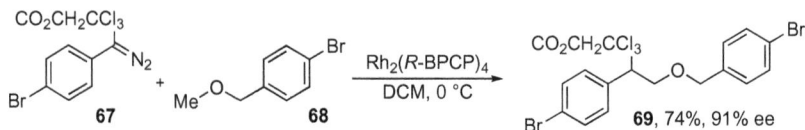

Late-stage C-H functionalization reaction

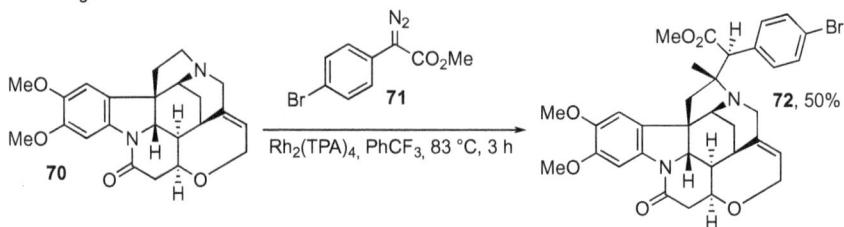

Scheme 7.23. Precious metal-catalyzed site-selective functionalization of methyl ethers (**68**) and late-stage C–H functionalization of C(sp^3)-H bond described by Davies *et al.*

and Fe(TPP)Cl catalyst for the C–H insertion reaction. In a first step, cyclohexane (**72**) was tested and the corresponding C–H insertion product (**73**) was obtained in 66% yield with a trace amount of dimeric carbene. No C–H insertion product was obtained when ethyl diazoacetate or dimethyl diazomalonate was used as a carbene precursor. When using tetrahydrofuran (**74**) as substrate, the C–H insertion reaction was exclusively observed in the β-position (**75**). A second ring-opening product, which occurs from the cleavage of the O–C bond, was observed as a minor reaction product (**76**) in 18% yield (Scheme 7.24).[37]

Additionally, competition experiments using a donor/acceptor diazo compound (**37**) were conducted. When mesitylene (**77**) is investigated as substrate, the C–H insertion reaction can occur at the C(sp^3) or at the C(sp^2) carbon—under the present reaction conditions—the C(sp^3)-H bond insertion reaction product (**78**) was detected as the major product (Scheme 7.25).[37]

The selectivity of the iron-catalyzed C(sp^3)-H insertion reaction of carbenes was further studied by Zhu *et al.* Studying the products of the reaction of *n*-hexane (**80**) and 2-methylpentane (**83**). The insertion reaction occurs preferentially at the higher substituted

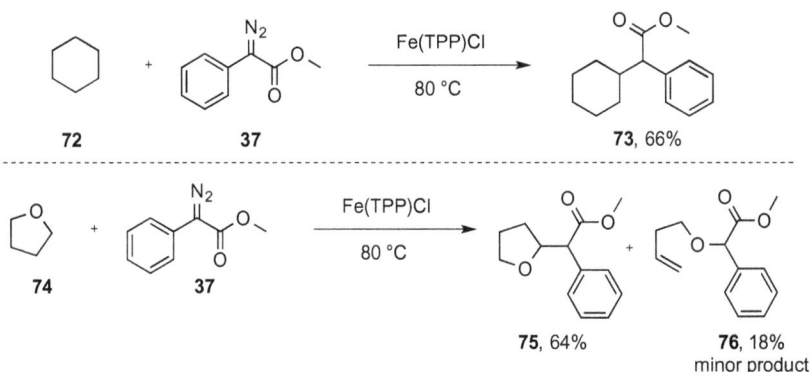

Scheme 7.24. Reaction of α-aryl-α-diazoesters with cyclohexane and THF described by Woo and coworkers.

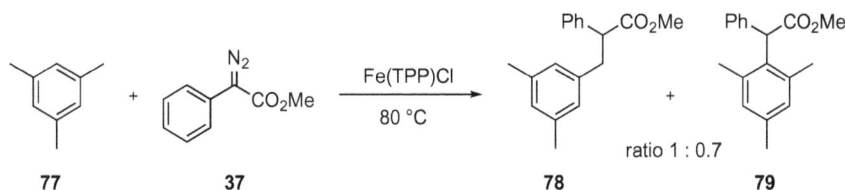

Scheme 7.25. Studies on the insertion reaction into C(sp³)-H- and C(sp²)-H-bonds using mesitylene **77**.

C–H bond. Using *n*-hexane (**80**) nearly no difference between the C2- and the C3-positon was detected, when changing to 2-metylpentane (**82**) a larger difference occurs. Interestingly, no insertion reaction into the primary C–H bond was observed in both reactions and the reactivity order of the C–H insertion reaction decreases from tertiary > secondary > allylic/benzylic >> primary C–H bonds (Scheme 7.26).[47]

An interesting approach for the intramolecular functionalization of C(sp³)-H bonds is described by Costas *et al.*—the group utilizes an electrophilic iron complex (**89**) and a Lewis acid for the reactivation of a diazoesters (**86**). Following this approach, an iron carbene intermediate is formed, which can undergo the intramolecular alkylation reaction to form a five-membered ring (**87**). The

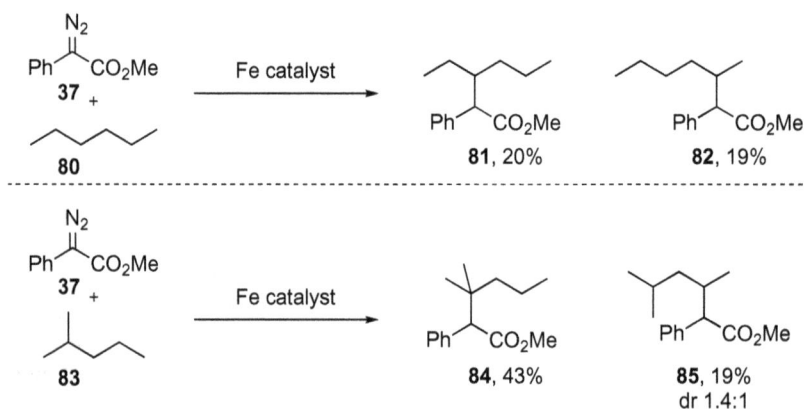

Scheme 7.26. Studies on the selectivity of iron-catalyzed C(sp³)-H insertion reactions.

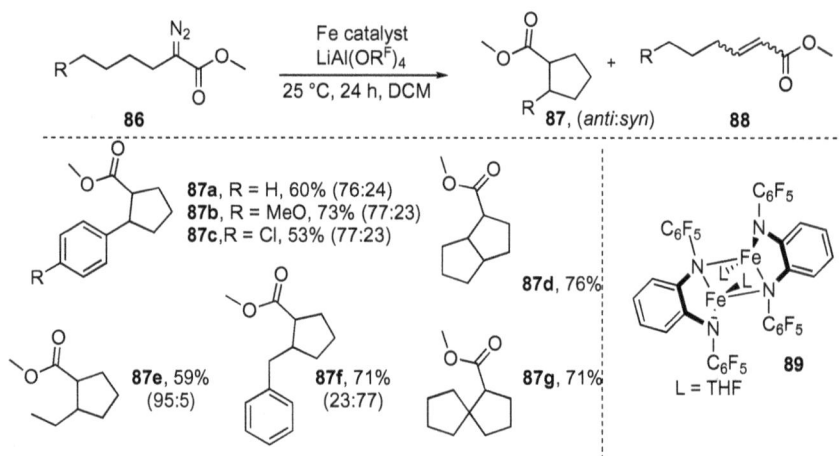

Scheme 7.27. Intramolecular alkylation reaction for the synthesis of five-membered rings (87) by Costas et al.

formation of α, β-unsaturated esters (88) was observed in all reactions and can occur from a β-hydride migration process. Using [Fe($_F$pda)-(THF)]$_2$ as a catalyst, the intramolecular C–H insertion reaction product was observed in good yields with formation of the α, β-unsaturated esters (88) in reduced yield. Following this strategy, bi- and spirocyclic systems were synthesized (Scheme 7.27).[48]

The White group explored another example of intramolecular cyclization *via* C–H functionalization reactions. The approach involves an iron phthalocyanine complex that acts as the carbene transfer catalyst. The carbene complex is generated from an acceptor/acceptor diazo compound (**90**). Using the high electrophilic sulfonate ester diazo compound (**90**) and adding noncoordinating counterions to increase the electrophilicity of the iron complex appeared to be a good strategy. The intramolecular C–H functionalization products (**91**) were isolated in moderate to good yields with broad functional group tolerance (Scheme 7.28).[49]

The enantio-, region-, and chemoselective functionalization of C(sp³)-H bond can be realized by the direct evolution of enzymes. Following this approach, the Arnold group investigated iron heme proteins in C(sp³)-H insertion reactions. In a first step (**78**), heme proteins as cytochrome P450, cytochrome *c*, and globulin homologues were studied. A variant of cytochrome P411 with an axial serine ligand showed a total turnover number (TTN) of 13 and was

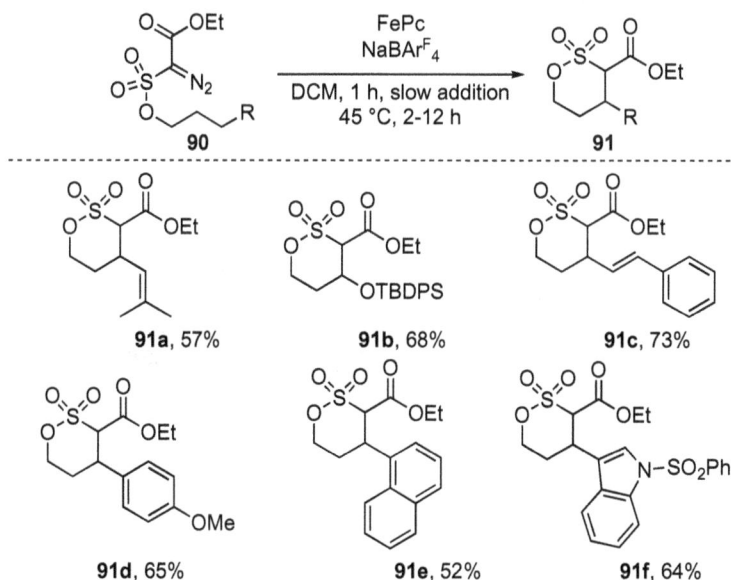

Scheme 7.28. Intramolecular cyclisation *via* C–H functionalization for the synthesis of sulfonate esters (**91**).

further engineered using directed evolution. The final variant—
P411-CHF enzyme—showed a TTN of 2020 and an enantiomeric
ratio of 96.7:3.3 was observed. With this P411 variant in hand, C–H
alkylation reactions of ethyl diazoacetate (**8**) with benzylic substrates
(**92**) were investigated and a great variance of different substrates
underwent the C–H insertion reaction. In further studies, allylic and
propargylic substrates and alkyl amines were investigated leading to
good TTN and high enantiomeric ratios (Scheme 7.29).[50]

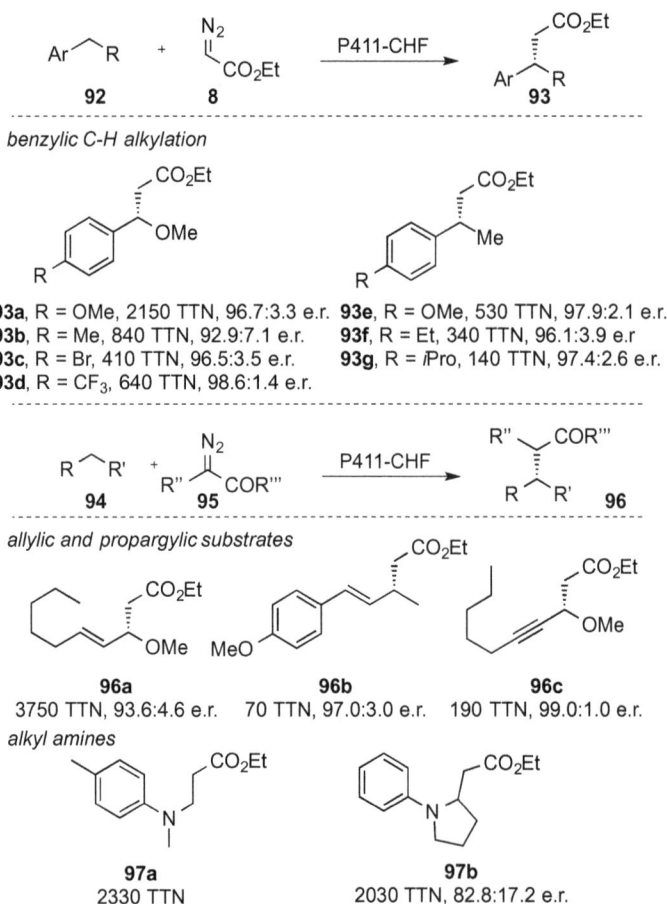

93a, R = OMe, 2150 TTN, 96.7:3.3 e.r. **93e**, R = OMe, 530 TTN, 97.9:2.1 e.r.
93b, R = Me, 840 TTN, 92.9:7.1 e.r. **93f**, R = Et, 340 TTN, 96.1:3.9 e.r
93c, R = Br, 410 TTN, 96.5:3.5 e.r. **93g**, R = iPro, 140 TTN, 97.4:2.6 e.r.
93d, R = CF₃, 640 TTN, 98.6:1.4 e.r.

96a
3750 TTN, 93.6:4.6 e.r.   **96b**
70 TTN, 97.0:3.0 e.r.   **96c**
190 TTN, 99.0:1.0 e.r.

97a
2330 TTN

97b
2030 TTN, 82.8:17.2 e.r.

Scheme 7.29. Benzylic C–H functionalization using engineered P411-CHF
enzymes; additional experience on allylic/propargylic substrates and alkyl amines
reported by Arnold *et al.*

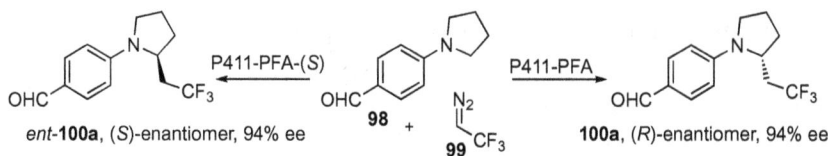

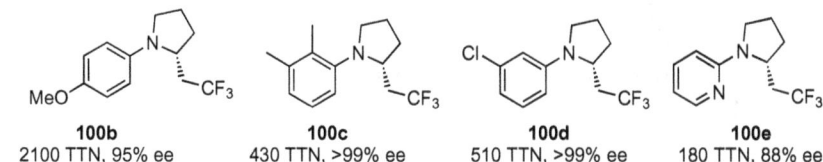

P411-PFA-catalyzed reaction products

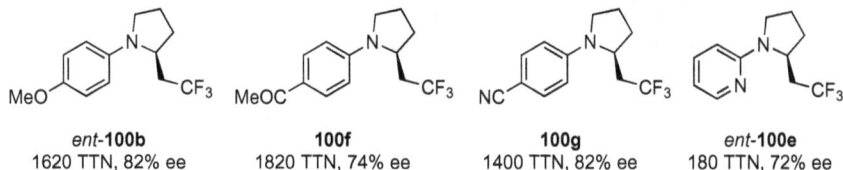

P411-PFA-(S)-catalyzed reaction products

Scheme 7.30. P411-PFA-catalyzed C–H trifluoroethylation reaction described by Arnold and coworkers.

Another interesting example of the enzymatic C–H functionalization of heterocycles using engineered cytochrome P450 enzymes, and a fluorocarbene precursor (**99**) was investigated in the Arnold group. Using a similar strategy for the mutation of the wildtype enzyme, a variant of cytochrome P411 showed high activity and good enantiomeric excess in the C–H functionalization reaction using 2,2,2-trifluoro-1-diazoethane (**99**) as carbene precursor. Further studies were conducted to achieve an enzyme to excess the other stereoisomer. Starting from a variant that provided the opposite absolute stereochemistry, direct evolution was performed leading to a variant that provides the inverse stereochemistry (Scheme 7.30).[51]

7.4 C–H Functionalization with Iron Nitrenes

The C–N bond is one of the most common hetero-carbon bonds and can be found in organic molecules ranging from natural products and drug molecules to material science. Owning this privileged roll, organic chemist are still exploring possibilities for the direct C–N bond formation reaction and expand the possible synthetic methodologies constantly.

During the past decade, the direct C–H amination reaction gained more and more interest and turned into a new research field. The application of azides as nitrene precursors is a common strategy and allows one of the most efficient reactions in terms of atoms and environmentally benign, since only nitrogen is formed as a by-product.[52]

The direct synthesis of *N*-heterocycles is still an open research field as most drug molecules and natural products contain these type of heterocycles. And only in 2008 the Driver group reported on their first studies for the iron-catalyzed, intramolecular C–H amidation reaction for the direct synthesis of benzimidazole derivatives. With this initial work on iron-catalyst nitrene transfer reaction, the Driver group pioneered the research field.[53]

7.4.1 *Aryl azides in nitrene transfer reactions*

As described by Driver *et al.* in 2008 benzimidazole derivatives (**103**) can be easily obtained by an intramolecular reaction of aryl azides (**101**) using FeBr$_2$ for the Lewis acid activation of the imine. The authors investigated a two-step procedure to generate the 2-azidoaryl imine (**102**) *in situ*. Following this approach, various benzimidazoles (**103**) were synthesized in moderate to high yield (Scheme 7.31).[53]

The synthesis of indole derivatives (**105**) using a similar approach was described by the Bolm group—using azido acrylates (**104**) in an intramolecular C–H amination reaction, the indole framework was furnished in one step. Iron(II) triflate was utilized as a readily available and robust catalyst for this transformation. A broad functional group tolerance on the ester group and the indole backbone was obtained (Scheme 7.32).[54]

The Driver group investigated an approach for the synthesis of functionalized indole heterocycles. *Ortho*-substituted aryl azides (**106**) underwent a tandem ethereal C–H bond amination [1,2]-shift

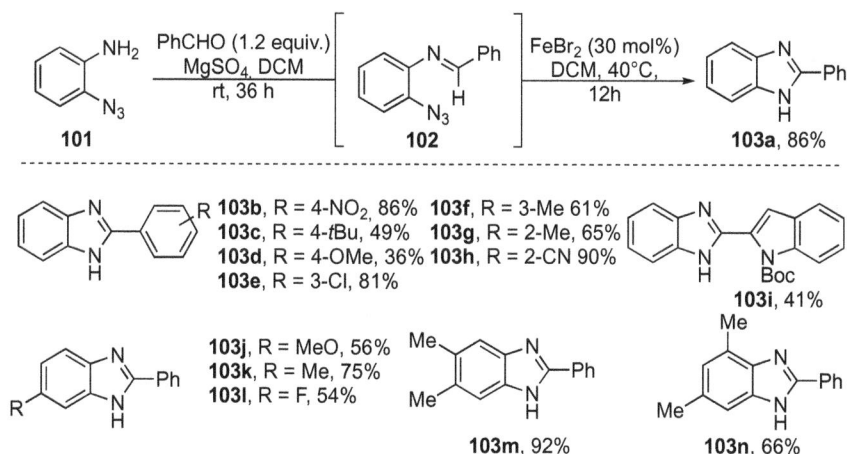

Scheme 7.31. Two-step procedure for the iron-catalyzed synthesis of benzimidazoles using aryl azides.

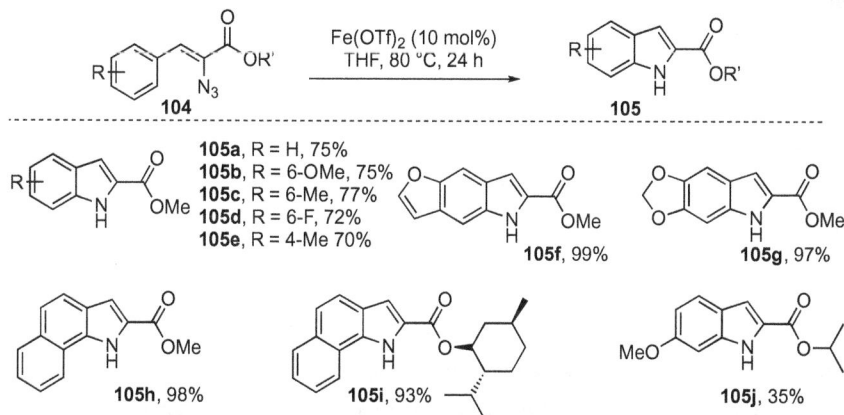

Scheme 7.32. Iron(III) triflate-catalyzed reaction for the synthesis of indoles by intramolecular C–H amination described by the Bolm group.

reaction to form 2,3-disubstituted and 3-substituted indoles (**108**). The reaction was carried out using FeBr$_2$ as a nitrene transfer catalyst and a variety of molecules containing the indole moiety were synthesized. Further, the migration group was studied in more

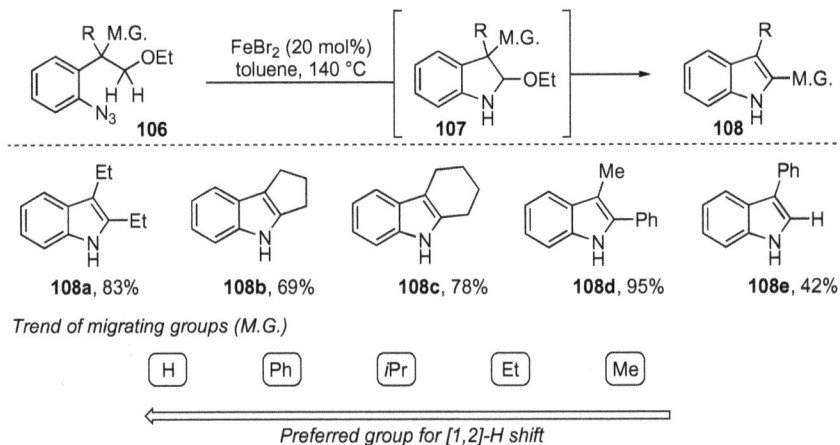

Scheme 7.33. Intramolecular C–H bond amination reaction following the approach of a tandem reaction of aryl azides (**106**); investigations of the migration group.

detail, resulting in a ranking of preferred groups for the [1,2]-H shift (Scheme 7.33).[55]

The reaction outcome of the reaction of styryl azides (**109**) with iron catalysts can be controlled by the iron catalyst as described by the Driver group. Either C–H bond functionalization or electrocyclization reactions were observed using FeBr$_2$ or electron-rich iron octaethylporphyrin chloride. As iron porphyrins are well known to catalyze N-atom transfer reactions *via* an one-electron mechanism, the authors investigated the radical-mediated C(sp^3)-H bond amination reaction in more detail. Only one single regioisomer was observed that indicated a long-lived radical species. A rotation step occurs to enable the radical recombination between the more substituted side and the nitrogen-centered radical (Scheme 7.34).[56]

The catalytic C–H bond amination reaction of toluene (**121**) using a high-spin iron imido complex (**127a**) was investigated by Betley *et al.*—a dipyrromethane ligand scaffold was synthesized and studied. The reaction of 1-azidoadamantane (**120**) with toluene (**121**) to form benzyl adamantylamine (**126**) was investigated as a model reaction.

The authors investigated the mechanism in further studies—a direct competition experiment was performed using toluene (**121**)

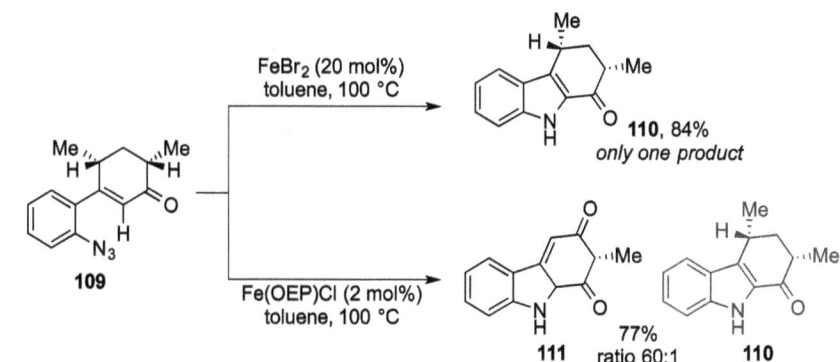

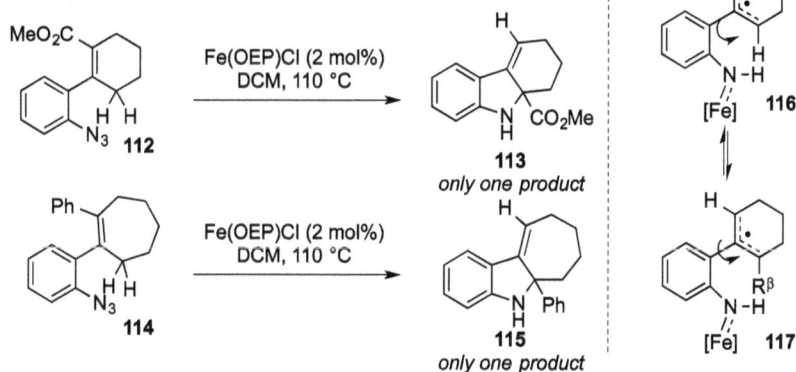

Scheme 7.34. C–H bond amination versus electrocyclization reaction—investigation on two iron-catalyzed reaction pathways; investigations on the chemoselectivity of the amination reaction.

and deutero toluene (*d*-121) and a KIE, k_H/k_D of 12.8 was observed. This KIE ratio can be attributed to a C–H bond-breaking step as the rate-determining step proceeding *via* a hydrogen atom abstraction.

The formation of the reactive formal iron(IV) imido complex (123) occurs from the reaction of the organic azide (120) with the iron(II) dipyrinato complex (122) under the liberation of nitrogen gas. Next, the imido species (123) undergoes a H-abstraction reaction with benzylic or allylic C–H bonds to form the iron(III) amido species (125) and the benzylic or allylic radical (124). The formation of the desired C–H amination product (126) is obtained by a radical–radical recombination step (Scheme 7.35).[57–59]

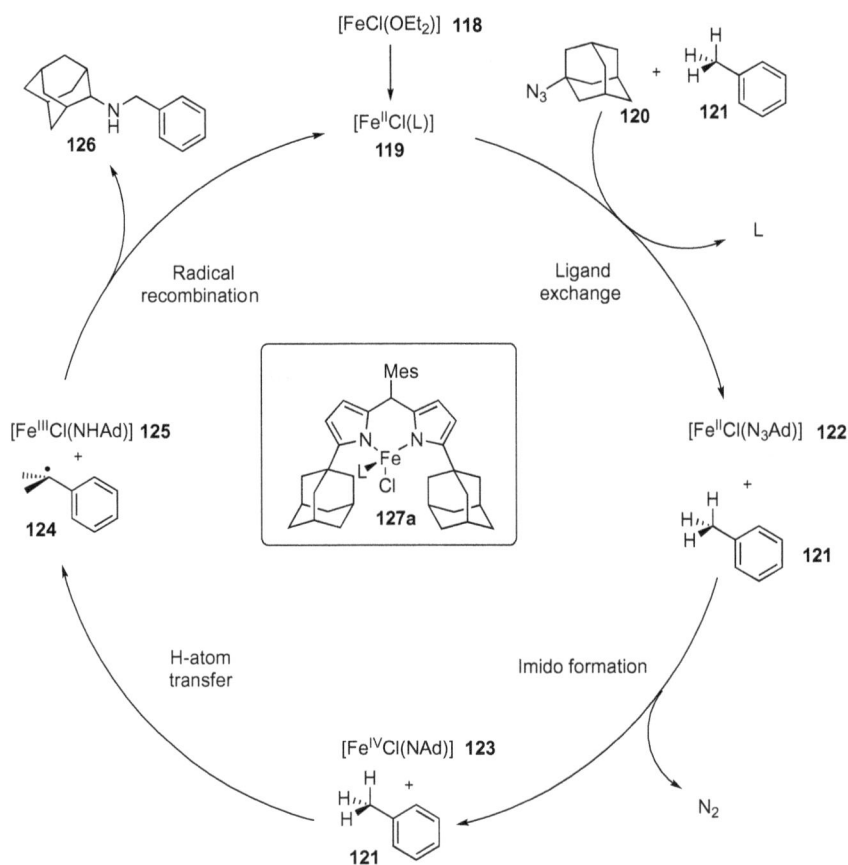

Scheme 7.35. Proposed catalytic cycle for the amidation of C–H bonds using organic azides investigated by the Betley group.

The Plietker group designed the nucleophile iron complex $Bu_4N[Fe(CO)_3(NO)]$—TBA[Fe]—for the intramolecular C–H amination reaction of α-azidobiaryls and (azidoaryl)alkenes (128). Following this approach, a variety of indole and carbazole heterocycles were synthesized. The transformation showed broad applicability and the corresponding heterocycles (129) were isolated in good to high yield (Scheme 7.36).[60]

In further studies, the Plietker group investigate the TBA[Fe]-catalyzed $C(sp^2)$-H bond amination reaction. In a first step,

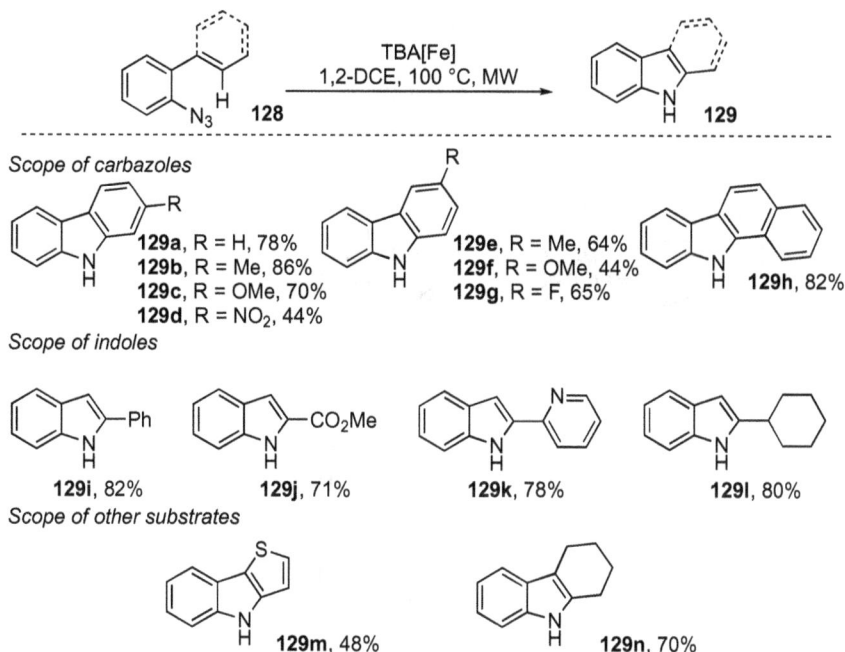

Scheme 7.36. TBA[Fe]-catalyzed intramolecular $C(sp^3)$-H amination reaction of alkylaryl azides (**128**) reported by the Plietker group.

ortho-tert-butylphenylazide (**130**) was investigated in an intramolecular reaction and the corresponding indoline (**131**) was observed in good yield. Next, a substrate scope was investigated and a broad variety of indoline heterocycles (**131**) was synthesized in moderate to good yield. Notably, substrates with more than one activated C–H bond react to a mixture of the corresponding indoline (**132**) and tetrahydroquinoline (**133**) (Scheme 7.37).[61]

The amination reaction of azides (**135**) to form benzylic amides (**136**) using iron catalysts (**137**) was described by Gallo *et al.* The synthesis of different glycoporphyrins was investigated as the saccharide moiety on the iron complex would allow the introduction of chiral, low-toxic scaffolds to the porphyrin framework. With the glycoporphyrins in hand, the reaction of ethylbenzene (**134**) with phenyl azides (**135**) was studied; the introduction of α-D-glucose

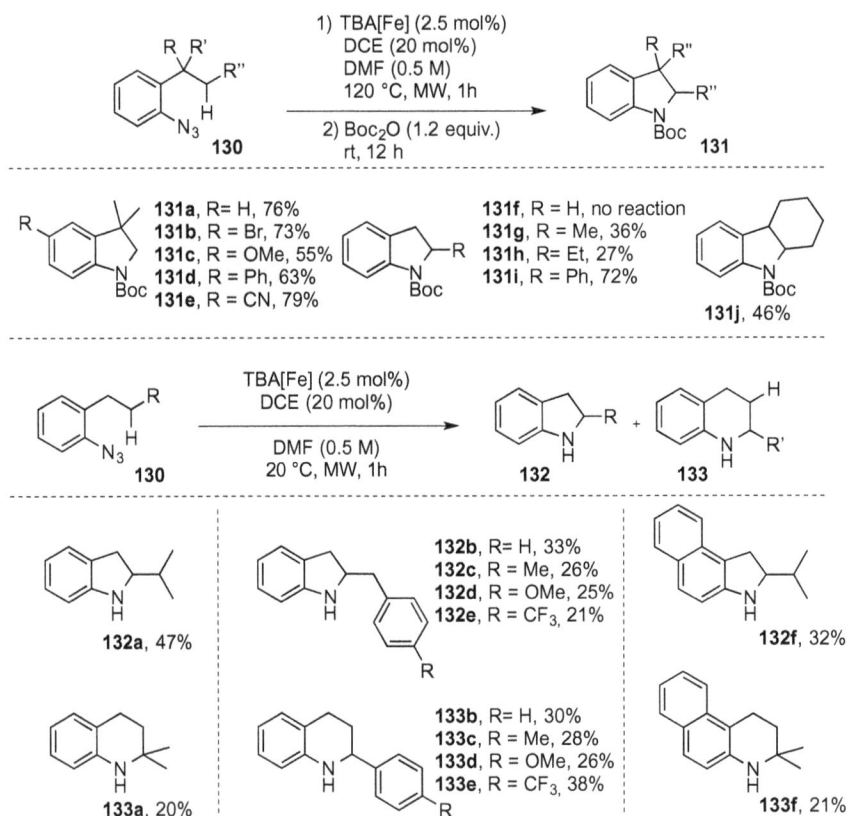

Scheme 7.37.   TBA[Fe]-catalyzed C(sp²)-H bond amination reaction reported by Plietker *et al.*

ligand showed high activity in this transformation. In further studies, a substrate scope was investigated demonstrating a high functional group tolerance and on both substrates and different allylic and aliphatic azides underwent the C–H functionalization reaction (Scheme 7.38).[62]

7.4.2 C–H functionalization with aliphatic iron nitrenes

The application of an iron-dipyrrinato catalyst enables the direct amination of aliphatic C–H bonds as described by Betley. The

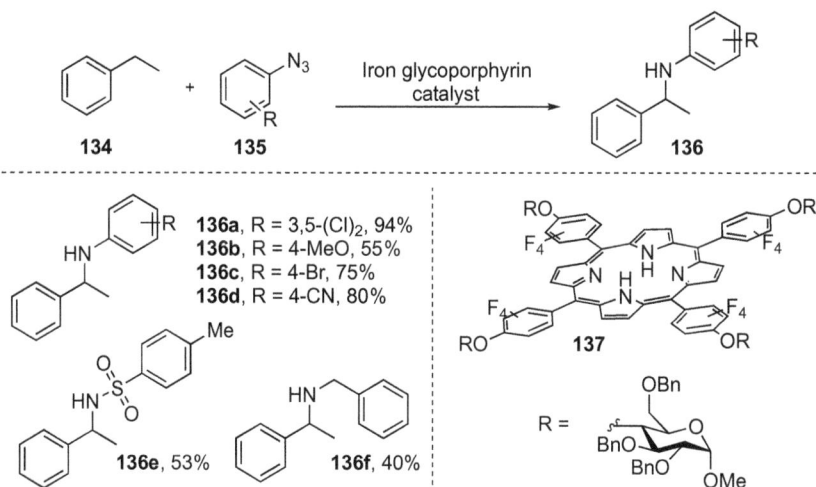

Scheme 7.38. Synthesis of benzylic amides catalyzed by an iron glycoporphyrin described by the Gallo group.

synthesis of saturated *N*-heterocycles is performed by an intramolecular reaction of organic azides (**138**). A broad scope on organic azides (**138**) was investigated and the corresponding five-membered *N*-heterocycles (**139**) were isolated in moderate to good yield. In a next step, the product distribution for azetidine, pyrrolidine, and piperidine formation using directing groups to affect the ring size was studied (Scheme 7.39).[63]

Moreover, the authors postulated a mechanism. The electronic structure of the high-spin Fe(III) complex (**127**) places substantial radical character on the iron and the nitrogen, which offers the possibility for two reaction pathways. An intramolecular H-atom abstraction could generate an alkyl radical and an Fe(II) amide (**141**), which undergoes radical recombination to form the desired *N*-heterocycle or a direct C–H bond insertion reaction takes place (Scheme 7.40).[63]

In further studies, the Betley group investigated the diastereoselectivity on the iron-catalyzed cyclization reaction to furnish 2,5-disubstituted pyrrolidines (**144/145**) from aliphatic azides (**143**). At first, 1-azido-1-aryl-hex-5-ene was studied using different high-spin

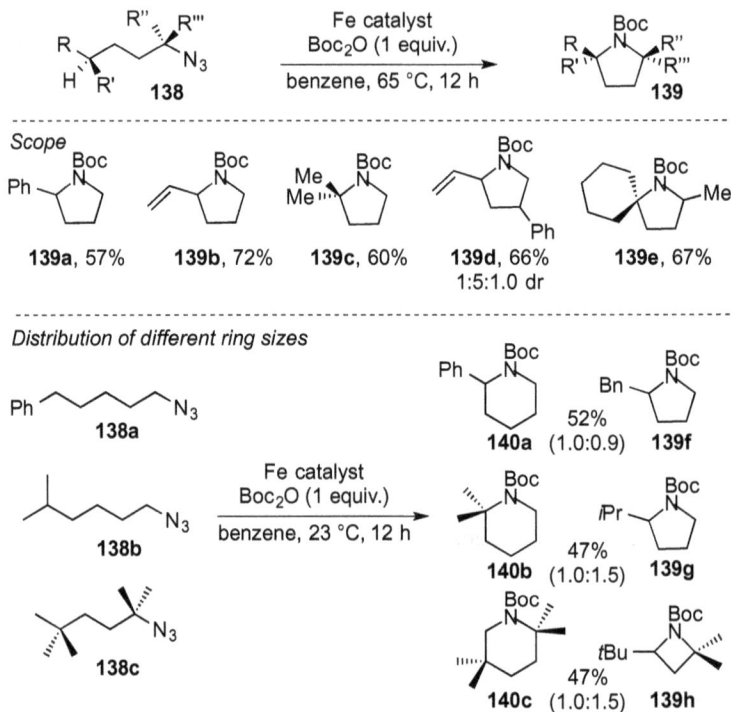

Scheme 7.39. Iron-catalyzed synthesis of pyrrolidine products; studies on the product distribution of azetidine, pyrrolidine, and piperidine formation described by Betley *et al.*

iron dipyrridin complexes—using DFT analysis a systematic modification was investigated leading to complexes with the structure $(^{Ad}L)Fe(OR)(THF)$. The investigated complexes showed significant higher diastereoselectivity (>20:1) than the previously used $(^{Ad}L)Fe(Cl)(OEt_2)$ complex. With the new iron complex in hand, a substrate scope was investigated; overall high diastereoselectivity was observed in moderate to good yields (Scheme 7.41).[64]

The van der Vlugt and de Bruin groups investigated a Fe(III) complex with a redox-active NNO ligand (**149**). The iron complex was formed using the iminosemiquinonato (ISQ) ligand and $FeCl_3$. In a next step, the synthesized ligand was studied in the intramolecular $C(sp^3)$-H amination reaction of aliphatic azides to furnish five-membered, saturated, N-protected heterocycles (**148**) (Scheme 7.42).[65]

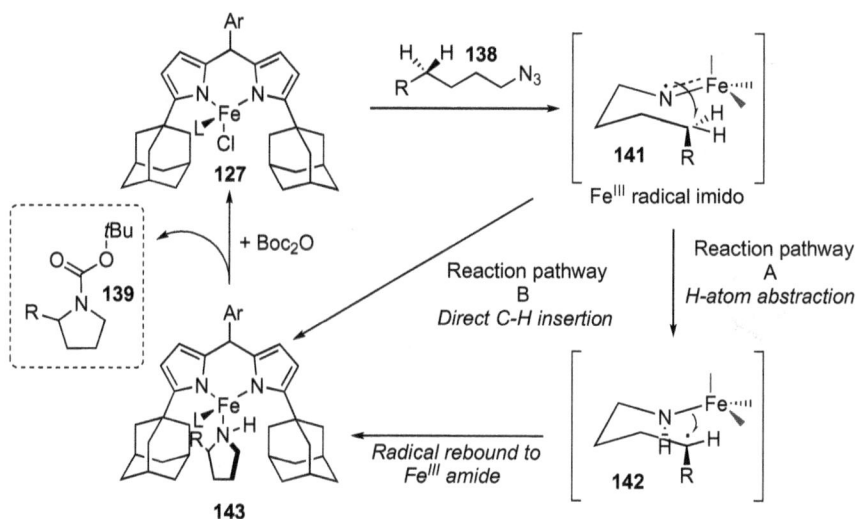

Scheme 7.40. Proposed reaction pathways for the iron-catalyzed synthesis of pyrrolidine described by Betley *et al.*

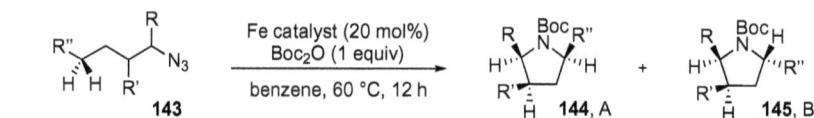

Investigated ligand systems

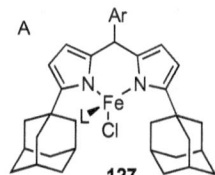

127

L = THF

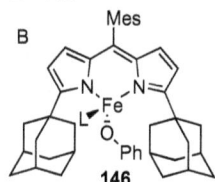

146

L = THF

Substrate scope

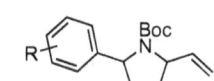

R = H	144a, A 72%, 3.9:1
	145a, B 77%, >20:1
R = 4-Cl	144b, A 63%, 3:1
	145b, B 75%, >20:1
R = 4-Me	144c, A 36%, 4:1
	145c, B 53%, 20:1
R = 3-Me	144d, A 47%, 4:1
	145d, B 55%, >20:1
R = 3-Br	144e, A 59%, 4:1
	145e, B 79%, >20:1
R = 2-Cl	144f, A 45%, >20:1
	145f, B 69%, >20:1
R = 2-Me	144g, A 27%, 4:1
	145g, B 50%, 11:1

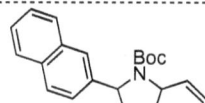

144h, A 22%, 3:1
145h, B 20%, 10:1

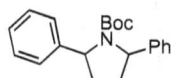

144i, A 34%, 5:1
145i, B 43%, 16:1

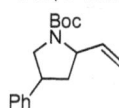

144j, A 66%, 1.5:1
145j, B 71%, 2:1

Scheme 7.41. Diastereoselective C–H amination reaction for the synthesis of disubstituted pyrrolidines reported by the Betley group.

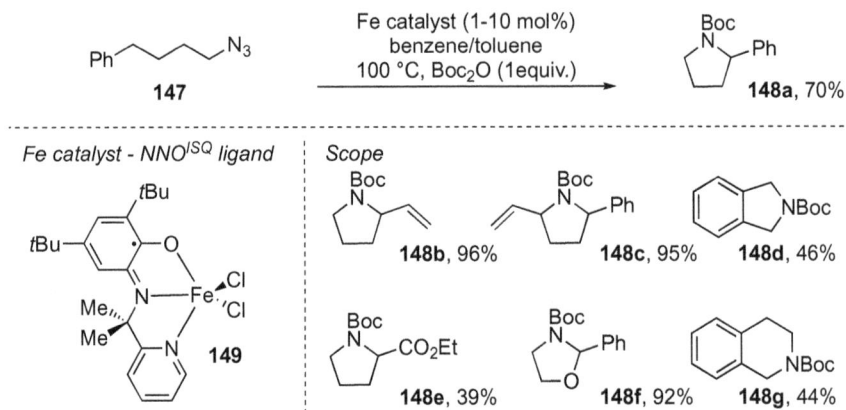

Scheme 7.42. C–H amination of aliphatic azides to N-heterocycles described by Vlugt and de Bruin.

Che *et al.* reported their investigations toward the intramolecular C–H amination reactions of alkyl azides (**150**) utilizing an iron(II) porphyrin-bearing axial N-heterocyclic carbene (NHC) ligand. A substrate scope was investigated, leading to a large number of five-membered, saturated N-heterocycles (**159**) including bi- and spirocyclic systems. It is noteworthy that the formation of piperidine analogs was observed. Moreover, the reactivity order of the C4-H bond identified as: 3° > 2° > benzylic, allylic >> 1° C–H bonds (Scheme 7.43).[66]

The application of simple $FeCl_2$ and β-diketiminate ligands was investigated by the Yu group for the synthesis of polysubstituted imidazolinones (**153**). Different substituted α-azidyl amides (**152**) were studied under the optimized reaction condition; isolating the imidazolinone (**153**) in good to excellent yield. Interestingly, no intramolecular C–H amination products formation was observed from 2-azido-2-methyl-1-(piperidin-1-yl) propan-1-one and 2-azido-2-methyl-1-(pyrrolidin-1-yl) propan-1-one. But in later investigation, addition of Boc_2O into the reaction tube can transform these substrates to the Boc-protected C–H insertion product **153m/0** with yields of 84% and 33% (Scheme 7.44).[67]

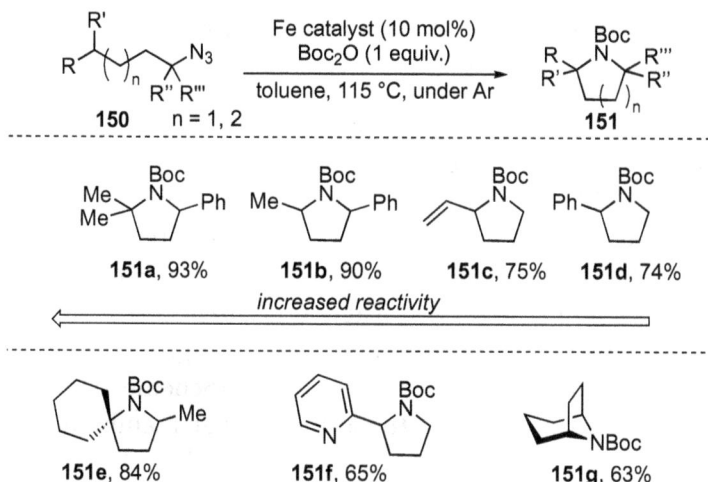

Scheme 7.43. Intramolecular, iron(II) porphyrin-catalyzed amination of alkyl azides (**150**) described by the Che group.

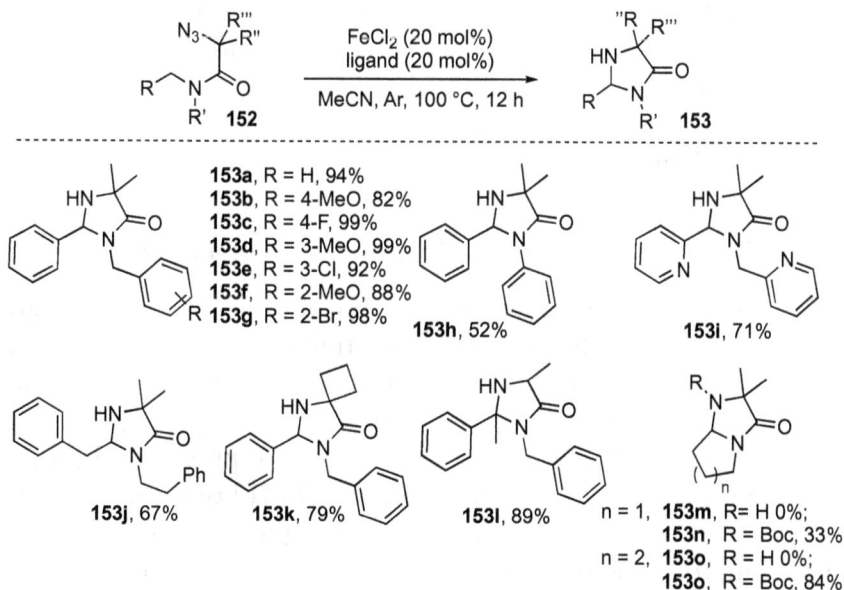

Scheme 7.44. Imidazoline synthesis described by Yu and coworkers using an iron-catalyzed, intramolecular C–H amination approach.

7.4.3 C–H functionalization with iron nitrenes from sulfonylazides

The nondirected amidation of aromatic C–H bonds was investigated by Loh and coworkers. The investigated studies utilize $FeBr_2$ and sulfonyl azides (155) as the nitrogen source for the direct functionalization of the $C(sp^2)$-H bond. A moderate to good regioselectivity was observed without the introduction of any directing group when using *m*-xylene. In the next step, both reaction partners were investigated in scope experiences leading to a good functional group tolerance and moderate to high yields and regioselectivities (Scheme 7.45).[68]

The Liu group studied the intramolecular amination of aliphatic C–H bonds of sulfamate esters (157)—their approach involves an *in situ* generated complex from $Fe(ClO_4)_2$ and bipyridine ligands. Moreover, aminopyridine and bipyridine were investigated as ligands, yet without any product formation. Following this strategy, nonactivated benzylic, allylic, propargylic, as well as primary and secondary aliphatic C–H bonds are aminated in moderate to good yield. This method provides an easy route to access complex bicyclic and spirocyclic heterocycles. Moreover, the authors performed functionalization reactions on natural product derivatives **158m/n** (Scheme 7.46).[69]

7.4.4 Enzymatic C–H functionalization with iron nitrenes

The enantioselective amination reaction of C–H bonds is of high interest for organic chemists, in nature transaminase or amino acid dehydrogenase enzymes can form those enantioselective aliphatic C-N bonds. It was found that the C–H bond cleavage is sensitive to bond strength and that weaker C–H bonds are preferentially aminated. Therefore, the amination of strong C–H bonds in the presence of weaker C–H bonds is often challenging. To overcome this limitation, the Arnold group investigated P411 variants for the regioselective amination reaction. In a first step, they studied the variant $P411_{BM3}$-CIS-T438S with moderate activity for the amination reagent sulfonylazide (159) but with high selectivity for the β-amination.

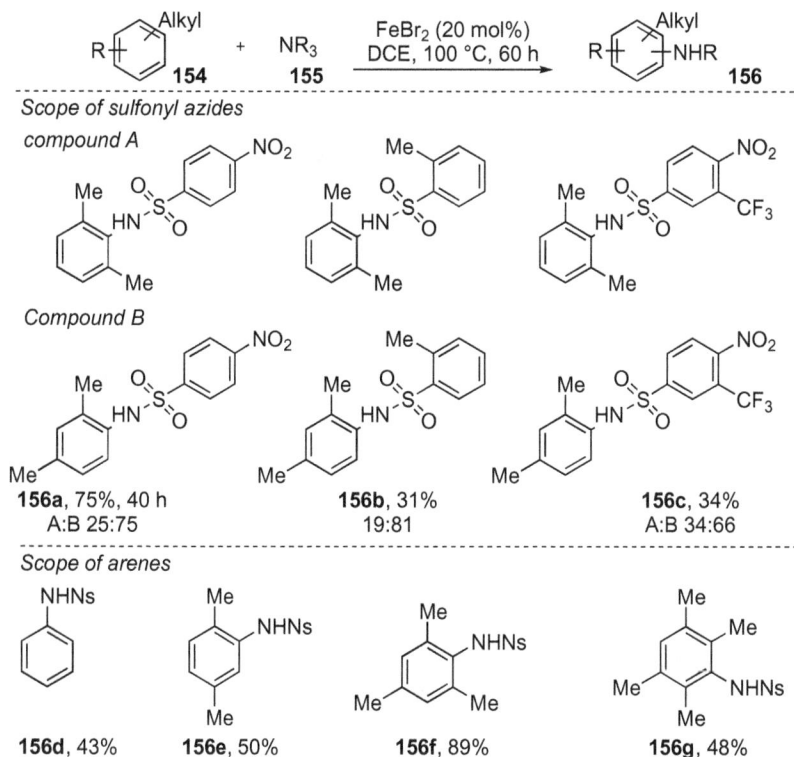

Scheme 7.45. General reaction and scope of the nondirected amination reaction of aromatic C–H bonds by Loh and coworkers.

Next positions in the active side were selected and screened by site-saturation mutagenesis resulting in a variant with 11 times higher activity. Moreover, a variant, which switches the selectivity from the β- to the α-amination, was found during the screening process. With those two variants in hand, different substrates were investigated for the intramolecular amidation of benzene sulfonylazide derivatives (**159**) resulting in high regio- and enantioselectivity for both enzymes (Scheme 7.47).[70]

In further studies, the Arnold group reported on nonheme iron enzymes catalyzed nitrene C–H insertion reactions. Their approach involves a screening of α-ketoglutarate-dependent iron

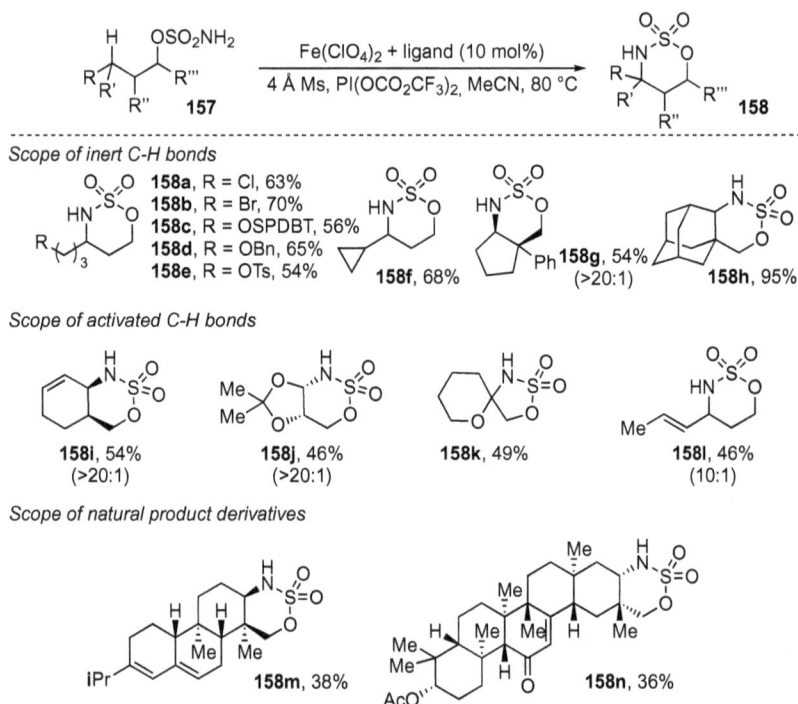

Scheme 7.46.   Intramolecular amidation reaction of aliphatic C–H bonds of sulfamate esters (**157**) described by Liu and coworkers, including the functionalization of natural product derivatives **158m/n**.

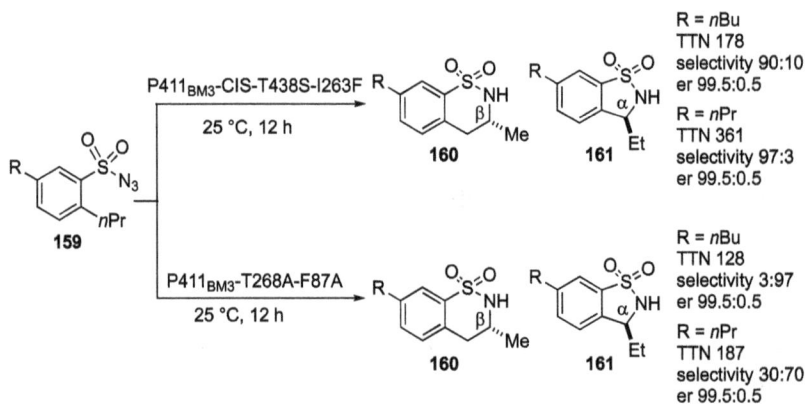

Scheme 7.47.   Intramolecular, regioselective α- and β-amination of benzene sulfonylazides using two engineered P411 variants.

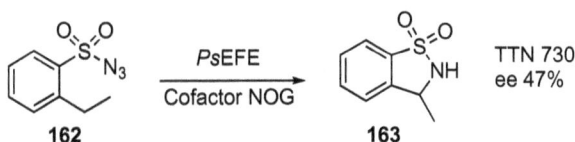

Scheme 7.48. Intramolecular, regioselective α- amination of benzenesulfonazides using a nonheme iron enzyme reported by the Arnold group.

dioxygenases—a family of enzymes with similar properties to the heme-containing cytochrome P450 family. Only the *Pseudomonas savastanoi* ethylene-forming enzyme (*Ps*EFE) as a relative of this enzyme family showed activity in the formation of aziridines from sulfonylazides. Therefore, *Ps*EFE was investigated in nitrene C–H insertion reactions using 2-ethylbenzene sulfonylazide (**162**). The *Ps*EFE showed good TTN, high chemoselectivity and moderate enantioselectivity and the corresponding α-amidation was obtained as the major product using *N*-oxalylglycine (NOG) as a cofactor (Scheme 7.48).[71]

The selective amidation reaction of indole heterocycles in the C2-position using a cytochrome P450 variant was described by Arnold *et al.* The variant P411-CSI (in whole cells) showed low yield and low selectivity in the reaction of tosyl azide (**165**) with 1-methylindole (**164**) as three products were observed; triazole from the cycloaddition reaction, C2-H functionalization product, and the reduction product. Using directed evolution for the site-saturation, mutagenesis resulted in the P411-IA enzyme for the direct C2-amination reaction of *N*-protected indoles. In further studies, the group investigated the formation of the triazole product (**167**) to achieve two P411 variants for the cycloaddition reaction (Scheme 7.49).[72]

7.5 Conclusion and Perspective

The implementation of iron complexes in the application of carbene and nitrene transfer reaction revolutionized the sustainability of these transformations. Especially, the generation of carbenes from diazo compounds—and nitrenes from azides—proved to be a powerful strategy as nitrogen gas is formed as the only by-product. A broad variety of iron complexes is known today and underlines

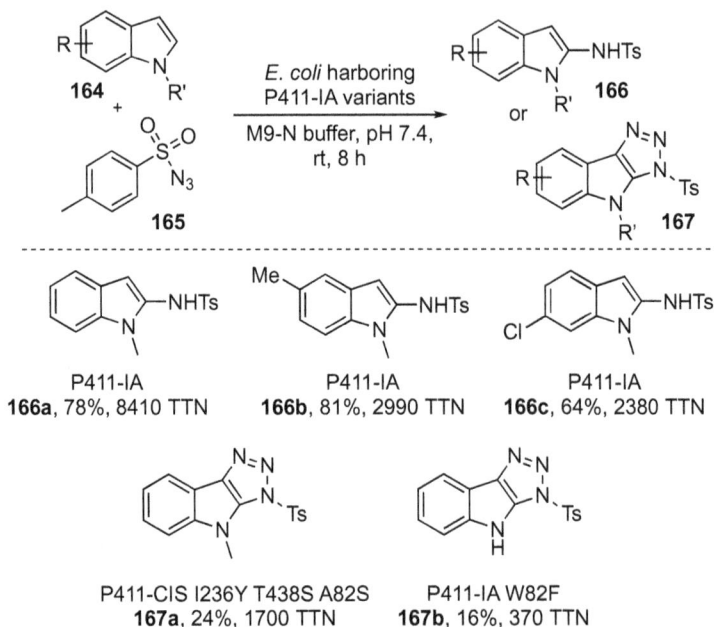

Scheme 7.49. C2-amination products and triazole products achieved by enzyme-catalyzed reactions reported by the Arnold group.

the possibility to adjust the reactivity of iron complexes. Today, a variety of iron-catalyzed carbene/nitrene transfer has been established and now provide organic chemists a toolbox for functionalization reactions via carbene or nitrene intermediates ranging from small molecule synthesis to natural product functionalization and heterocycles construction.

In light of enzyme-catalyzed carbene/nitrene transfer reaction, the directed evolution of wild type iron-containing enzymes, e.g., P450, is another "hot topic." Different enzymes have been engineered by now and show remarkable chemo-, regio, and stereoselectivity.

References

1. J. Norinder, A. Matsumoto, N. Yishikai and E. Nakamura, Iron-catalyzed direct arylation trough directed C–H bond activation, *J. Am. Chem. Soc.* **130**(18), 5858–5850 (2008).

2. M. Zhang, Y. Zhang, X. Jie, H. Zhao, G. Lim and W. Su, Recent advances in directed C–H functionalizations using monodentate nitrogen-based directing groups, *Org. Chem. Fron.* **1**, 843–895 (2014).

3. J. Wencel-Delord, T. Dröger, F. Liu and F. Glorius, Towads mild metal-catalyzed C–H activation, *Chem Soc. Rev.* **40**, 4740–4761 (2011).

4. C. Jia, T. Kitamura and Y. Fujiwara, Catalytic functionalization of arenes and alkanes via C–H bond activation, *ACC. Chem. Res.* **34**(8), 633–639 (2001).

5. D. M. Guptill and H. M. L. Davies, 2,2,2-trichloroethyl aryldiazoacetates as robust reagents for the enantioselective C–H functionaization of methyl ethers, *J. Am. Chem. Soc.* **136**(51), 177718–177721 (2014).

6. S. Pan and T. Shibata, Recent advances in iridum-catalyzd alkylation of C–H and N-H bonds, *ACS Catal.* **3**(4), 704–712 (2013).

7. S. E. Ammann, G. T. Rice and M. C. White, Terminal olefins to chromans, isochromans and pyrans via allylic C–H oxidation, *J. Am. Chem. Soc.* **136**(31), 10834–10837 (2014).

8. P. Gandeepan, T. Müller, D. Zell, G. Cera, S. Warratz and L. Ackermann, 3d tansition metals for C–H activation, *Chem. Rev.* **119**, 2192–2452 (2019).

9. E. Burbidge, G. R. Burbidge, A. W. Fowler and F. Hoyle, Synthesis of the elements in stars, *Rev. Mod. Phys.* **29**, 547–560 (1957).

10. R. Shang, L. Illies and E. Nakamura, Iron-catalyzed C–H bond activation, *Chem. Rev.* **117**(13), 9986–9139 (2017).

11. A. L. Felg and S. J. Lippard, Reactions of non-heme iron(II) centers with dioxygene in biology and chemistry, *Chem. Rev.* **94**, 759–805 (1994).

12. H. Meerwein, H. Rathjen and H. Werne, Die Methylierung von RH-Verbindungen mittels Diazomeethan unter Mitwirkung des Lichtes, *Ber. Dtsch. Chem. Ges.* **24**, 136–137 (1942).

13. W. von E. Doering, L. Knox and M. Jones, Notes. Reaction of methylene with diethyl ether and tetrahydrofuran, *J. Org. Chem.* **24**(1), 136–137 (1959).

14. H. M. L. Davies and J. R. Manning, Catalytic C–H functionalization by metal carbenoid and nitrenoid insertion, *Nature*, **451**, 417–424 (2008).

15. A. Ford, M. Miel, A. Ring, C. N. Slattery, A. R. Maguire and M. A. McKervey, Modern organic synthesis with α-doazocarbonyl compounds, *Chem. Rev.* **115**(18), 9981–10080 (2015).

16. M. P. Doyle, R. Duffy, M. Ratnikov and L. Zhou, Catalytic carbene insertion into C–H bonds, *Chem. Rev.* **110**(2), 704–724 (2010).

17. C. Empel and R. M. Koenigs, Sustainable carbene transfer reactions with iron and light, *Synlett*, **30**(17), 1929–1934 (2019).

18. H. M. L. Davies and D. Morton, Guiding principles for site selective and stereoselective intramolecular C–H functionalization by donor/acceptor rhodium carbenes, *Chem. Soc. Rev.* **40**, 1857–1869 (2011).

19. W. J. Seitz, A. K. Saha and M. M. Hossain, Iron Lewis acid catalyzed cyclopropanation reaction of ethyl diazoacetate and olefins, *Organometallics*, **12**(7), 2604–2608 (1993).

20. I. Bauer and H.-J. Knölker, Iron catalysis in organic synthesis, *Chem. Rev.* **115**(9), 3170–3387 (2015).

21. B. Plietker, *Iron Catalysis in Organic Chemistry: Reactions and Applications*, 2nd ed., Wiley-VCH, Weinheim (2008).

22. J. R. Wolf, C. G. Hamaker, J.-P. Djukic, T. Kodadek and L. K. Woo, Shape and stereoselective cyclopropanation of alkenes catalyzed by iron porphyrins, *J. Am. Chem. Soc.* **117**(36), 9194–9199 (1995).

23. T.-S. Lai, F.-Y. Chan, P.-K. So, D.-L. Ma, K.Y. Wong and C.-M. Che, Alkene cyclopropanation catalyzed by Halterman iron porphyrin: Participation of organic based axial ligands, *Dalton Trans.* 4845–4851 (2006).

24. I. Aviv and Z. Gross, Corrole-based applications, *Chem. Commun.* **20**, 1987–1999 (2007).

25. B. Morandi and E. M. Carreira, Iron-catalyzed cyclopropanation with trifluoroethylamine hydrochloride and olefines in aqueous media: In situ generation of trifluoromethyl diazomethane, *Angew. Chem. Int. Ed.* **49**, 938–941 (2010).

26. P. Le Maux, S. Juillard and G. Simonneaux, Asymmetric synthesis of trifluoromethylphenyl cyclopropanes catalyzed by chiral metalloporphyrins, *Synthesis*, **10**, 1701–1704 (2006).

27. Y. Cai, S.-F. Zhu, G.-P. Wang and Q.-L. Zhou, Iron-catalyzed C–H functionalization of indoles, *Adv. Synth. Catal.* **353**, 2939–2944 (2011).

28. B. Wang, I. G. Howard, J. W. Pope, E. D. Conte and Y. Deng, Bis(imino)pyridine iron complexes for catalytic carbene transfer reactions, *Chem. Sci.* **10**, 7958–7963 (2019).

29. R. L. Khade and Y. Zhang, C–H insertions by iron porphyrin carbene: Basis mechanism and origin of substrate selectivity, *Chem. Eur. J.* **23**, 17654–17658 (2017).

30. E. Nakamura and N. Yoshikai, Low-valent iron-catalyzed C-C bond dormation-addition, substitution, and C–H bond activation, *J. Org. Chem.* **75**(18), 6061–6067 (2010).

31. S. Ishii and P. Helquist, Intramolecular C–H insertion reaction of iron carbene complexes as a general method for synthesis of bicyclo[n.3.0] alkanones, *Synlett*, **4**, 508–510 (1997).

32. S. Ishii, S. Zhao, G. Nehta, C. J. Knors and P. Helquist, Intramolecular C–H insertion reactions of (h^5-cyclopentadienyl)dicabonyliron carbene complexes: Scope od the reaction and application to the synthesos of (±)-sterpurene and (±)-pentalene, *J. Org. Chem.* **66**, 3449–3458 (2001).

33. S. Ishii, S. Zhao and P. Helquist, Stereochemical probes of intramolecular C–H insertion reaction of iron-carbene complexes, *J. Am. Chem. Soc.* **122**, 5897–5898 (2000).

34. V. Postils, M. Rodriguez, G. Sabenya, A. Conde, M. Mar Diaz-Requejo, P. J. Pérez, M. Costas, M. Solà and J. M. Luis, Mechanism of the selective Fe-catalyzed arene carbon-hydrogen bond functionalization, *ACS Catal.* **8**, 4313–4322 (2018).

35. A. Conde, G. Sabenya, M. Rodriguez, V. Postil, J. M. Luis, M. Mar Diaz-Requejo, M. Costas and P. J. Pérez, Iron and manganese catalysts for the selective functionalization of arene $C(sp^2)$–H bonds by carbene insertion. *Angew. Chem Int. Ed.* **55**, 6530–6534 (2016).

36. A. Padwa, D. J. Austin, A. T. Price, M. A. Semones, M. P. Doyle, M. N. Protopopova, W. P. Winchester and A. Tran, Ligand effects on dirhodium(II) carbene reactivities. Highly effective switch between competitive carbenoid transformations, *J. Am. Chem. Soc.* **115**(19), 8669–8680 (1993).

37. H. M. Mbuvi and L. K. Woo, Catalytic C–H insertions using iron(III) Porphyrin complexes, *Organometallics*, **27**, 637–645 (2008).

38. J.-M. Yang, Y. Cai, S.-F. Zhu and Q.-L. Zhou, Iron-catalyzed arylation of α-aryl-α -diazoesters, *Org. Biomol. Chem.* **14**, 5516–5519 (2016).

39. T. Itahara, M. Ikeda and T. Sakakibara, Alkenylation of 1-acylindoles with olefins bearing electron-withdrawing substituents and palladium acetate, *J. Che. Soc. Pekin Trans.* **1**, 1361–1363 (1983).

40. A. H. Sandtrov, Transition metal-catalyzed C–H activation of indoles, *Adv. Synth. Catal.* **357**, 2403–2435 (2015).

41. J. A. Leich, Y. Bhonoah and C. G. Frost, Beyond C2 and C3: Transition-metal-catalyzed C–H functionalization of indoles, *ACS Catal.* **7**, 5618–5627 (2017).

42. K. J. Hock, A. Knorrscheidt, R. Hommelsheim, J. Ho, M. J. Weissenborn and R. M. Koenigs, Tryptamine synthesis by iron porphyrin catalyzed C–H functionalization of indoles with diazoacetonitrile, *Angew. Chem. Int. Ed.* **58**, 3630–3634 (2019).

43. D. V. Vargas, A. Tinoco, V. Tyagi and R. Fasan, Myoglobin-catalyzed C–H functionalization of unprotected indoles, *Angew. Chem. Int. Ed.* **57**, 9911–9915 (2018).

44. O. F. Brandenberg, K. Chen and F. H. Arnold, Directed evolution of a cytochrome P450 carbene transferase for selective functionalization of cyclic compounds, *J. Am. Chem. Soc.* **141**, 8989–8995 (2019).

45. B. Wang, D. Qui, Y. Zhang and J. Wang, Recent advantages in $C(sp^3)$-H bond functionalization via metal-carbene insertion, *Beilstein J. Org. Chem.* **12**, 796–804 (2016).

46. J. He, L. G. Harmann, H. M. L. Dabies and R. E. J. Beckwith, Late-stage C–H functionalization of complex alkaloids and drug molecules via intramolecular rhodium-carbenoid insertion, *Nat. Commun.* **6**, 5943–5952 (2015).

47. Q.-Q. Chen, J.-M. Yang, H. Xu and S.-F. Zhu, Iron-catalyzed carbenoid insertion into $C(sp^3)$-H bonds, *Synlett* **28**, 1327–1330 (2017).

48. A. Hernán-Gómez, M. Rodriges, T. Parella and M. Costas, Electrophilic iron catalyst paired with a lithium cation enables selective functionalization of non-activated aliphatic C–H bonds via metallocarbene intermediates, *Angew. Chem. Int. Ed.* **58**, 13904–13911 (2019).

49. J. R. Griffin, C. I. Wendell, J. A. Garwin and M. C. White, Catalytic $C(sp^3)$-H alkylation via an iron carbene intermediate, *J. Am. Chem. Soc.* **139**, 13624–13627 (2017).

50. R. K. Zhang, K. Chen, X. Huang, L. Wohlschlager, H. Renata and F. H. Arnold, Enzymatic assembly of carbon-carbon bonds via iron-catalysed sp^3 C–H functionalization, *Nature* **565**, 67–72 (2019).

51. J. Zhang, X. Huang, R. K. Zhang and F. H. Arnold, Enantiodivergent α-amino C–H fluoroalkylation catalysed by engineered cytochrome P450s, *J. Am. Chem. Soc.* **141**, 9798–9802 (2019).

52. B. Plietker and A. Röske, Recent advantages in Fe-catalyzed C–H aminations using azides as nitrene precursors, *Catal. Sci. Technol.* **9**, 4188–4197 (2019).

53. M. Shen and T. G. Driver, Iron(II) bromide-catalyzed synthesis of benzimodazoles from aryl azides, *Org. Lett.* **10**(15), 3367–3370 (2008).

54. J. Bonnamour and C. Bolm, Iron(II) triflates as a catalyst for the synthesis of indoles by intramolecular C–H amidation, *Org. Lett.* **13**(8), 2012–2014 (2011).

55. Q. Nguyen, T. Nguyen and T. G. Driver, Iron(II) bromide-catalyzed intramolecular C–H bond amination [1,2]-shift tandem reactions of aryl azides, *J. Am. Chem. Soc.* **135**, 620–623 (2013).

56. C. Kong, N. Jana, C. Jones and T. G. Driver, Conrtol of the chemoselectivity of metal *N*-Aryl nitrene reaktivity: C–H bond amination versus electrocyclisation, *J. Am. Chem. Soc.* **138**, 13271–13280 (2016).

57. M. J. T. Wilding, D. A. Iovan, A. T. Wrobel, J. T. Lukens, S. N. MacMillan, K. M. Lancaster and T. A. Betley, Direct comparison of C–H bond amination efficacy through manipilation of nitrogen-valence centered redox: Imido versus iminyl, *J. Am. Chem. Soc.* **139**, 14757–14766 (2017).

58. E. R. King, E. T. Hennessy and T. A. Betley, Catalytic C–H bond amination from high-spin iron imido complexes, *J. Am. Chem. Soc.* **133**, 4917–4923 (2011).

59. P. Wang and L. Deng, Recent Advantages in Iron-catalyzed C–H bond amination *via* iron imido intermediate, *Chin. J. Chem.* **36**, 1222–1240 (2018).

60. I. T. Alt and B. Plietker, Iron-catalyzed intramolecular C(sp^2)-H amination, *Angew. Chem. Int. Ed.* **55**, 1519–1522 (2016).

61. I. T. Alt, C. Guttroff and B. Plitker, Iron-catalyzed intramolecular aminations of C(sp^3)-H bonds in alkylaryl azides, *Angew. Chem. Int. Ed.* **56**, 10582–10586 (2017).

62. G. Tseberlidis, P. Zardi, A. Caselli, D. Cancogni, M. Fusari, L. Lay and E. Gallo, Glycoporphyrin catalysts for efficient C–H bond aminations by organic azides, *Organometallics*, **34**, 3774–3781 (2015).

63. E. T. Hennessy and T. A. Betley, Complex N-heterocycle synthesis via iron-catalyzed direct C–H bond amination, *Science* **340**(6132), 591–595 (2013).

64. D. A. Iovan, M. J. T. Wilding, Y. Baek, E. T. Hennessy and T. A. Betley, Diastereoselective C–H bond amination for disubstituted pyrrolidines, *Angew. Chem. Int. Ed.* **56**, 15599–15602 (2017).

65. B. Bagh, D. L. J. Broere, V. Sinha, P. F. Kuijpers, N. P. van Leest, B. de Bruin, S. Demeshko, M. A. Siegler and J. I. van der Vlugt, Catalytic synthesis of N-heterocycles via direct C(sp³)-H amination using an air-stable iron(III) species with a redox-active ligand, *J. Am. Chem. Soc.* **139**, 5117–5124 (2017).

66. K.-P. Shinf, Y. Liu, B. Cao, X.Y. Chang, T. You and C.-M. Che, N-heterocyclic carbene iron(iii) porphyrin-catalyzed intramolecular C(sp³)-H amidation of alkyl azides, *Angew. Chem. Int. Ed.* **57**, 11947–11951 (2018).

67. X. Zhao, S. Liang, X. Fan, T. Yang and W. Yu, Iron-Catalyzed Intramolecular C–H amidation of α-azodyl amides, *Org. Lett.* **21**, 1559–1563 (2019).

68. Y. Ding, S.-Y. Zhang, Y.-C. Chen, S.-X. Fan, J.-S. Tian and T.-P. Loh, Regioselective C–H Amidation of (alkyl)arenes by iron(II) catalysis, *Org. Lett.* **21**, 2736–2739 (2019).

69. W. Liu, D. Zhong, C.-L. Yu, Y. Zhang, D. Wu, Y.-L. Feng, H. Cong, X. Lu and W.-B. Liu, Iron-catalyzed intramolecular amination of aliphatic C–H bonds of sulfamate esters with high reactivity and chemoselectivity, *Org. Lett.* **21**, 2673–2678 (2019).

70. T. K. Hyster, C. C. Farwell, A. R. Buller, J. A. McIntosh and F. H. Arnold, Enzyme-controlled nitrogen-atom transfer enables regiodivergent C–H amination, *J. Am. Chem. Soc.* **136**, 15505–15508 (2014).

71. N. W. Goldberg, A. M. Knight, R. K. Zhang and F. H. Arnold, Nitrene transfer catalyzed by a non-heme iron enzyme and enhanced by non-native small molecule cofactors, *ChemRxiv. Preprint* (2019). doi.:10.26434/chemrxiv.10062044.v1.

72. O. F. Brandenberg, D. C. Miller, U. Markel, A. Ouald Chaib and F. H. Arnold, Engineering chemoselectivity in hemoprotein-catalyzed indole amination, *ACS Catal.* **9**, 8271–8275 (2019).

Chapter 8

Iron Catalysis in Metal-Ion Batteries

A. Gomez-Martin and J. Ramirez-Rico*

*Institute of Materials Science, Condensed Matter Physics,
Universidad de Sevilla, 41012 Spain*

*jrr@us.es

Metal-ion batteries are an essential link in the reliable implementation of renewable energies and in the development of electric vehicles to reduce our dependence on fossil fuels and green-house emissions. The manufacturing of the state-of-the-art graphite anode for lithium-ion batteries (LIBs) requires high-temperature treatments and long processing times resulting in high costs and energy consumption. The prior addition of an iron catalyst to a non-graphitizing carbon precursor has been shown to promote *in situ* graphitization during heat treatment at moderate temperatures (below 1,000 °C), resulting in considerable energy savings for graphite synthesis.

This chapter reviews on the use of iron as catalyst for inducing the graphitization of carbon materials for application as anodes for metal-ion battery. The phenomena of catalytic graphitization of carbon materials by an Fe catalyst and main parameters influencing final graphitic structure are reviewed. Then, the chapter discusses about the electrochemical investigations of these materials as anodes for metal-ion batteries and how synthesis features affect the lithium-ion intercalation compared to state-of-the-art electrodes.

8.1 Introduction

The global energy consumption has continuously increased over the last decade. Fossil fuels (including coal, natural gas, and oil) still remain as the world's main energy sources, accounting for 85.2% of global energy consumption, despite entailing serious environmental risks. For that reason and the predicted short-term shortage of fossil fuels, there is a growing interest to gradually move toward renewable energies in the near future. Although today's energy consumption from renewable energies only accounts for 3.6% of the total (data from 2017), they would supply energy in a sustainable and effective manner if the strong supply dependence on atmospheric conditions is overcome. In this sense, energy storage systems are not only playing an important role to solve the intermittency of supply and balance the electrical energy from high demand and low demand periods, but also pushing the development of electric vehicles.

Among the vast choice of systems based on either chemical or physical processes capable of storing electrical energy, three major technologies are nowadays in the front line: fuel cells, batteries, and supercapacitors.[1]

8.2 Lithium-Ion Batteries

The first successful commercialization of rechargeable lithium-ion batteries (LIBs) was developed by Sony in 1991,[2] using a coke-based carbon anode and lithium cobalt oxide ($LiCoO_2$) cathode. Since then, this technology has overwhelmingly revolutionized the energy storage devices market for portable electronic devices, and more recently, for large-scale applications such as grid storage, electric and hybrid vehicles.

The so-called *rocking chair* or lithium-ion battery is based on the reversible insertion/deinsertion (uptake and release, respectively) of charge carriers into the structures of two host electrodes. The use of lithium insertion compounds emerged as a breakthrough solution to the safety concerns related to the use of metallic lithium as anode material in primary lithium metal rechargeable batteries in

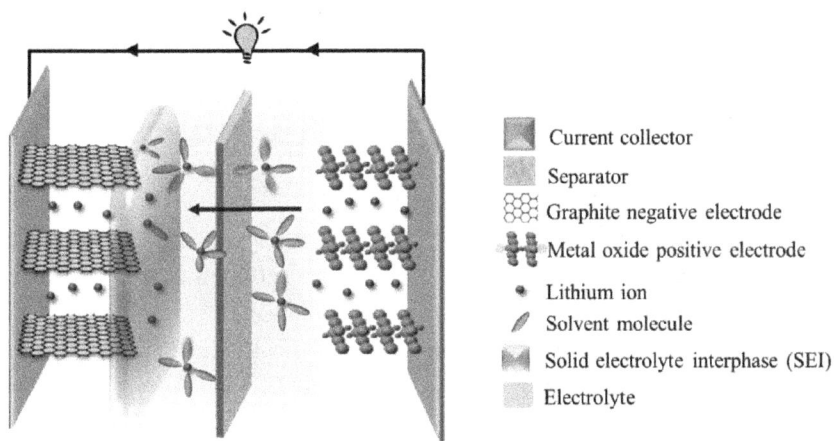

Fig. 8.1. Schematic illustration of the working principle of a lithium-ion cell (during charge). The cell reaction, based on graphite as negative electrode and transition metal oxide as positive electrode, is shown in the inset.

the mid 1990s: the nonuniform deposition of metallic lithium upon charge/discharge processes when using liquid organic electrolytes causes dendrite formation[3] and a high risk of cell ignition when the dendrites grow from the anode and passes through the separator to the cathode side, resulting in a short circuit.

A LIB cell is mainly formed by four components (outlined in Fig. 8.1): a negative electrode (anode), a positive electrode (cathode) containing lithium in the structure, a porous separator, and an electrolyte. During charge, lithium ions from the cathode are transferred across the electrolyte and inserted or intercalated into the anode. Simultaneously, electrons transfer through the external circuit in the same direction. The reverse behavior happens during discharge process and lithium ions are extracted from the negative electrode and reversibly inserted back into the cathode structure.

The two electrodes, anode and cathode, are separated by a porous membrane made of polyethylene or polypropylene,[4] soaked with an electrolyte containing a lithium conducting salt (often lithium hexafluorophosphate—$LiPF_6$) dissolved in a mixture of organic, carbonate-based solvents to increase the mobility and conductivity of lithium ions. The solid electrolyte interface (SEI) is a protective layer formed on the negative electrode surface (at a potential ~ 0.8 V vs. Li/Li$^+$) as a result of electrolyte decomposition during the first few charge/discharge cycles.[5]

In addition to active components, both the anode and the cathode electrodes contain different inactive compounds, including a polymeric binder, which provides a linkage between active material particles, conductive additives, and current collectors. Copper foil often acts as a current collector for the anode material, whereas aluminum foil is used for the cathode due to its ability to form a passive film, which makes the electrolyte/aluminum interface stable even at potentials higher than 4V vs. Li/Li$^+$. Aluminum, however, is not used for the anode side despite being lighter and cheaper because it forms alloys compounds with lithium at potentials ≈ 0.2 V vs. Li/Li$^+$.[6,7]

Even though, the general concept of LIBs has been slightly modified, ongoing intensive research efforts have been devoted either in the past decade to the development of different active (anode and cathode) and inactive materials (electrolyte, separator, and binders) or in the development of alternative metal-ion batteries such as sodium-ion batteries (SIBs) and potassium-ion batteries, among others. The working principle and battery components of sodium and potassium-ion batteries are similar to that of LIBs, but with a different charge carrier ion (Na$^+$/K$^+$ rather than Li$^+$): two host electrodes (anode and cathode) reversibly take up and release Na$^+$/K$^+$ ions during charge/discharge processes. Future perspectives rely on improving practical aspects such as higher specific and volumetric densities, better cycling performance, lower costs, and battery sustainability.

Cathode materials, although being decisive for cell performance, are not discussed in this chapter. However, the reader can refer to the extensive literature.[8-10] Layered transition metal oxides such as lithium nickel cobalt aluminum oxide (NCA; LiNi$_{0.8}$Co$_{0.15}$Al$_{0.05}$O$_2$) and lithium nickel manganese cobalt oxide (NMC; LiNi$_{1-x}$Mn$_y$Co$_z$O$_2$) represent the state-of-the-art cathode materials.

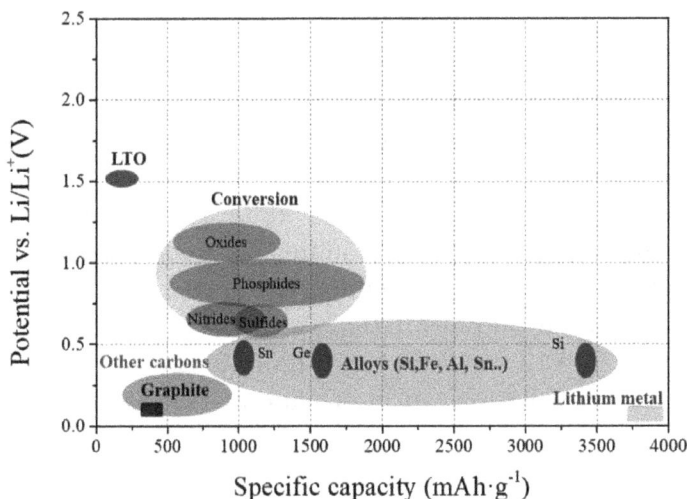

Fig. 8.2. Overview of active anode materials for LIBs represented in terms of working potential vs. Li/Li$^+$ and the corresponding specific capacity. Includes data from Ref. 12.

From a general point of view, anode materials for LIBs can be largely classified into three categories depending on the type of reaction/storage mechanism involved: insertion/intercalation, alloying, and conversion anodes. Figure 8.2 shows an overview of different proposed anodes as a function of corresponding working potential and theoretical specific capacities. Different approaches have been employed for improving the performance of LIB anode materials and a short discussion is presented below. As stated before, metallic lithium (specific capacity, $Q \approx 3,862$ mAh·g^{-1}) is not used in practice due to the associated high safety risks, although is considered as the most promising one for next-generation lithium-metal batteries.[11]

Insertion anodes are based on the reversible electrochemical insertion of lithium ions (guest species) into the available crystal lattice sites of the host material. When the host material has a well-defined layered lattice structure and the lithium ions are stored within its interlayer spacing, this insertion is termed an intercalation reaction. Among intercalation-type negative electrodes, graphite is the state-of-the-art anode material since the early commercialization of LIBs due to its moderate gravimetric capacity (372 mAh·g^{-1}), low operating potential (<0.2 V vs. Li/Li$^+$), good cycling stability, low cost as well as low-voltage hysteresis, and hence high voltage and energy efficiencies.

Conversion anodes are based on the reversible redox reactions between lithium ions and transition metal cations. During a conversion reaction, transition metal compounds M_aX_b, where M refers to the metallic cation (mostly Mn, Fe, Co, Ni) and X to the anion specie (S, N, P, O, F) are reduced by lithium ions during charge processes leading to the formation of nanometric-sized metal crystals embedded into a thermodynamically stable lithium binary matrix (Li_nX, being n the formal valence state of X). Achievable capacity of conversion reaction-type anode materials ranges from 350 mAh·g^{-1} for Cu_2S to 1,800 mAh·g^{-1} for MnP_4.[13] Although these type of anodes can deliver higher capacities compared to insertion/intercalation anodes, they have some drawbacks such as poor electronic conductivity and relative large volume expansion, which results in loss of electrode integrity and causes rapid capacity fading upon cycling.

The last type of anode materials, alloying anodes, are based on the formation of a Li-alloy compound. During lithiation, the starting crystalline structure undergoes an amorphization process to form a Li_nX alloy. Examples of alloying anodes are the Group-IV elements such as Si, Ge, and Sn (theoretical gravimetric capacities 4,200, 1,384, 994 mAh·g^{-1}). Among alloying anodes, Si appears the most promising and competitive future choice due to its outstanding theoretical gravimetric capacity, as high as 4,200 mAh·g^{-1} (theoretically up to 4.4 lithium atoms per silicon unit), low cost, and natural abundance.[14] Unfortunately, Si itself suffers from electrode pulverization and capacity fading upon cycling as a result of the huge volume expansion it undergoes during the process of lithium uptake/extraction (up to 300% for Si anodes). Strategies aimed at mitigating this effect and improving cell performance are similar to those employed for conversion reaction anodes: synthesizing composites containing an inactive material that buffers volume expansion (amorphous carbon,[15] graphite,[16] encapsulation in core-shell structures[17]), design of nanostructures[18,19] and size reduction.[20]

The current market (data from 2016) is mainly dominated by carbon materials: natural graphites (46%), synthetic graphites (43%), and amorphous carbons (7%).[21] Lithium titanate (LTO) and Si-based anodes only account for 2% each of the total market. Given that the resources to synthetize natural and synthetic graphites may be limited, the development of carbon materials from sustainable resources, such as biomass, has received considerable attention in the last five years.[22] However, the direct carbonization of biomass resources leads to amorphous structures, which suffer from a large irreversible capacity in the first charge/discharge cycles and a large voltage hysteresis on their potential profile, resulting in low-energy densities.

8.3 Fe-Catalysis in Carbon Anodes for LIBs

Carbon materials have industrial interest in a wide variety of advanced fields,[23–25] including their application as electrodes in energy storage systems such as LIBs, SIBs, and electric double-layer capacitors (EDLCs).[26–28] A variety of carbon allotropes can be found depending on their morphology, structure, and type of chemical hybridization, such as graphite, graphene, diamond, carbon nanotubes (CNTs), fullerenes, and amorphous carbons.[29] While graphite currently represents the material of choice as anode for LIBs, amorphous carbons are preferred as anodes for SIBs, and electrodes for EDLCs.

Graphite is a layered allotrope formed of stacked graphene sheets, which themselves are layers of sp^2-hybridized carbon atoms arranged in a honeycomb-like planar structure.[30] The three covalent bonds at an angle of 120° in the same plane form hexagonal rings with a carbon–carbon bond length of 1.42 Å. In graphite, graphene sheets are stacked with an interlayer spacing of 3.354 Å, weakly bonded by the interaction between perpendicular delocalized π-electron orbital clouds and van der Waals forces. The most fundamental and thermodynamically stable crystallographic unit cell in graphite is hexagonal with an –ABABAB– regular stacking (Fig. 8.3). The second layer (labeled as B) is displaced by 2/3 and 1/3 of the first layer (labeled as A) along the axes a_1 and a_2, respectively.[28] The

(A)

(B)

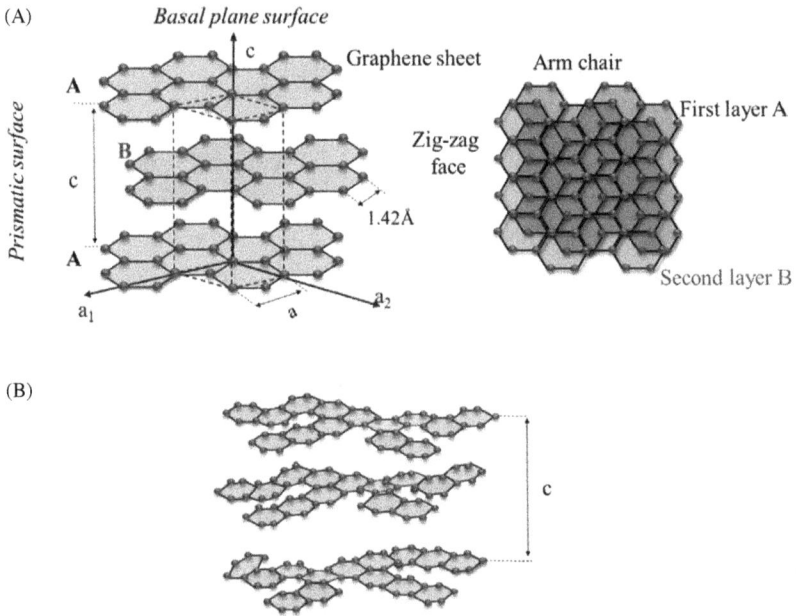

Fig. 8.3. Schematic illustration of (A) crystal structure of hexagonal graphite and the unit cell; (B) crystallographic structure of turbostratic carbon with stacking disorder. Modified and redrawn from Refs. 3 and 31.

structure of graphite is composed by two types of surfaces, prismatic (edges) and basal, which influence the physicochemical properties of graphitic materials depending on the direction.

Such a perfectly stacked layered structure is characteristic of natural graphite, although highly ordered synthetic graphite can be obtained by heat-treating some carbon precursors at temperatures in excess of 2,500 °C. Most carbon materials, however, lack the long-range order of natural graphite due to random stacking between graphene layers, and are referred to as amorphous or turbostratic carbons. Depending on their structure, amorphous carbons can convert to graphite by heating at very high temperatures in a process called *graphitization*. Amorphous carbons that can be graphitized upon heat-treatment are called *soft*, while those that cannot are *hard* carbons.

The electrochemical behavior and properties of the lithium insertion into carbon materials (either graphite, soft and hard carbons) such as potential profile, specific capacity, and insertion mechanism and kinetics strongly depend on the material structural features, including crystallinity, morphology, surface properties, and functional groups.[26]

As stated before, graphitic materials possess two main kind of surfaces, prismatic (edge) and basal surfaces. The intercalation of lithium ions into the graphitic structure is well-known to mainly take place *via* the prismatic surfaces between adjacent stacked graphene sheets or trough defects located at the basal surfaces.[3,32] During the charge process and the electrochemical reduction of the graphitic anode, lithium ions intercalate into the graphite structure and form lithiated graphitic intercalation compounds (GICs), Li_xC_n.

At ambient pressure, the maximum intercalation stage for highly crystalline materials corresponds to one lithium atom per six carbon atoms (in Li_xC_6 $x = 1$) with a theoretical gravimetric capacity of 372 $mAh \cdot g^{-1}$. Upon intercalation, the stacking order of the graphene layers shifts from ABAB (hexagonal graphite) to AAAA stacking. The interlayer distance is increased from 3.35 to 3.70 Å (increase of 10.3% for LiC_6) leading to a Li–Li distance of 4.3 Å.[33]

The storage mechanism involved in graphitic structures can be described by the *staging phenomena*. This process occurs in the potential range from ≈ 0.22 to 0 V (vs. Li/Li^+) and is labeled with respect to the stage index, s (s = I, II, III, IV), which represents the number of empty neighboring graphene layers between one lithiated layer along the c-axis. Stages higher than s = IV were also reported in the literature, but there are still some discrepancies regarding their composition.[34] Figure 8.4 shows a scheme of the staging mechanism and the corresponding ideal galvanostatic (constant current) charge potential profile as a function of the composition and cyclic voltammetry plot of the reduction of graphite to form the maximum intercalation stage, LiC_6.

Stage III corresponds to one occupied layer for every three graphene layers ($x = 0.2$ in Li_xC_6). In stage II, alternate graphene layers are occupied. This stage can be divided into s = II ($x = 0.5$ in Li_xC_6)

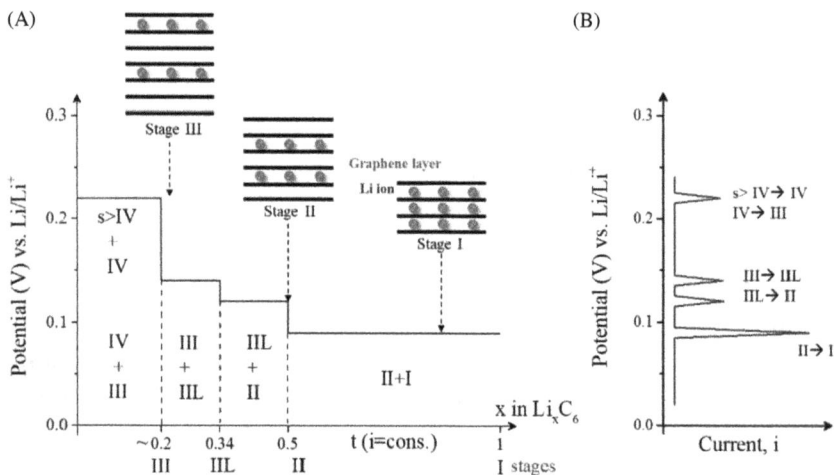

Fig. 8.4. (A) Schematic illustration of stage formation during electrochemical intercalation of lithium ions into graphite structure and galvanostatic charge curve. (B) Schematic voltammetry curve. Figure from Ref. 32. Reproduced with permission. Copyright 2011 Wiley-VCH.

and s = II-L ($x = 0.34$ in Li_xC_6) due to different lithium packing densities. Finally, in stage I (maximum intercalation stage, $x = 1$ in Li_xC_6) every interlayer gap is occupied by intercalated guest ions. The potential profile of the graphite anode exhibits different plateaus in galvanostatic experiments (voltage remains constant for a certain time) and sharp peaks in voltammetry, indicating the transition between different staging phases at potential below ~ 0.2 V vs. Li/Li+. The transition between stages IV and III occurs at ca. 0.22 V vs. Li/Li+, from stages III to II-L at ca. 0.14 V vs. Li/Li+, from stages II-L to II at ca. 0.12 V vs. Li/Li+, and from stages II to I at ca. 0.09 V vs. Li/Li+.[3,28]

Non graphitizing and *graphitizing* carbons exhibit different electrochemical properties with regard to the potential profile, Coulombic efficiency, and reversible capacity as anode for LIBs. Dahn et al.[35,36] proved that both *graphitizing* (soft) and *non graphitizing* (hard) carbons heat-treated at temperatures below 800–900 °C are considered *high specific charge* carbons. This means that they can achieve reversible capacities between 400 and 2,000 mAh·g⁻¹ (1.2 ≤

$x \leq 5$ in Li_xC_6), depending on the processing parameters and starting precursor.[3] Their potential profile exhibits a large hysteresis as the lithium insertion occurs at potential close to 0 V vs. Li/Li^+, whereas lithium de-insertion occurs at much more positive potentials. The hysteresis and high capacities are usually related to the content of hydrogen-functional groups. Alternatively, such a high capacity has also been related to different mechanisms in the literature of hard carbons: the formation of covalent Li_2 molecules occupying nearest neighbor sites in the carbon structure,[37] the accommodation of lithium at the zigzag and arm-chair faces between two adjacent crystallites,[38] the clustering of lithium ions in nanopores,[39] and storage of lithium ions in graphite surface edges.[3] Dahn *et al.*[35,36] proposed the adsorption of lithium ions on both sides of single graphene sheets, arranged like falling cards, and the storage in micropores.

Above this temperature range, the electrochemical behavior depends on the nature of the carbon structure. At treatment temperatures up to 1,000 °C, hydrogen is progressively released from the carbon precursor. In general, carbon (either from hard or soft precursors) treated at temperatures above 1,000 °C delivers low-specific capacities (with x values between ≈ 0.5 and 0.7 in Li_xC_6). However, after very high-temperature treatments, the development of graphitic-ordered regions leads to an increased specific capacity close 300 mAh·g^{-1} at 2,600 °C for soft carbons. In contrast, an increase in the temperature for hard carbons leads to a drastic reduction in the specific capacity ($Q \approx 125$ mAh·g^{-1}, $x \approx 0.4$ in Li_xC_6).[40]

Fromm *et al.*[41] also investigated the correlation between the reversible specific capacity and the carbonization temperature of biomass sources as representative hard carbons compared to petroleum coke as soft carbon reference material at temperatures between 800 and 2,800 °C. In terms of the specific discharge capacity, their work reported a general trend: a decrease in capacity from 800 to $\approx 1,800$–2,000 °C, attributed to structural changes and the reduction of remaining surface functional groups, followed by an increase in capacity from temperatures rising above 2,000 °C, attributed to an enhanced crystalline order.

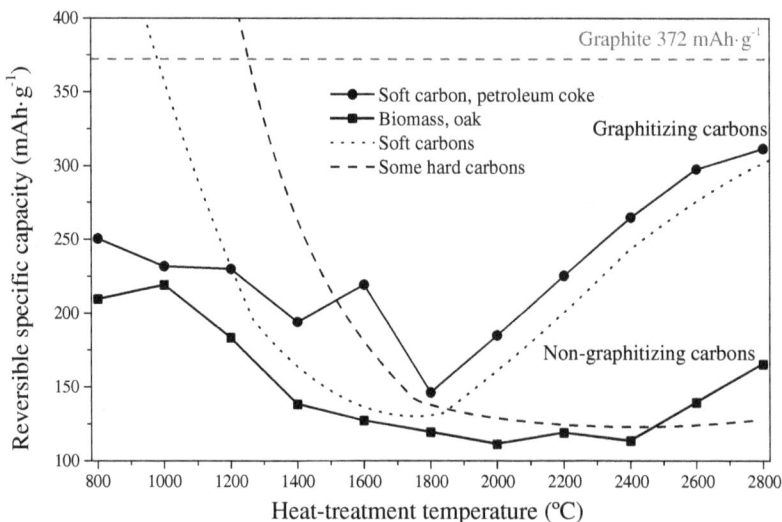

Fig. 8.5. Evolution of the reversible specific capacity as a function of HTT for graphitizing and non graphitizing carbons without catalyst. Contains data from Refs. 40 and 41.

The evolution of the specific capacities for both hard and soft carbons without catalysts as a function of the heat-treatment (HTT) temperature is illustrated in Fig. 8.5 from findings of the work of Dahn *et al.*[40] and Fromm *et al.*[41] As can be clearly seen from this graph, soft carbons need temperatures above 2,800 °C to achieve reversible capacities of $Q \approx 310$ mAh·g^{-1}, close to that of theoretical graphite. This carries substantial economic costs and high-energy consumption, so there is a current need to find more sustainable and cheaper methods to synthetize graphitic anode materials.

8.3.1 *Carbonization and graphitization processes*

Most carbon materials are produced by thermal decomposition of carbon-rich organic precursors such as biomass resources, polymers, or cokes under an inert atmosphere (Ar or N_2) up to temperatures ranging between 800 and 1,500 °C. Indeed, the possibility of transforming a variety of waste biomass resources into functional carbon materials has been the subject of extensive research efforts because of the possible social, economic, and environmental benefits.[42]

The term "*biomass*" refers to all organic matter from plants or plant-based materials derived from nature, including woody biomass, food and agricultural residues, among others. Main components of biomass resources are cellulose (40%–50%), hemicellulose (20%–30%) and lignin (15%–20%), and for this reason, plant-based biomass resources are often termed lignocellulosic materials.[43]

The average annual biomass production in the European Union can be estimated to be close to 1,500 million tons per year, considering only agricultural and forestry sources.[44] Thus, the use of biomass resources for industrial applications can offer a way of recycling and a safe method of waste disposal. At the same time, it is believed that carbon materials from biomass resources could pave the way for the development of more sustainable energy storage systems by using renewable and green chemistry concepts.[45,46]

The thermal decomposition of carbon-rich organic precursors is termed "*carbonization*" since it involves an increase in the relative carbon content in the final product. During carbonization, organic precursors undergo pyrolysis, cyclization, aromatization, polycondensation, and carbonization processes until an inorganic carbon scaffold is obtained. In the literature, carbonization is also occasionally referred to as pyrolysis as both processes are in fact overlapped with each other upon heating. However, the former normally refers to the use of higher temperatures (above 800 °C) and lower heating rates than pyrolysis, with the aim of obtaining a higher relative carbon content in the final material.[47]

A general scheme of the carbonization and graphitization processes for different organic precursors can be seen in Fig. 8.6, showing the main gas products during reaction and the structural evolution of the resulting carbon scaffold. Upon heating biomass resources, the physically adsorbed water is first released at about 150 °C. Low molecular weight aliphatic molecules and aromatic hydrocarbons are subsequently released. Cyclization, aromatization, and polycondensation of aromatic molecules progressively happens, accompanied with the release of oxygen and hydrogen (CO_2, CO, CH_4). The actual

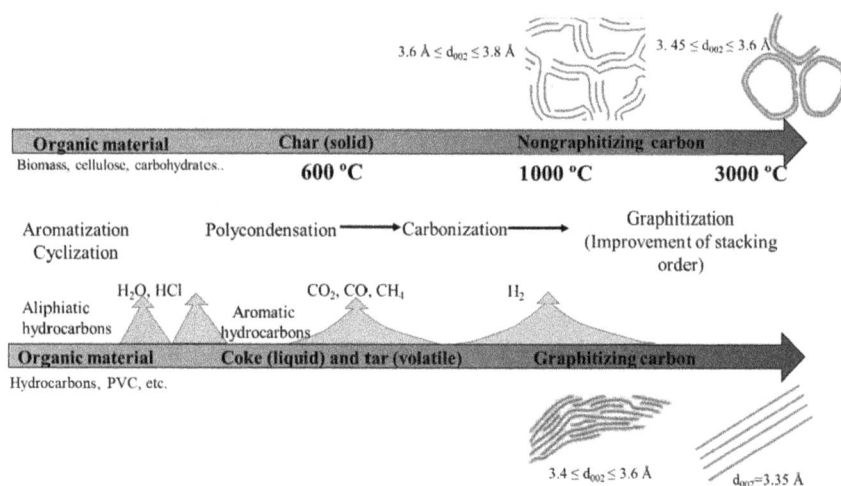

Fig. 8.6. Schematic illustration of carbonization and graphitization processes depending on the starting precursor (graphitizing and nongraphitizing carbons). Modified and redrawn from Refs. 49 and 50.

temperature at which decomposition occurs in biomass resources is determined by the thermal stability of their components. Hemicellulose first degrades at 220–315 °C, followed by cellulose (320–400 °C) and finally lignin (220–700 °C). Most of the overall weight loss during the carbonization of biomass resources occurs between 220 and 600 °C, leaving a residue with a carbon yield of 15%–30%, depending on the starting precursor composition.[48] At temperatures higher than 1,000 °C, the concentration of remaining functional groups and heteroatoms gradually decreases (mostly hydrogen content). The carbon content in the resulting precursor after carbonization can reach 100 wt% at treatment temperatures above 1,500 °C.[30]

According to the state of the precursor during heating, the carbonization can be classified into either gas-phase, liquid-phase, or solid-phase carbonization. In the specific case of solid-phase carbonization, no remarkable changes in morphology are reported.[30] The resulting carbon microstructure depends then not only on the starting precursor, but also on the maximum treatment temperature.

After carbonization, the structure of the resulting carbon strongly depends on the starting precursor and it is often classified into two categories, *graphitizing* (also referred to as soft carbons) and *nongraphitizing* (also referred to as hard carbons), based on its ability to convert to graphite upon heat-treatment. This clear distinction was first drawn by Rosalind Franklin in 1951[51] through powder X-ray diffraction (XRD) analysis of carbons derived from carbonization of different organic materials. Her results proved that carbons exhibited different behavior when heated at temperatures over 2,500 °C depending on the starting precursor. Indeed, hard carbons exhibit a strong cross-linking and random orientation between crystallites after carbonization, which precludes the development of a truly graphitic structure upon further heat-treatment. By contrast, soft carbons exhibit much weaker cross-linking and higher degree of alignment between neighboring carbon crystallites, thus facilitating the graphitization upon heat-treatment.

Some works suggest that the occurrence of an intermediate liquid state during carbonization makes the difference in the ability to graphitize of soft carbons precursors such as pitches and cokes.[30] In this case, the resulting carbon melts and does not maintain the starting precursor morphology.[52] In contrast, if the carbon residue remains solid during carbonization and does not pass through a liquid state, closed voids formed by gas release are unable to collapse into a denser crystalline structure. Examples of *nongraphitizing* carbons include those derived from biomass resources,[53] polyvinylidene chloride,[54] sugar,[51] or cellulose.[55] In contrast, polyvinyl chloride,[51] petroleum coke,[41] pitch[56] or polyvinyl acetate[57] are examples of *graphitizing* carbons.

Franklin proposed two different structural models for the two classes. In her view, randomly oriented graphitic nanodomains are bound by amorphous regions that act as cross-linking groups in hard carbons. Later on, Harris *et al.*[58,59] suggested that *nongraphitizing* carbons may contain curved carbon sheets with a structure similar to that of fullerene, being the curvature a result of topological defects such as nonhexagonal rings and saddle points.

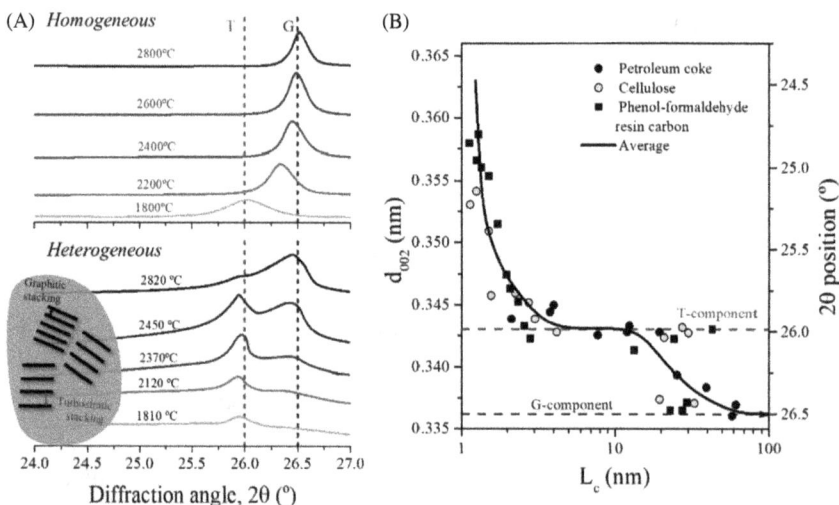

Fig. 8.7. (A) XRD patterns of petroleum coke (homogeneous graphitization, data from Ref. 41) and charcoal (heterogeneous graphitization, data from Ref 30. (B) Relation between d_{002}, L_c and (002) reflection position for both nongraphitizing carbons (cellulose and phenol-formaldehyde resin) and graphitizing carbons (petroleum coke). Contains data from Ref. 55.

The effect of HTT on the structural development of carbonaceous materials has been extensively evaluated by measuring the average interplanar distance (d_{002}) and the crystallite dimensions along the basal (L_a; length) and stacking directions (L_c; thickness). The dependences of d_{002}, L_a, and L_c on HTT is shown in Fig. 8.7,[30,41,53,60] including values reported for biomass-derived carbons (e.g., corn straw,[61] corn cob,[62] mangosteen shell,[63] cellulose,[64] date palm[65] and kelp).[66] According to the literature, *graphitizing* and *nongraphitizing* carbons exhibit crystallites sizes L_a and L_c in the ≈ 1 nm range at temperatures around 1,000 °C. However, the effect of HTT clearly highlights their different nature from 1,500 °C onward: while L_a and L_c reach values around 100 nm at 3,000 °C for *graphitizing* carbons, *nongraphitizing* carbons only reach values of about 10 and 4 nm, indicating poor structural development in the stacking direction. With regard to the interplanar distance, *nongraphitizing* carbons only reach a minimum value of $d_{002} \approx 0.35$ nm at 3,000 °C, whereas *graphitizing* carbons exhibit a value close to the theoretical value of ideal graphite ($d_{002} \approx 0.335$ nm).

During heat treatment at temperatures above 2,000 °C, some carbon precursors can undergo a homogeneous and continuous graphitization process, while others undergo a heterogeneous process (also called as multiphase graphitization) with a transition between turbostratic and graphitic phases.

In XRD analysis, homogeneous graphitization results in a progressive shift in the (002) reflection toward the position of ideal graphite ($2\theta \approx 26.5°$) with increasing HTT, as is shown in petroleum coke[41] in Fig. 8.7A. In contrast, XRD patterns of carbon materials that undergo heterogeneous graphitization show two overlapping reflections: the T-component corresponding to turbostratic order with an interplanar distance of about $d_{002} \approx 0.343$ nm and the G-component with an interlayer spacing similar to that of graphite ($d_{002} \approx 0.335$ nm).

With increasing treatment temperature, soft carbon precursors only exhibit the G-component (Fig. 8.7B), while hard carbons exhibit both reflections with relative intensities varying with the temperature. For instance, multiphase graphitization, showing both turbostratic and graphitic contributions, has been observed for sugar-derived carbons[67] as well as wood-derived carbons.[41]

8.3.2 *Catalytic graphitization*

An effective approach to accelerate the development of graphitic domains in *nongraphitizing* carbons at moderate temperatures (below 1,000 °C) is the use of catalysts to induce graphitization. *Catalytic graphitization* involves the addition of some foreign species to the carbon precursor that, upon heating, can induce the precipitation of ordered carbon regions and thus a decrease in the activation energy for crystallization.

This process received much attention in the 1980s[31,68] due to the possibility of synthesizing graphite at temperatures far below what was conventionally required (3,000 °C), resulting in significant cost and energy savings. Similar approaches were later extensively used to grow CNTs by chemical vapor deposition (CVD).[69] However, despite the large body of work devoted to employ this method for the synthesis of carbon materials, the mechanisms involved in catalytic graphitization are still not well understood and limited progress has been made in this direction. Currently, attention is focused on the application of catalytically graphitized carbon in different advanced fields such as electrodes for supercapacitors devices,[70-72] as anode materials for lithium-ion batteries[73] and others.[74,75]

Hard carbon precursor materials, including phenolic resins, xerogels, cellulose, lignin, and biomass resources, have been successfully graphitized in the presence of catalysts in a solid-state process.[76-78] Some previous publications on catalytic graphitization of *nongraphitizing* carbons have explored the effect of a large variety of possible catalysts, such as the works from Ōya *et al.*[68] (10 wt% catalyst addition to phenol formaldehyde; HTT $\approx$ 2,600–3,000 °C), Yokokawa *et al.*[79] (5 wt% catalyst addition to furfural alcohol; HTT $\approx$ 1,400–2,300 °C), Weisweiler *et al.*[80] (unreported catalyst amount, HTT higher than the melting point of each metal), Maldonado *et al.*[77] (catalyst addition 1 wt% to resorcinol and formaldehyde, HTT $\approx$ 500–1,800 °C), Wang *et al.*[81] (1–5 wt% catalyst addition to pitch; HTT $\approx$ 1,000–2,800 °C), Shoujun *et al.*[82] (2–9 wt% catalyst addition to furan resin; HTT $\approx$ 2,100–2,200 °C), Sevilla *et al.*[83] (~ 3 mmol catalyst per gram of carbon from silica xerogel; HTT $\approx$ 900 °C), and Zhao *et al.*[84] (0.009–0.019 wt% catalyst addition to glucose; HTT $\approx$ 800–1,000 °C).

Among the different catalysts employed, VI–VIII transition metals, and especially iron (Fe), cobalt (Co), and nickel (Ni), have stood out for their good graphitization efficiency.[77,85-89] Metal transition oxides such as Fe_2O_3 and Fe_3O_4 and alloy compounds such as Fe-Ni and Co-Ni have also proven to be effective catalysts.[90-92]

In this regard, of particular interest are previous studies from Maldonado *et al.*,[77] Yan *et al.*,[85] Thambiliyagodage *et al.*[86] and Sevilla *et al.*[87-89] evaluating the graphitization effectiveness of different transition metals with the same concentration on a certain carbon precursor. The evolution of the interplanar distance d_{002} and Raman

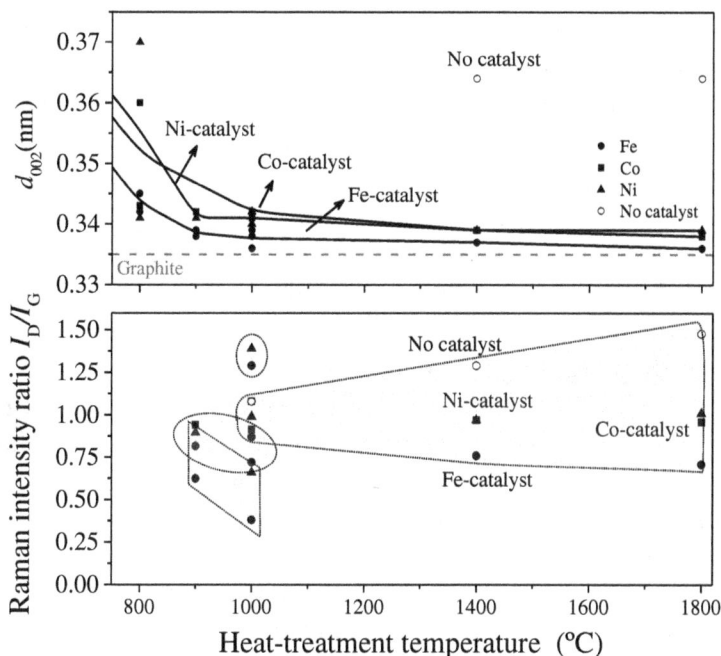

Fig. 8.8. Dependence of interplanar distance (d_{002}) (top panel) and Raman intensity ratio between D and G bands (I_D/I_G; bottom panel) with HTT for carbons treated with transition metal catalysts. The differences between I_D/I_G ratio for the same precursor are visually shown with surrounding dotted areas to guide the eye. Contains data from Refs. 77, 85–89.

intensity ratio between *D* and *G* bands (I_D/I_G), which is often an indicator of the extent of the graphitization, is shown in Fig. 8.8. All these works agree that the highest catalytic activity is often achieved using Fe as a catalyst, followed by Co and then Ni, as can be also inferred from the lower interplanar distances close to the theoretical of graphite (dotted gray line; Fig. 8.8—top panel) and the lower I_D/I_G ratio obtained regardless of the raw precursor and the catalyst concentration (Fig. 8.8—bottom panel). It is believed that the effectiveness shown by group VIII transition metals stems from their partially occupied d-orbitals valence shell (only occupied 6–8 electrons in their d-shell orbitals) and the number of free electron vacancies.[85]

Numerous papers have confirmed that the extent of graphitization within the carbon material can be enhanced through an increase in the starting catalyst concentration.[78,93–95] For instance, Chen et al.[96] reported graphitization percentages at 800 °C for cotton of 12%, 39%, 51%, and 46% using iron nitrate catalytic solution concentrations of 0.1, 0.3, 0.5, and 1.0 M, respectively. Previous studies have also highlighted how relevant is the catalyst addition process and the catalyst particle size toward an effective graphitization, but in a much less extent than the catalyst concentration.[93] A homogeneous distribution of fine nano-sized catalyst particles can be obtained by heating after impregnation with inorganic- and organic-based solutions containing the metal acting as catalyst. Coprecipitation of nitrate salts and hydrothermal impregnation are recent alternative approaches to introduce catalyst into the carbon precursor.[94,97]

The size of the catalyst particles was found to influence the precipitated graphite microstructures and might be the main responsible of inducing a multiphase catalytic graphitization. Different works have shown that the size of the metal particles is intimately tied to the carbonization temperature and increases due to a coarsening effect upon heating.[98,99] If the particles are finely divided in the order of a few nanometers in size (e.g., $\approx$ 20 nm according to Oya et al.[100] in the case of a Ni catalyst), onion-like graphitic layers surrounding catalyst particles precipitate showing a turbostratic component in the XRD pattern (T-component; Fig. 8.9A). As HTT increases when the catalyst concentration is high, metal particles become progressively bigger as a result of the diffusion process. Particle size distributions at different processing temperatures from Ref. 100 are shown in Fig. 8.9B. In such a case, at higher temperatures, three-dimensional bulk graphite crystals precipitate (from catalyst sizes above 80 nm in the case of Ni), showing a graphite component in the XRD pattern (G-component, Figs. 8.9A, B).[100]

In catalytically graphitized carbons, metal transition particles are often subsequently removed by room temperature stirring in concentrated nitric or hydrochloric acid[101–103] to introduce emerging porosity with a view toward EDLC applications. However, the higher the catalyst loading and catalyst particle size, the lower the surface

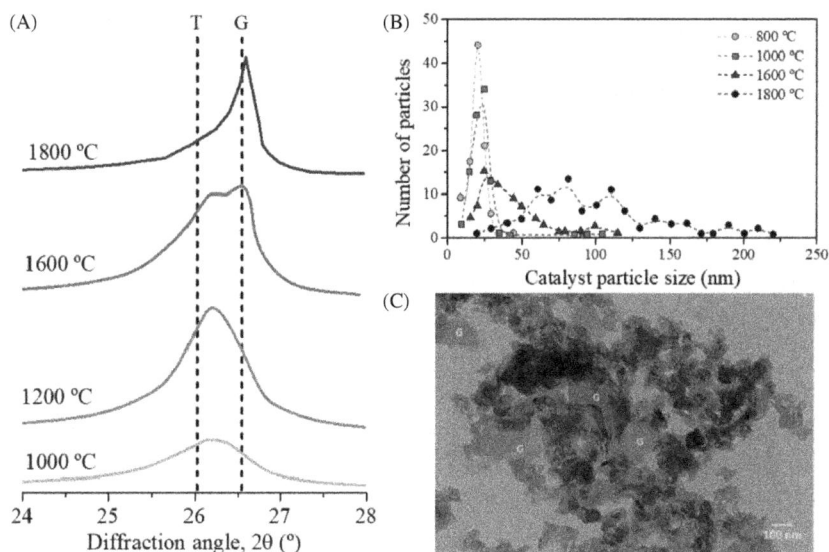

Fig. 8.9. (A) Changes of (002) diffraction profiles of the catalyzed phenol formaldehyde carbon (30 wt% catalyst), data from Ref. 100. (B) Size distributions of metal particles as a function of temperature and (C) TEM image of catalytically graphitized beech-derived carbon at 1600 °C where it can be observed the precipitation of three-dimensional graphite crystals (labeled as *G*). TEM image is from Ref. 99, used with permission. Copyright 2014, Elsevier.

area after acid etching treatment to remove remaining catalyst particles afterward.[78,94] Yan *et al.*[94] found an increase in the surface area of Kraft lignin-derived carbon at 1,000 °C for Fe loadings ranging from 0 to 10 wt%, point from which it then drastically decreased.

According to numerous experiments on catalytic graphitization, the peak temperature of the heat-treatment process is a key parameter controlling graphitization.[101,104] According to *in situ* XRD experiments from Hoekstra *et al.*,[105] the onset of the graphitization by Fe occurs from 715 °C, while Ni and Co require temperatures above 800 °C. From this temperature on, several authors have reported a monotonically increase in the degree of graphitization with increasing HTT. Most works are limited to low treatment temperatures and give no clear correlations between microstructural parameters and HTT,[78,96] but Ramirez-Rico *et al.*[106] reported an increase in the degree of crystallinity by using Fe from 27% to 63% as treatment temperature increases from 850 to 1,600 °C, as calculated from Raman measurements. The increase in crystallinity in that work was more pronounced in the 1,100–1,400 °C temperature range. A similar trend was also reported by Kakunuri et al.,[107] showing the largest extent of graphitization occurring between 900 and 1,200 °C. So far, there is no clear evidence of a saturation point in the degree of crystallinity obtained from catalytic graphitization, though further studies considering a wider temperature range could clarify this factor.

Other parameters that have a somewhat influence are the heating rate and the holding time at the peak temperature. Bitencourt et al.[93] evaluated the influence of both parameters on the graphitization degree of a *Novolac* resin using Fe as a catalyst. For a fixed impregnation concentration of 2 wt% of ferrocene, the graphitization degree at 1,400 °C was estimated to be 27.0%, 28.4%, and 32.6% for respective residence times of 6 min, 5 h, and 10 h. Alternatively, the effect of the heating rate (1 °C–4 °C·min^{-1}) used during the carbonization of the resins was also analyzed, with an optimum value of 3 °C·min^{-1}. Skowroński et al.[108] graphitized a phenolic-resin carbon using Fe as a catalyst up to 1,000 °C using two residence times; 20 h and 100 h. Such a wide time difference gave clear evidence of enhanced crystalline ordering with increasing residence time. Although previous works revealed that the catalytic graphitization may be directly improved with increasing the residence time, the impregnation method, and finding the optimum heating rate, these parameters seem to be secondary. They provide a crystallinity enhancement to a much lesser extent than others, such as catalyst concentration or carbonization temperature.

There are two main proposed mechanism responsible for the catalytic graphitization and the conversion of solid amorphous carbon precursors to graphitic materials trough heat-treatment.[31,81,94,109] The most cited mechanism is dissolution–precipitation, whereby amorphous carbon first dissolves into metal particles, forming near-eutectic liquid droplets (M_xC_y, where M stands for the metal atom), which continuously saturate and precipitate graphite crystals due to the free enthalpy difference between amorphous and graphitic carbon. The other possible mechanism involves the formation–decomposition of an intermediate metal carbide compound that, with increasing/decreasing temperature, can decompose to metal particles and graphite. Although catalytic graphitization has been extensively used for many years, there are still remained open questions regarding why some authors detect the presence of metal carbides after cooling down, while others do not, and whether the carbide takes an active part in the graphitization.[86]

In the representative case of using Fe as a catalyst and lignocellulosic or polymeric carbon sources, many experimental works agree that the carbonization process first leads to the formation of iron oxides, Fe_2O_3/Fe_3O_4, at low processing temperatures (<500 °C), which further carbothermally reduce to FeO and finally to metallic Fe at 600 °C (ferrite phase α, body-centered cubic).[77,96,110,111] The same trend is observed when using Fe_2O_3 as starting catalyst agent.[112] From previous *in situ* XRD experiments upon heating, iron carbide (cementite phase, Fe_3C) and graphite reflections become discernible at temperatures from 700 °C.[110] By *in situ* transmission electron microscopy (TEM) experiments at 640 °C, liquid metal-C were observed randomly moving through amorphous carbon areas, precipitating graphitic regions along their path. The propagation of the catalyst particle within the solid carbon structure is believed to be facilitated by the high thermal expansion coefficient of Fe, Co, and Ni, which forces the graphitic shell surrounding catalyst particles to break out upon increasing temperature, allowing catalysts to exit and continue moving.[105,113] Alternatively, according to Huo *et al.*,[114] the coalescence of catalyst particles upon heating can occur through defective carbon areas or by Fe core ejection due to contraction strain.

By further heating at temperatures above 700 °C, the relative intensity of the graphite peak increases with HTT. In addition, ferrite phase is progressively converted to austenite, γ-Fe (face-centered cubic) with increasing temperature.[110] From *ex situ* XRD experiments, it can be expected a decrease in the carbide concentration after cooling down from temperatures above 1,000 °C.[115]

Some works have reported the presence of cementite carbide phase (Fe_3C) after cooling down from temperatures below 1,000 °C,[111,115–117] while others have not.[94] Interestingly, Neeli *et al.*[117] recently found that an increase in drying time after impregnation of cellulose with iron nitrate solution resulted in higher yield of oxide phases (Fe_2O_3) and cementite formation upon heating. Thompson *et al.*[78] Yan *et al.*[94] and Zhou *et al.*[111] reported an increase in the relative fraction of the cementite phase after cooling down with increasing the Fe loading within the carbon matrix. This apparently contradicts what is expected on the basis of the binary Fe-C phase diagram, as will be shown below.

Some works argue that the precipitation of curved graphitic nanostructures surrounding encapsulated catalyst particles as well as the formation of carbide phases might be rather related to the cooling process.[118] Since the solubility of carbon in the molten metal catalyst particles decreases with decreasing temperature, the excess of carbon can precipitate around catalyst particles core to reduce the surface energy upon cooling.[30,117,119] Liu *et al.*[104] evaluated the effect of the cooling rates with values of 2 and 20 °C·min^{-1}, but no clear influence was reported. Saenger *et al.*[120] proved by *in situ* XRD experiments upon heating/cooling treatments that the graphite formation by using a Ni catalyst occurs upon heating at ≈ 730 °C. These findings are in agreement with those *in situ* XRD experiments upon heating reported by Hoekstra *et al.*[110] using an Fe catalyst. Still, Saenger *et al.*[120] also found that graphite peak intensity reached during cooling was stronger than during heating. This fact was attributed to possible continuous precipitation of dissolved carbon,[109] changes in sample morphology and/or detection geometry, resulting in the same amount of graphitic carbon having a stronger XRD signal,[120] Little attention has been paid to the cooling process and there are not still clear evidences of its influence.

According to the binary Fe-C phase diagram (Fig. 8.10), the solubility of carbon into α-Fe particles at 727 °C is limited to 0.022 wt%, turning to 2.14 wt% into γ-Fe (1,147 °C) due to the higher number of available interstitial positions. However, Krivoruchko *et al.*[109] reported the formation of unusual liquid near-eutectic Fe-C

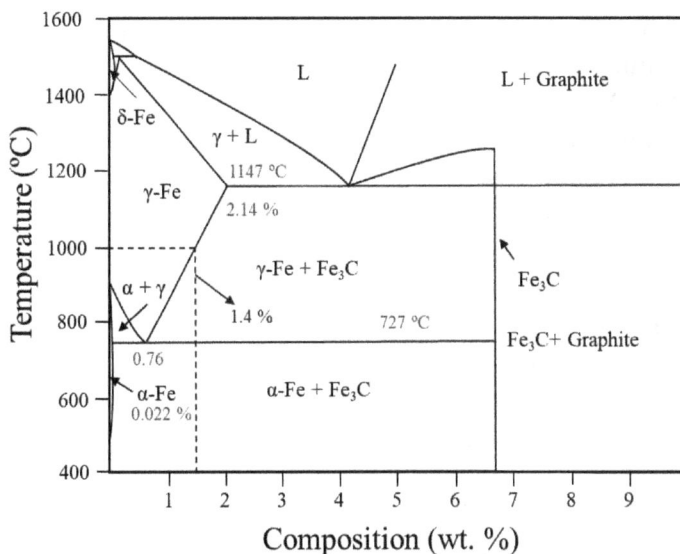

Fig. 8.10. Fe-C phase diagram. Figure reprinted from Ref. 117. Used with permission. Copyright 2018, Elsevier.

droplets at anomalously low temperatures (640 °C) with a suggested composition of around 50 at. %. This liquid state cannot be explained by the corresponding equilibrium phase diagram, so this phenomenon is often attributed to size effects.[121,122]

Neeli *et al.*[117] proposed that, when the catalyst particles reach a supersaturation state, additional carbon can be either incorporated into a new intermediate phase Fe_3C with a bulk carbon content of 6.7 wt%,[118] or precipitate as crystalline graphitic structures. For a dissolved carbon content higher than 6.7 wt%, all Fe particles should be converted to Fe_3C. Theoretically, at temperatures above 727 °C the decomposition rate of Fe_3C is higher than its formation rate, thus the unstable carbide form could decompose into graphite and metal particles.[94] This agrees with the apparent less carbide relative composition contribution at higher carbonization temperatures.[115]

Other works argue that the catalytic activity of transitions metals may be rather related to the number of free electron vacancies in the d-shell orbitals.[31,85] Both Fe, Co, and Ni are in group VIII with corresponding electron configurations of Fe: [Ar] $3d^6 4s^2$, Co: [Ar] $3d^7 4s^2$, Ni: [Ar] $3d^8 4s^2$, resulting in two to four free electron vacancies. The energy levels of their configuration would slightly change with the acceptance of electrons from carbon, thus enabling the dissolution of carbon and the formation of strong covalent bonds between metal and carbon atoms. According to the d-electron configuration, the catalytic activity of transition metals should theoretically have the order of Fe > Co > Ni, in agreement with some experimental works.[77,85–89]

The process of catalytic graphitization by transition metals has recently aroused interest not only from the point of view of increasing the degree of structural order within the carbon material, which in turn increases the electronic conductivity with increasing the carbonization temperature,[123] but also because it is an effective way to introduce emerging meso-porosity when the remaining catalyst nanoparticles are removed by acid etching after HTT, as illustrated in Fig. 8.11. This causes an increase in the material surface area and a better electrochemical performance for supercapacitor applications compared to nontreated carbons. The resulting pore structure and surface area can be tuned by selecting the starting carbon source, the metal catalyst loading,[102,124] and tuning heat-treatment parameters. This approach has been recently addressed by several authors for synthetizing alternative electrode carbon materials for supercapacitor applications.[101,116,125,126]

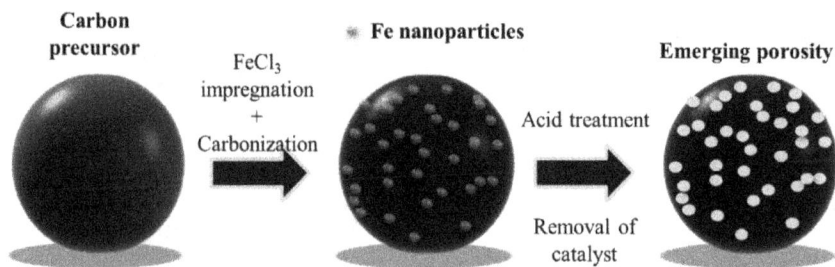

Fig. 8.11. Schematic illustration of catalytic graphitization process used for introducing porosity into carbon materials.

8.4 Fe-Catalyzed Graphitic Carbons in Metal-Ion Batteries

In the past literature, some works have already investigated the electrochemical properties of graphitized carbon precursors by the addition of transition metals such as Fe,[73,90,107,108,127–131] Co[132] and Ni[133–135] as anodes for LIBs. Comparison between different works is difficult due to the different synthesis parameters (starting precursors, routes to introduce the catalyst into the carbon scaffold, catalyst size and amount, HTT, as well as heating rate and holding time), as well as electrochemical characterization routines (different electrode composition, mass loading, porosity, and cycling conditions). Nonetheless, the effect of HTT on the achievable reversible capacity of all carbon anode materials graphitized by means of Fe is summarized in Fig. 8.12. As can be seen, most of these works are limited at very low processing temperatures (below 1,200 °C). Notwithstanding that, samples obtained at low temperatures typically exhibit potential profiles characteristic of disordered carbons regardless of the starting precursor and catalyst concentration.[90,129] The high

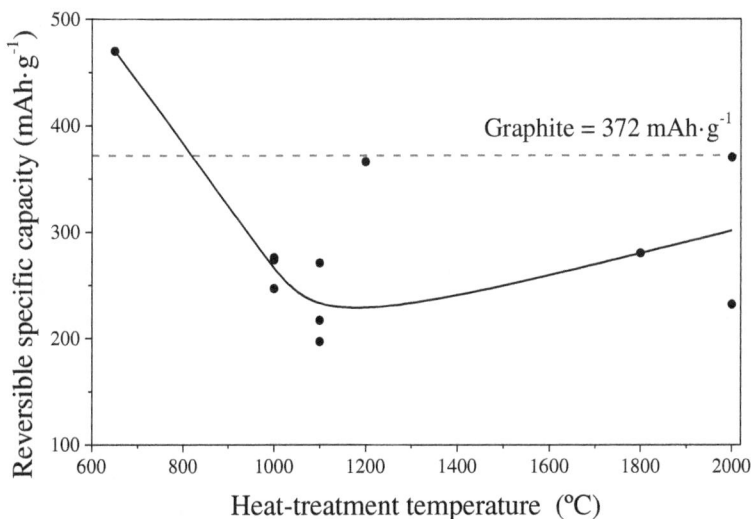

Fig. 8.12. Evolution of the reversible specific capacity as a function of treatment-temperature of carbons catalytically graphite by iron. Data from Refs. 73, 90, 107, 108, 127–131.

reversible capacities achieved within the temperature range below 800 °C can stem from additional storage due to remaining functional groups. According to works on this topic,[73,128] the catalyst concentration and heat-treatment peak temperature are the two most influential parameters on the reversible capacity as anodes for LIBs. Although the catalyst concentration influence is not explicitly shown in Fig. 8.12, a moderately increasing trend in reversible capacity can be discerned with increasing the temperature, related to an increased structural order and graphitization level.

The catalytic graphitization of carbon-rich precursors has emerged as a fairly new route to develop graphitic anodes for LIBs since 2007. Skowroński et al.[129] first proved that the heat-treatment of disordered carbon spheres with Fe powder (C:Fe volume ratio of 1:1) at 1,000 °C allowed to improve the reversible capacity as anode for LIBs from 74 to 250 mAh·g[-1] due to the partial electrode graphitization. Later on, Skowroński et al.[108] graphitized carbon spheres using Fe powder (C:Fe weight ratio of 1:11) up to 1,000 °C using two residence times of 20 h and 100 h. Interestingly, even though an increased residence time gave clear evidence of enhanced crystalline ordering within the structure, both carbons exhibited similar reversible capacities when tested as anode for LIBs (274 and 275 mAh·g[-1], respectively). This work together with the promising results paved the way for further research. Obrovac et al.[127] reported for a graphite derived from glucose-derived hard carbon and Fe powder (C:Fe molar ratio of 87.5:12.5) heated up to 1,200 °C a reversible capacity as high as 366 mAh·g[-1]. Gaikwad et al.[130] reported a reversible capacity of 192 mAh·g[-1] by a resorcinol formaldehyde-derived carbon treated up to 1,100 °C with 5 wt.% of iron acetate solution.

Generally, the higher the HTT and catalyst concentration, the better the reversible capacity. Zhao et al.[128] synthetized Fe-graphite composites with starting concentrations of 10 and 17 wt% of Fe by ball milling of graphite powder followed by annealing at 1,100 and 2,000 °C. Such a work first highlighted the influence of the catalyst concentration and the peak temperature on electrochemical properties. At a temperature of 1,100 °C, no difference was reported in terms of the reversible capacity for the two different amounts of catalyst. However, this work reported a clear increase in the reversible capacity from 232 to 370 mAh·g⁻¹ when increasing the catalyst concentration from 10 to 17 wt%, and an increase in reversible capacity from 197 to 370 mAh·g⁻¹ as the treatment temperature raises from 1,100 °C to 2,000 °C for a fixed Fe concentration of 17 wt%. Meanwhile, Gaikwad *et al.*[131] investigated the role or iron loading on the degree of graphitization of carbon xerogels treated at 1,100 °C by incorporating 0, 2, 5, and 10 wt% HTT of iron acetate as a catalyst solution. In this case, the reversible capacity increased from 232 to 264 mAh·g⁻¹ when increasing the starting catalytic solution concentration from 0 to 10 wt%.

Gomez-Martin *et al.*[73] evaluated microstructural and electrochemical properties as anode for LIBs of graphitized hard carbon materials from biomass resources by means of an Fe catalyst (fixed concentration of ≈ 12 wt%) at temperatures ranging between 850 and 2,000 °C. Authors made use of natural resources as starting sources to improve the possible environmental friendliness and sustainability of batteries. A comparison in terms of reversible capacity of Fe-graphitized carbons to those of soft and hard carbons obtained without a catalyst from Ref. 73 is shown in Fig. 8.13. Without the

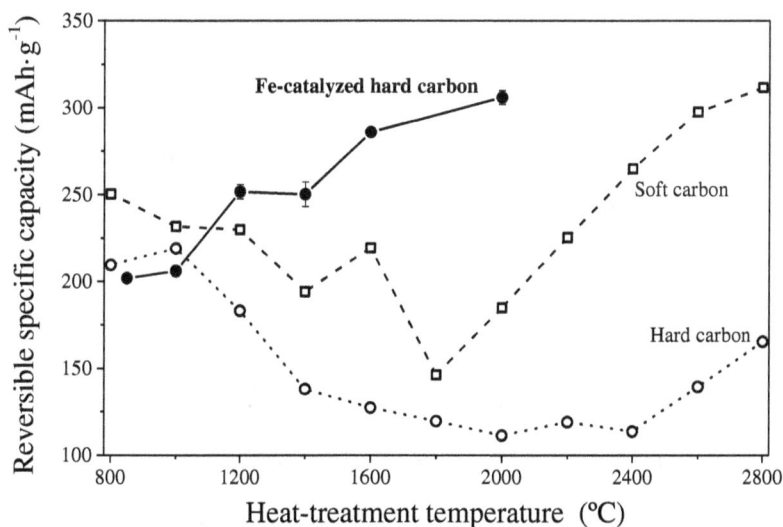

Fig. 8.13. Evolution of the reversible specific capacity (specific current of 37.2 mA·g⁻¹) as a function of carbonization temperature. Figure adapted from Ref. 73.

effect of any catalyst, the trend is clear as reported in Fig. 8.5: a decrease up to ≈ 2,000 °C, followed by a slight enhancement with increasing temperature. A maximum specific discharge capacity of around 160 mAh·g^{-1} at 2,800 °C is achieved in the case of hard carbons (in which the development of graphitic domains is rather limited), whereas a more pronounced increase is achieved in soft carbons, reaching specific capacities close to 310 mAh·g^{-1} at 2,800 °C.

The capacity of Fe-treated hard carbons, however, monotonically increased from ≈ 202 mAh·g^{-1} at 800 °C to ≈ 310 mAh·g^{-1} at 2,000 °C due to the enhanced crystalline order.[73] Results at the highest temperatures were thus comparable to those of commercially used synthetic graphite derived from a petroleum coke precursor at higher temperatures. Furthermore, by the use of Fe as a catalyst at a carbonization temperature of 2,000 °C, capacities up to two times higher than that of noncatalyzed hard carbons could be reached. As can be seen in the potential profile shown in Fig. 8.14, without

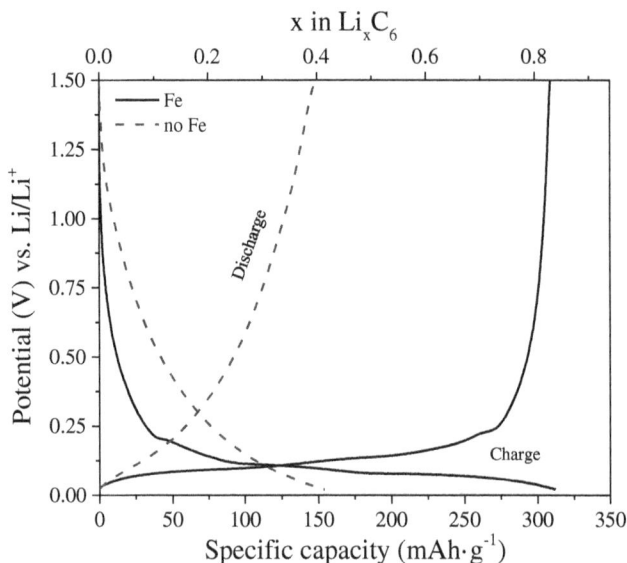

Fig. 8.14. Representative comparative of charge/discharge voltage profiles at a specific current of 37.2 mA·g^{-1}) of medium-density fiberboard (MDF) carbon heated up to 2,000 °C with and without Fe. As can be seen, the staging phenomena is not observed for noncatalyzed carbon sample. Figure adapted from Ref. 73.

catalyst staging phenomena was not apparently taking place due to the limited structural order, thus resulting in a capacity as low as » 150 mAh·g^{-1}. However, the potential profile of Fe-catalyzed carbons exhibited three voltage plateaus, showing the transition between different lithium–graphite intercalation compounds characteristic of highly graphitized carbon.

The capacities reported by carbons graphitized by an Fe catalyst are not far from meeting commercial requirements in terms of specific capacities. Although the theoretical specific capacity of graphite is 372 mAh·g^{-1}, values of 310–350 mAh·g^{-1} are often achieved in commercial cells.[136] Apart from reversible capacities, Gomez-Martin *et al.*[73] reported that the rate capability became worse at high current densities as the structural order increased in catalytically graphitized carbons. This is because the lithium intercalation through the graphitic sheets in the more graphitic samples have slower kinetics than the adsorption mechanism.

Although advances have been made to understand electrochemical properties of such materials, there is still a long road for improvement in this field. The main drawback of these materials still to be overcome in future works, and that make them not commercially competitive for now, is the low coulombic efficiency in the first charge/discharge cycles. An extremely low coulombic efficiency due to the formation of the SEI on the anode surface means that the amount of active lithium in the cathode material in a full-cell configuration may be fully irreversibly consumed in the first cycles. The coulombic efficiencies in the first cycle reported in the literature of Fe-catalyzed carbons range between 32 and 87%,[73,90,107,108,127–131] which are much lower to that of synthetic graphitic anodes used in commercial LIBs (between 80% and 95%). In this regard, Gomez-Martin *et al.*[73] found a clear linear correlation between the coulombic efficiency and the basal-plane surface area, as a higher surface area resulted in more electrolyte decomposition and SEI formation. The high surface area shown by these materials is related to the removal of iron particles by acid washing after heat-treatment process. Kakunuri *et al.*,[107] Gomez-Martin *et al.*,[73] Zhao *et al.*[128] and Gaikwad *et al.*[131] have reported an increase in the first-cycle coulombic efficiency by increasing the carbonization temperature and the catalyst loading. This can be explained by the coarsening effect of Fe nanoparticles by either increasing the catalyst loading or the temperature, which resulted in enlarged pore dimensions when the particles are removed, causing an overall decrease in the surface area. Although these aforementioned works have helped to understand the dependence of this parameter, it is important that future works make an effort to improve it.

The use of an Fe catalyst has also received considerable attention as an intermediate stage to synthetize conversion–reaction anodes. Some authors have reported the development of graphitic carbon-encapsulated α-Fe_2O_3[90,137,138] and Fe_4N[139] anodes with high reversible capacity (up to 1,000 mAh·g⁻¹) and improved cycling stability. It is believed that the graphitic layer surrounding metal transition particles could increase the electrical conductivity of the composite and buffer the volume expansion related to the lithiation enhancing the electrode structural stability upon cycling.

Furthermore, recent studies have also investigated the use of graphitized carbon materials by transition metals at low processing temperatures as anodes for SIBs[132,140] and potassium-ion batteries.[141] Although the structural order induced by the metal catalyst would not be desirable to favor sodium intercalation, the use of lower temperatures allows to introduce heteroatoms such as N and O species from nitrate-based catalytic solutions. These incorporated functional groups have proven to increase the sodium-ion adsorption capability.

8.5 Concluding Remarks

Catalytic graphitization is a novel approach that allows the production of highly crystalline materials from nongraphitizable or hard carbon sources. Among the transition metals usually employed as catalysts, iron has the greatest effect in promoting graphitization at moderate temperatures, resulting in materials that are close to the graphite standard while resulting in lowered production costs thus paving the way for more environmentally friendly anodes for metal-ion batteries, and in particular for lithium-ion. Catalytic graphitization is also an effective way of incorporating iron nanoparticles into a carbon matrix that can be later transformed into oxides or nitrides that incorporate additional capacity by means of faradic reactions at the anode (conversion electrodes) paving the way for hybrid lithium-ion batteries.

References

1. M. Winter and R. J. Brodd, What are batteries, fuel cells, and super-capacitors?, *Chem. Rev.* **104**, 4245–4270 (2004).
2. T. Nagaura and K. Tazawa, Lithium ion rechargeable battery, *Prog. Batteries. Sol. Cells* **9**, (1990).
3. M. Winter, J. O. Besenhard, M. E. Spahr and P. Novák, Insertion electrode materials for rechargeable lithium batteries, *Adv. Mater.* **10**, 725–763 (1998).
4. R. Schmuch, R. Wagner, G. Hörpel, T. Placke and M. Winter, Performance and cost of materials for lithium-based rechargeable automotive batteries, *Nat. Energy* **3**, 267–278 (2018).
5. M. Winter, The solid electrolyte interphase – The most important and the least understood solid electrolyte in rechargeable li batteries, *Z. Phys. Chem.* **223**, 1395–1406 (2009).
6. S.-T. Myung, Y. Hitoshi and Y.-K. Sun, Electrochemical behavior and passivation of current collectors in lithium-ion batteries, *J. Mater. Chem.* **21**, 9891–9911 (2011).
7. S. Theivaprakasam, G. Girard, P. Howlett, M. Forsyth, S. Mitra and D. MacFarlane, Passivation behaviour of aluminium current collector in ionic liquid alkyl carbonate (hybrid) electrolytes, *Npj Mater. Degrad.* **2**, 13 (2018).
8. M. S. Whittingham, Lithium batteries and cathode materials, *Chem. Rev.* **104**, 4271–4302 (2004).
9. D. Andre, S.-J. Kim, P. Lamp, S. F. Lux, F. Maglia, O. Paschos and B. Stiaszny, Future generations of cathode materials: An automotive industry perspective, *J. Mater. Chem. A* **3**, 6709–6732 (2015).
10. A. Manthiram, B. Song and W. Li, A perspective on nickel-rich layered oxide cathodes for lithium-ion batteries, *Energy Storage Mater.* **6**, 125–139 (2017).

11. H. Yang, C. Guo, A. Naveed, J. Lei, J. Yang, Y. Nuli and J. Wang, Recent progress and perspective on lithium metal anode protection, *Energy Storage Mater.* **14**, 199–221 (2018).

12. S. Goriparti, E. Miele, F. De Angelis, E. Di Fabrizio, R. Proietti Zaccaria and C. Capiglia, Review on recent progress of nanostructured anode materials for Li-ion batteries, *J. Power Sources* **257**, 421–443 (2014).

13. J. Cabana, L. Monconduit, D. Larcher and M. R. Palacín, Beyond intercalation-based li-ion batteries: The state of the art and challenges of electrode materials reacting through conversion reactions, *Adv. Mater.* **22**, E170–E192 (2010).

14. X. Zuo, J. Zhu, P. Müller-Buschbaum and Y.-J. Cheng, Silicon based lithium-ion battery anodes: A chronicle perspective review, *Nano Energy* **31**, 113–143 (2017).

15. J. Jang, I. Kang, K.-W. Yi and Y. W. Cho, Highly conducting fibrous carbon-coated silicon alloy anode for lithium ion batteries, *Appl. Surf. Sci.* **454**, 277–283 (2018).

16. J. Qin, S. Zhu, C. Feng, N. Zhao, C. Shi, E.-Z. Liu, F. He, L. Ma, J. Li and C. He, In-situ space-confined catalysis for fabricating 3D mesoporous graphene and their capacitive properties, *Appl. Surf. Sci.* **433**, 568–574 (2018).

17. H. Tao, L.-Z. Fan, W.-L. Song, M. Wu, X. He and X. Qu, Hollow core–shell structured Si/C nanocomposites as high-performance anode materials for lithium-ion batteries, *Nanoscale* **6**, 3138–3142 (2014).

18. M. R. Zamfir, H. T. Nguyen, E. Moyen, Y. H. Lee and D. Pribat, Silicon nanowires for Li-based battery anodes: A review, *J. Mater. Chem. A* **1**, 9566–9586 (2013).

19. W. Wang, L. Gu, H. Qian, M. Zhao, X. Ding, X. Peng, J. Sha and Y. Wang, Carbon-coated silicon nanotube arrays on carbon cloth as a hybrid anode for lithium-ion batteries, *J. Power Sources* **307**, 410–415 (2016).

20. X. H. Liu, L. Zhong, S. Huang, S. X. Mao, T. Zhu and J. Y. Huang, Size-dependent fracture of silicon nanoparticles during lithiation, *ACS Nano* **6**, 1522–1531 (2012).

21. C. Pillot, The rechargeable battery market and main trends 2016–2025, *Avicenne Energy* (2017).

22. D. Larcher and J. M. Tarascon, Towards greener and more sustainable batteries for electrical energy storage, *Nat. Chem.* **7**, 19 (2014).

23. S. Samad, K. S. Loh, W. Y. Wong, T. K. Lee, J. Sunarso, S. T. Chong and W. R. Wan Daud, Carbon and non-carbon support materials for platinum-based catalysts in fuel cells, *Int. J. Hydrogen Energy* **43**, 7823–7854 (2018).

24. C. Zhang, R. Kong, X. Wang, Y. Xu, F. Wang, W. Ren, Y. Wang, F. Su and J.-X. Jiang, Porous carbons derived from hypercrosslinked porous polymers for gas adsorption and energy storage, *Carbon* **114**, 608–618 (2017).

25. R. Alshehri, A. M. Ilyas, A. Hasan, A. Arnaout, F. Ahmed and A. Memic, Carbon nanotubes in biomedical applications: Factors, mechanisms, and remedies of toxicity, *J. Med. Chem.* **59**, 8149–8167 (2016).

26. S. Flandrois and B. Simon, Carbon materials for lithium-ion rechargeable batteries, *Carbon* **37**, 165–180 (1999).

27. E. Frackowiak, Carbon materials for supercapacitor application, *PCCP* **9**, 1774–1785 (2007).

28. F. Beguin and E. Frackowiak, eds., *Carbons for Electrochemical Energy Storage and Conversion Systems.* CRC Press, US (2009).

29. P. S. Karthik, A. L. Himaja and S. P. Singh, Carbon-allotropes: Synthesis methods, applications and future perspectives, *Carbon Lett.* **15**, 219–237 (2014).

30. M. Inagaki and F. Kang, eds., *Materials Science and Engineering of Carbon: Fundamentals.* Butterworth-Heinemann (2014).

31. A. Ōya and H. Marsh, Phenomena of catalytic graphitization, *J. Mater. Sci.* **17**, 309–322 (1982).

32. J. O. Besenhard, ed. *Handbook of Battery Materials.* Wiley VCH (1998).

33. R. C. Boehm and A. Banerjee, Theoretical study of lithium intercalated graphite, *J. Chem. Phys.* **96**, 1150–1157 (1992).

34. M. Wakihara and O. Yamamoto, eds., *Lithium Ion Batteries.* Kodanshda/Wiley-VCH (1998).

35. J. R. Dahn, W. Xing and Y. Gao, The "falling cards model" for the structure of microporous carbons, *Carbon* **35**, 825–830 (1997).

36. E. Buiel and J. R. Dahn, Li-insertion in hard carbon anode materials for Li-ion batteries, *Electrochim. Acta* **45**, 121–130 (1999).

37. K. Sato, M. Noguchi, A. Demachi, N. Oki and M. Endo, A Mechanism of Lithium Storage in Disordered Carbons, *Science* **264**, 556–558 (1994).

38. E. Peled, C. Menachem, D. Bar-Tow and A. Melman, Improved graphite anode for lithium-ion batteries chemically: Bonded solid electrolyte interface and nanochannel formation, *J. Electrochem. Soc.* **143**, L4–L7 (1996).

39. H. Fujimoto, A. Mabuchi, K. Tokumitsu and T. Kasuh, Irreversible capacity of lithium secondary battery using meso-carbon micro beads as anode material, *J. Power Sources* **54**, 440–443 (1995).

40. J. R. Dahn, T. Zheng, Y. Liu and J. S. Xue, Mechanisms for lithium insertion in carbonaceous materials, *Science* **270**, 590–593 (1995).

41. O. Fromm, A. Heckmann, U. C. Rodehorst, J. Frerichs, D. Becker, M. Winter and T. Placke, Carbons from biomass precursors as anode materials for lithium ion batteries: New insights into carbonization and graphitization behavior and into their correlation to electrochemical performance, *Carbon* **128**, 147–163 (2018).

42. J. Deng, M. Li and Y. Wang, Biomass-derived carbon: Synthesis and applications in energy storage and conversion, *Green Chem.* **18**, 4824–4854 (2016).

43. W. De Jong, ed. *Biomass as a Sustainable Energy Source for the Future.* John Wiley & Sons (2014).

44. A. Camia and N. Robert, eds., *Biomass Production, Supply, Uses and Flows in the European Union.* Publications Office of the European Union (2018).

45. J. Wang, P. Nie, B. Ding, S. Dong, X. Hao, H. Dou and X. Zhang, Biomass derived carbon for energy storage devices, *J. Mater. Chem. A* **5**, 2411–2428 (2017).

46. Y. Yao and F. Wu, Naturally derived nanostructured materials from biomass for rechargeable lithium/sodium batteries, *Nano Energy* **17**, 91–103 (2015).

47. V. Dhyani and T. Bhaskar, A comprehensive review on the pyrolysis of lignocellulosic biomass, *Renew. Energy* **129**, 695–716 (2018).

48. A. Demirbaş, Relationships between lignin contents and fixed carbon contents of biomass samples, *Energy Convers. Manage.* **44**, 1481–1486 (2003).

49. D. Saurel, B. Orayech, B. Xiao, D. Carriazo, X. Li and T. Rojo, From charge storage mechanism to performance: A roadmap toward high specific energy sodium-ion batteries through carbon anode optimization, *Adv. Energy Mater.* **8**, 1703268 (2018).

50. S. Somiya, ed. *Handbook of Advanced Ceramics.* Academic Press (2013).

51. R. E. Franklin and J. T. Randall, Crystallite growth in graphitizing and non-graphitizing carbons, *Proc. Royal Soc. A* **209**, 196–218 (1951).

52. J. J. Kipling, J. N. Sherwood, P. V. Shooter and N. R. Thompson, Factors influencing the graphitization of polymer carbons, *Carbon* **1**, 315–320 (1964).

53. X. Dou, I. Hasa, D. Saurel, C. Vaalma, L. Wu, D. Buchholz, D. Bresser, S. Komaba and S. Passerini, Hard carbons for sodium-ion batteries: Structure, analysis, sustainability, and electrochemistry, *Mater. Today* **23**, 87–104 (2019).

54. L. L. Ban, D. Crawford and H. Marsh, Lattice-resolution electron microscopy in structural studies of non-graphitizing carbons from polyvinylidene chloride (PVDC), *J. Appl. Crystallogr.* **8**, 415–420 (1975).

55. H. Honda, K. Kobayashi and S. Sugawara, X-ray characteristics of non-graphitizing-type carbon, *Carbon* **6**, 517–523 (1968).

56. R. Franklin, The structure of graphitic carbons, *Acta Crystallogr.* **4**, 253–261 (1951).

57. N. Sonobe, T. Kyotani and A. Tomita, Carbonization of polyfurfuryl alcohol and polyvinyl acetate between the lamellae of montmorillonite, *Carbon* **28**, 483–488 (1990).

58. P. J. F. Harris, Fullerene-related structure of commercial glassy carbons, *Philos. Mag.* **84**, 3159–3167 (2004).

59. P. J. F. Harris, New perspectives on the structure of graphitic carbons, *Crit. Rev. Solid State Mater. Sci.* **30**, 235–253 (2005).

60. F. G. Emmerich, Evolution with heat treatment of crystallinity in carbons, *Carbon* **33**, 1709–1715 (1995).

61. Y.-E. Zhu, H. Gu, Y.-N. Chen, D. Yang, J. Wei and Z. Zhou, Hard carbon derived from corn straw piths as anode materials for sodium ion batteries, *Ionics* **24**, 1075–1081 (2018).

62. P. Liu, Y. Li, Y.-S. Hu, H. Li, L. Chen and X. Huang, A waste biomass derived hard carbon as a high-performance anode material for sodium-ion batteries, *J. Mater. Chem. A* **4**, 13046–13052 (2016).

63. K. Wang, Y. Jin, S. Sun, Y. Huang, J. Peng, J. Luo, Q. Zhang, Y. Qiu, C. Fang and J. Han, Low-cost and high-performance hard carbon anode materials for sodium-ion batteries, *ACS Omega* **2**, 1687–1695 (2017).

64. S. Qiu, L. Xiao, M. L. Sushko, K. S. Han, Y. Shao, M. Yan, X. Liang, L. Mai, J. Feng, Y. Cao, X. Ai, H. Yang and J. Liu, Manipulating adsorption -insertion mechanisms in nanostructured carbon materials for high-efficiency sodium ion storage, *Adv. Energy Mater.* **7**, 1700403 (2017).

65. I. Izanzar, M. Dahbi, M. Kiso, S. Doubaji, S. Komaba and I. Saadoune, Hard carbons issued from date palm as efficient anode materials for sodium-ion batteries, *Carbon* **137**, 165–173 (2018).

66. P. Wang, X. Zhu, Q. Wang, X. Xu, X. Zhou and J. Bao, Kelp-derived hard carbons as advanced anode materials for sodium-ion batteries, *J. Mater. Chem. A* **5**, 5761–5769 (2017).

67. R. Franklin, The interpretation of diffuse X-ray diagrams of carbon, *Acta Crystallogr.* **3**, 107–121 (1950).

68. A. Ōya and S. Ōtani, Catalytic graphitization of carbons by various metals, *Carbon* **17**, 131–137 (1979).

69. Y. Homma, Y. Kobayashi, T. Ogino, D. Takagi, R. Ito, Y. J. Jung and P. M. Ajayan, Role of transition metal catalysts in single-walled carbon nanotube growth in chemical vapor deposition, *J. Phys. Chem. B* **107**, 12161–12164 (2003).

70. K. Wang, Y. Cao, X. Wang, P. R. Kharel, W. Gibbons, B. Luo, Z. Gu, Q. Fan and L. Metzger, Nickel catalytic graphitized porous carbon as electrode material for high performance supercapacitors, *Energy* **101**, 9–15 (2016).

71. X. Zhang, K. Zhang, H. Li, Q. Wang, L. e. Jin and Q. Cao, Synthesis of porous graphitic carbon from biomass by one-step method And its role in the electrode for supercapacitor, *J. Appl. Electrochem.* **48**, 415–426 (2018).

72. B. Chang, Y. Guo, Y. Li, H. Yin, S. Zhang, B. Yang and X. Dong, Graphitized hierarchical porous carbon nanospheres: Simultaneous activation/graphitization and superior supercapacitance performance, *J. Mater. Chem. A* **3**, 9565–9577 (2015).

73. A. Gomez-Martin, J. Martinez-Fernandez, M. Ruttert, A. Heckmann, M. Winter, T. Placke and J. Ramirez-Rico, Iron-catalyzed graphitic

carbon materials from biomass resources as anodes for lithium-ion batteries, *Chem Sus Chem* **11**, 2776–2787 (2018).

74. A. Abdelwahab, J. Castelo-Quibén, J. Vivo-Vilches, M. Pérez-Cadenas, F. Maldonado-Hódar, F. Carrasco-Marín and A. Pérez-Cadenas, Electrodes based on carbon aerogels partially graphitized by doping with transition metals for oxygen reduction reaction, *Nanomaterials* **8**, 266 (2018).

75. L. Chen, H. Wang, H. Wei, Z. Guo, M. A. Khan, D. P. Young and J. Zhu, Carbon monolith with embedded mesopores and nanoparticles as a novel adsorbent for water treatment, *RSC Adv.* **5**, 42540–42547 (2015).

76. G. Hasegawa, K. Kanamori and K. Nakanishi, Facile preparation of macroporous graphitized carbon monoliths from iron-containing resorcinol–formaldehyde gels, *Mater. Lett.* **76**, 1–4 (2012).

77. F. J. Maldonado-Hódar, C. Moreno-Castilla, J. Rivera-Utrilla, Y. Hanzawa and Y. Yamada, Catalytic graphitization of carbon aerogels by transition metals, *Langmuir* **16**, 4367–4373 (2000).

78. E. Thompson, A. F. Danks, L. Bourgeois and Z. Schnepp, Iron-catalyzed graphitization of biomass, *Green Chem.* **17**, 551–556 (2015).

79. C. Yokokawa, K. Hosokawa and Y. Takegami, Low temperature catalytic graphitization of hard carbon, *Carbon* **4**, 459–465 (1966).

80. W. Weisweiler, N. Subramanian and B. Terwiesch, Catalytic influence of metal melts on the graphitization of monolithic glasslike carbon, *Carbon* **9**, 755–761 (1971).

81. R. Wang, G. Lu, W. Qiao and J. Yu, Catalytic graphitization of coal-based carbon materials with light rare earth elements, *Langmuir* **32**, 8583–8592 (2016).

82. S. Yi, Z. Fan, C. Wu and J. Chen, Catalytic graphitization of furan resin carbon by yttrium, *Carbon* **46**, 378–380 (2008).

83. M. Sevilla and A. B. Fuertes, Catalytic graphitization of templated mesoporous carbons, *Carbon* **44**, 468–474 (2006).

84. L. Zhao, X. Zhao, L. T. Burke, J. C. Bennett, R. A. Dunlap and M. N. Obrovac, Voronoi-tessellated graphite produced by low-temperature catalytic graphitization from renewable resources, *Chem Sus Chem* **10**, 3409–3418 (2017).

85. Q. Yan, J. Li, X. Zhang, E. B. Hassan, C. Wang, J. Zhang and Z. Cai, Catalytic graphitization of kraft lignin to graphene-based structures with four different transitional metals, *J. Nanopart. Res.* **20**, 223 (2018).
86. C. J. Thambiliyagodage, S. Ulrich, P. T. Araujo and M. G. Bakker, Catalytic graphitization in nanocast carbon monoliths by iron, cobalt and nickel nanoparticles, *Carbon* **134**, 452–463 (2018).
87. M. Sevilla and A. B. Fuertes, Fabrication of porous carbon monoliths with a graphitic framework, *Carbon* **56**, 155–166 (2013).
88. M. Sevilla, C. Sanchís, T. Valdés-Solís, E. Morallón and A. B. Fuertes, Synthesis of graphitic carbon nanostructures from sawdust and their application as electrocatalyst supports, *J. Phys. Chem. C* **111**, 9749–9756 (2007).
89. M. Sevilla, C. Salinas Martínez-de Lecea, T. Valdés-Solís, E. Morallón and A. B. Fuertes, Solid-phase synthesis of graphitic carbon nano-structures from iron and cobalt gluconates and their utilization as electrocatalyst supports, *PCCP.* **10**, 1433–1442 (2008).
90. F. Wu, R. Huang, D. Mu, B. Wu and Y. Chen, Controlled synthesis of graphitic carbon-encapsulated $\alpha\text{-}Fe_2O_3$ nanocomposite via low-tem-perature catalytic graphitization of biomass and its lithium storage property, *Electrochim. Acta* **187**, 508–516 (2016).
91. H.-h. Zhou, Q.-l. Peng, Z.-h. Huang, Q. Yu, J.-h. Chen and Y.-f. Kuang, Catalytic graphitization of PAN-based carbon fibers with electrodepos-ited Ni-Fe alloy, *T. Nonferr. Metal Soc.* **21**, 581–587 (2011).
92. J. Wang, X. Deng, H. Zhang, H. Duan, F. Cheng and S. Zhang, Low-Temperature catalytic graphitization of phenolic resin using a co-ni bimetallic catalyst, *Interceram* **65**, 24–27 (2016).
93. C. S. Bitencourt, A. P. Luz, C. Pagliosa and V. C. Pandolfelli, Role of catalytic agents and processing parameters in the graphitization pro-cess of a carbon-based refractory binder, *Ceram. Int.* **41**, 13320–13330 (2015).
94. Q. Yan, J. Li, X. Zhang, J. Zhang and Z. Cai, In situ formation of graphene-encapsulated iron nanoparticles in carbon frames through catalytic graphitization of kraft lignin, *Nanomater. Nanotechno.* **8**, 188–192 (2018).
95. S. Kubo, Y. Uraki and Y. Sano, Catalytic graphitization of hardwood acetic acid lignin with nickel acetate, *J. Wood. Sci.* **49**, 188–192 (2003).

96. L. Chen, T. Ji, L. Mu, Y. Shi, L. Brisbin, Z. Guo, M. A. Khan, D. P. Young and J. Zhu, Facile synthesis of mesoporous carbon nanocomposites from natural biomass for efficient dye adsorption and selective heavy metal removal, *RSC Adv.* **6**, 2259–2269 (2016).

97. M. Xie, J. Yang, J. Liang, X. Guo and W. Ding, In situ hydrothermal deposition as an efficient catalyst supporting method towards low-temperature graphitization of amorphous carbon, *Carbon* **77**, 215–225 (2014).

98. R. Fu, T. F. Baumann, S. Cronin, G. Dresselhaus, M. S. Dresselhaus and J. H. Satcher, Formation of Graphitic Structures in Cobalt- and Nickel-Doped Carbon Aerogels, *Langmuir* **21**, 2647–2651 (2005).

99. A. Gutierrez-Pardo, J. Ramirez-Rico, A. R. de Arellano-Lopez and J. Martinez-Fernandez, Characterization of porous graphitic monoliths from pyrolyzed wood, *J. Mater. Sci.* **49**, 7688–7696 (2014).

100. A. Ōya and S. Ōtani, Influences of particle size of metal on catalytic graphitization of non-graphitizing carbons, *Carbon* **19**, 391–400 (1981).

101. A. Gutiérrez-Pardo, J. Ramírez-Rico, R. Cabezas-Rodríguez and J. Martínez-Fernández, Effect of catalytic graphitization on the electrochemical behavior of wood derived carbons for use in supercapacitors, *J. Power Sources* **278**, 18–26 (2015).

102. C. Ma, E. Cao, J. Li, Q. Fan, L. Wu, Y. Song and J. Shi, Synthesis of mesoporous ribbon-shaped graphitic carbon nanofibers with superior performance as efficient supercapacitor electrodes, *Electrochim. Acta* **292**, 364–373 (2018).

103. Y. Li, W. Ou-Yang, X. Xu, M. Wang, S. Hou, T. Lu, Y. Yao and L. Pan, Micro-/mesoporous carbon nanofibers embedded with ordered carbon for flexible supercapacitors, *Electrochim. Acta* **271**, 591–598 (2018).

104. Y. Liu, Q. Liu, J. Gu, D. Kang, F. Zhou, W. Zhang, Y. Wu and D. Zhang, Highly porous graphitic materials prepared by catalytic graphitization, *Carbon* **64**, 132–140 (2013).

105. J. Hoekstra, A. M. Beale, F. Soulimani, M. Versluijs-Helder, J. W. Geus and L. W. Jenneskens, Base metal catalyzed graphitization of cellulose: A combined raman spectroscopy, temperature-dependent X-ray diffraction and high-resolution transmission electron microscopy study, *J. Phys. Chem. C* **119**, 10653–10661 (2015).

106. J. Ramirez-Rico, A. Gutierrez-Pardo, J. Martinez-Fernandez, V. V. Popov and T. S. Orlova, Thermal conductivity of Fe graphitized wood derived carbon, *Mater. Des.* **99**, 528–534 (2016).

107. M. Kakunuri, S. Kali and C. S. Sharma, Catalytic graphitization of resorcinol-formaldehyde xerogel and its effect on lithium ion intercalation, *J. Anal. Appl. Pyrolysis* **117**, 317–324 (2016).

108. J. M. Skowroński and K. Knofczyński, Catalytically graphitized glass-like carbon examined as anode for lithium-ion cell performing at high charge/discharge rates, *J. Power Sources* **194**, 81–87 (2009).

109. O. P. Krivoruchko and V. I. Zaikovskii, A new phenomenon involving the formation of liquid mobile metal–carbon particles in the low-temperature catalytic graphitisation of amorphous carbon by metallic Fe, Co and Ni, *Mendeleev Commun.* **8**, 97–99 (1998).

110. J. Hoekstra, A. M. Beale, F. Soulimani, M. Versluijs-Helder, D. van de Kleut, J. M. Koelewijn, J. W. Geus and L. W. Jenneskens, The effect of iron catalyzed graphitization on the textural properties of carbonized cellulose: Magnetically separable graphitic carbon bodies for catalysis and remediation, *Carbon* **107**, 248–260 (2016).

111. W. Zhou, Y. Yu, X. Xiong and S. Zhou, Fabrication of α-Fe/Fe3C/Woodceramic Nanocomposite with Its Improved Microwave Absorption and Mechanical Properties, *Materials* **11**, 878 (2018).

112. K. Lotz, A. Wütscher, H. Düdder, C. M. Berger, C. Russo, K. Mukherjee, G. Schwaab, M. Havenith and M. Muhler, Tuning the properties of iron-doped porous graphitic carbon synthesized by hydrothermal carbonization of cellulose and subsequent pyrolysis, *ACS Omega* **4**, 4448–4460 (2019).

113. H. Rastegar, M. Bavand-vandchali, A. Nemati and F. Golestani-Fard, Catalytic graphitization behavior of phenolic resins by addition of in situ formed nano-Fe particles, *Physica E* **101**, 50–61 (2018).

114. J. Huo, H. Song, X. Chen, S. Zhao and C. Xu, Structural transformation of carbon-encapsulated iron nanoparticles during heat treatment at 1,000 °C, *Mater. Chem. Phys.* **101**, 221–227 (2007).

115. S. H. Park, S. M. Jo, D. Y. Kim, W. S. Lee and B. C. Kim, Effects of iron catalyst on the formation of crystalline domain during carbonization of electrospun acrylic nanofiber, *Synth. Met.* **150**, 265–270 (2005).

116. X. Zhang, H. Li, K. Zhang, Q. Wang, B. Qin, Q. Cao and L. E. Jin, Strategy for preparing porous graphitic carbon for supercapacitor: Balance on porous structure and graphitization degree, *J. Electrochem. Soc.* **165**, A2084–A2092 (2018).

117. S. T. Neeli and H. Ramsurn, Synthesis and formation mechanism of iron nanoparticles in graphitized carbon matrices using biochar from biomass model compounds as a support, *Carbon* **134**, 480–490 (2018).

118. F. Ding, A. Rosén, E. E. B. Campbell, L. K. L. Falk and K. Bolton, Graphitic encapsulation of catalyst particles in carbon nanotube production, *J. Phys. Chem. B* **110**, 7666–7670 (2006).

119. W. Lian, H. Song, X. Chen, L. Li, J. Huo, M. Zhao and G. Wang, The transformation of acetylene black into onion-like hollow carbon nanoparticles at 1000 °C using an iron catalyst, *Carbon* **46**, 525–530 (2008).

120. K. L. Saenger, J. C. Tsang, A. A. Bol, J. O. Chu, A. Grill and C. Lavoie, In situ x-ray diffraction study of graphitic carbon formed during heating and cooling of amorphous-C/Ni bilayers, *Appl. Phys. Lett.* **96**, 153105 (2010).

121. M. Yudasaka, K. Tasaka, R. Kikuchi, Y. Ohki, S. Yoshimura and E. Ota, Influence of chemical bond of carbon on Ni catalyzed graphitization, *J. Appl. Phys.* **81**, 7623–7629 (1997).

122. S. Glatzel, Z. Schnepp and C. Giordano, From paper to structured carbon electrodes by inkjet printing, *Angew. Chem. Int. Ed.* **52**, 2355–2358 (2013).

123. M. T. Johnson and K. T. Faber, Catalytic graphitization of three-dimensional wood-derived porous scaffolds, *J. Mater. Res.* **26**, 18–25 (2011).

124. A. Abdelwahab, J. Castelo-Quibén, M. Pérez-Cadenas, F. J. Maldonado-Hódar, F. Carrasco-Marín and A. F. Pérez-Cadenas, Insight of the effect of graphitic cluster in the performance of carbon aerogels doped with nickel as electrodes for supercapacitors, *Carbon* **139**, 888–895 (2018).

125. S. Zhang, Y. Su, S. Zhu, H. Zhang and Q. Zhang, Effects of pretreatment and FeCl3 preload of rice husk on synthesis of magnetic carbon composites by pyrolysis for supercapacitor application, *J. Anal. Appl. Pyrolysis* **135**, 22–31 (2018).

126. Y. Yu, J. Du, L. Liu, G. Wang, H. Zhang and A. Chen, Hierarchical porous nitrogen-doped partial graphitized carbon monoliths for supercapacitor, *J. Nanopart. Res.* **19**, 119 (2017).
127. M. N. Obrovac, X. Zhao, L. T. Burke and R. A. Dunlap, Reversible lithium insertion in catalytically graphitized sugar carbon, *Electrochem. Commun.* **60**, 221–224 (2015).
128. X. Zhao, Y. Yao, A. E. George, R. A. Dunlap and M. N. Obrovac, Electrochemistry of catalytically graphitized ball milled carbon in li batteries, *J. Electrochem. Soc.* **163**, A858–A66 (2016).
129. J. M. Skowroński, K. Knofczyński and M. Inagaki, Changes in electrochemical insertion of lithium into glass-like carbon affected by catalytic graphitization at 1000 C, *Solid State Ionics* **178**, 137–144 (2007).
130. M. Gaikwad, M. Kakunuri and C. S. Sharma, Catalytically graphitized nanostructured carbon xerogels as high performance anode material for lithium ion battery, *ECS Trans.* **85**, 1–10 (2018).
131. M. M. Gaikwad, M. Kakunuri and C. S. Sharma, Enhanced catalytic graphitization of resorcinol formaldehyde derived carbon xerogel to improve its anodic performance for lithium ion battery, *Mater. Today Commun.* **20**, 100569 (2019).
132. N. Wang, Q. Liu, B. Sun, J. Gu, B. Yu, W. Zhang and D. Zhang, N-doped catalytic graphitized hard carbon for high-performance lithium/sodium-ion batteries, *Sci. Rep.* **8**, 9934 (2018).
133. Z. Yang, H. Guo, X. Li, Z. Wang, J. Wang, Y. Wang, Z. Yan and D. Zhang, Graphitic carbon balanced between high plateau capacity and high rate capability for lithium ion capacitors. *J. Mater. Chem. A* **5**, 15302–15309 (2017).
134. Z. Yang, H. Guo, F. Li, X. Li, Z. Wang, L. Cui and J. Wang, Cooperation of nitrogen-doping and catalysis to improve the Li-ion storage performance of lignin-based hard carbon, *J. Energy Chem.* **27**, 1390–1396 (2018).
135. H. Ouyang, Q. Gong, C. Li, J. Huang and Z. Xu, Porphyra derived hierarchical porous carbon with high graphitization for ultra-stable lithium-ion batteries, *Mater. Lett.* **235**, 111–115 (2019).
136. M. Endo, C. Kim, K. Nishimura, T. Fujino and K. Miyashita, Recent development of carbon materials for Li ion batteries, *Carbon* **38**, 183–197 (2000).

137. M. Li, H. Du, L. Kuai, K. Huang, Y. Xia and B. Geng, Scalable dry production process of a superior 3D net-like carbon-based iron oxide anode material for lithium-ion batteries, *Angew. Chem. Int. Ed.* **56**, 12649–12653 (2017).

138. Y. Yan, H. Tang, F. Wu, Z. Xie, S. Xu, D. Qu, R. Wang, F. Wu, M. Pan and D. Qu, Facile synthesis of Fe_2O_3@graphite nanoparticle composite as the anode for Lithium ion batteries with high cyclic stability, *Electrochim. Acta* **253**, 104–113 (2017).

139. D. Zhang, G. Li, M. Yu, J. Fan, B. Li and L. Li, Facile synthesis of $Fe_4N/Fe_2O_3/Fe$/porous N-doped carbon nanosheet as high-performance anode for lithium-ion batteries, *J. Power Sources* **384**, 34–41 (2018).

140. X. Liu, Y. Zhu, N. Liu, M. Chen, C. Wang and X. Wang, synthesis of hard/soft carbon hybrids with heteroatom doping for enhanced sodium storage, *Chemistry Select* **4**, 3551–3558 (2019).

141. K. Han, Z. Liu, P. Li, Q. Yu, W. Wang, C.-Y. Lao, D. He, W. Zhao, G. Suo, H. Guo, L. Song, M. Qin and X. Qu, High-throughput fabrication of 3D N-doped graphenic framework coupled with Fe_3C@porous graphite carbon for ultrastable potassium ion storage, *Energy Storage Mater.* **22**, 185–193 (2019).

Chapter 9

Iron-Catalysis in Environmental Remediation

Noelia Losada-García and Jose M. Palomo*

*Department of Biocatalysis, Institute of Catalysis (CSIC),
Marie Curie 2, Cantoblanco, 28049 Madrid, Spain*
*josempalomo@icp.csic.es

This chapter shows the application of different iron nanostructured
materials as successful catalyst for environmental elimination of different important toxic organic molecules. Within these toxic organic
molecules are organic pollutants, chlorinated and heavy metals,
among others. The zero-valent iron (ZVI) nanoparticles are one of
the main candidates for the remediation of toxic compounds. This
chapter is focused on specifically the remediation of aromatic pollutants such as Bisphenol-A (BPA), chlorinated organic compounds,
and heavy metals.

9.1 Introduction

On earth there are different types of metals. They can be found in
their elemental form or forming mineral structures. Among all metals, iron (Fe) is one of the most abundant elements.[1] Throughout
history this element has played an important role in the development of civilization, but in recent decades the role in remediation

has been discovered due to the interest caused by global warming and therefore the care of the environment.

Recent and historical industrial and agricultural activities have led to numerous sites with elevated contaminant concentrations in soils, sediments, surface, and groundwater. In 2007, the European Environment Agency (EEA) estimated 250,000 contaminated sites where remediation was required.[2] Widespread contaminants are trace elements, metalloids, and aromatic, polyaromatic, and chlorinated organic compounds. During the last decades, Fe oxides were drawn into focus of the development of new remediation technologies due to their sorptive and reactive character.[3] *In situ* Fe-based treatment methods are potentially cost-effective remediation options (Table 9.1).[4]

One of the most interesting and successful applications of iron nanoparticles (NPs) has been the removal of contaminants like organic pollutants, metals, and metalloids.[5,6] From the different iron NPs, the zero-valent iron (ZVI) NPs have been the most successfully used in this area, followed by the magnetite NPs.[7-9] One important problem in the application of these NPs on remediation is their tendency to aggregate, which limits their dispersibility and mobility in effluents.[10] Therefore, most of the actual strategies are based on the synthesis of iron NPs supported on different materials.[11-14]

From this, the objective of this chapter is to provide the reader with new advances in the use of Fe-based catalysts for remediation.

Table 9.1. Different Fe-based remediation technologies: stage of development and mode of application.[4]

Technology	Stage of development	*In situ* or *ex situ*	Remedial mechanism
Assisted natural remediation	Laboratory and pilot fields trials	*In situ*	Contaminant immobilization
Chemical reduction via addition of Fe (II)-containing solutions	Commercial systems available	*In situ* or *ex situ*	Reductive precipitation
Permeable reactive barriers	Commercial systems available	*In situ*	Adsorption or degradation in barrier

We will focus on the elimination of important pollutants as Bisphenol-A (BPA), different organic chlorinated compounds, and heavy metals.

9.2 Degradation of BPA

BPA (2,2-bis (4-hydroxyphenyl) propane) is widely used in the synthesis of polymers that include phenolic resins, epoxy resins, polycarbonates, polyesters, and polyacrylates.[15] BPA-based polymers are important raw materials for the manufacture of various products for daily consumption, such as water bottles, food container coatings, sports equipment, including safety equipment, eyeglasses, computer and cell phone casings, and water and beverage bottles. BPA is also consumed as a resin in dental fillings, coatings on cans, powder paints, and additives in thermal paper.[16]

The BPA can leach from these products and end up in wastewater. BPA concentrations reported in surface water range between 0.016 and 0.5 mg/L; however, levels >10 mg/L BPA have also been detected in old landfill leachate.[17]

In the last 20 years, the presence of certain chemical substances in drinking water sources has attracted scientific and public interest due to its possible harmful impacts on the environment and health. In the case of endocrine disruptors, they are related to the increase in cases of breast cancer, infertility, low sperm count, genital deformities, obesity, early puberty, and diabetes, as well as alarming mutations in wildlife.[16,18] Therefore, the presence of BPA in the environment has aroused great concern among the public and regulatory agencies.

At present, there are only a few methods developed for the removal of BPA from aqueous solutions. For example, adsorption, biochemical oxidation, and wet chemical oxidation and others,[19,20] but all of these methods have suffered from some disadvantages such as long reaction times, high cost, and low efficiency. In addition, many of the processes use high temperature/pressure or precious metal catalysts, which are too expensive for practical application and widespread adoption. These low concentrations render their effective removal by conventional biological, physical and chemical methods difficult and costly. Biological processes have proven to be rather ineffective in the degradation of BPA.[21]

Among these methods, advanced oxidation processes have been explored and found to be a promising mechanism. In particular, Fenton reaction technology follows a radical reaction and can explain for the degradation of most refractory organic contaminants. It also benefits from the fact that no UV or visible light is needed to initiate the reaction; this is especially important for groundwater systems.[22]

Owing to the superior chemical properties and large surface area, nanoscale ZVI (nZVI) Fe (0) has been successfully used for the degradation of a wide range of contaminants in aqueous environment in the last 5 years, including dyes,[23] heavy metals,[24] chlorinated and brominated entities,[25] and nitrobenzene.[26]

Magnetic nanoparticles (MNPs) of Fe_3O_4 have been used as a heterogeneous catalyst by Huang *et al.*[27] They have attracted great interest in heterogeneous Fenton system due to their advantages such as reusability and intrinsic peroxidase-like activity in which Fe_3O_4 MNPs catalyze H_2O_2 to produce hydroxyl radicals to decompose organic pollutants.[28–30] According to previous researches, the peroxidase-like activity originates mainly from iron ions on the surface of Fe_3O_4 MNPs.[31] It was reported that organic pollutants could be partially degraded in the H_2O_2 + Fe_3O_4 MNPs system.[28,30,32] However, the aggregation of Fe_3O_4 MNPs during the reaction would reduce their surface activity and dispersibility in aqueous solution, thus reducing the catalytic activity.[28,29]

It is well known that ultrasonic in aqueous solution will decrease mass transfer limitations and provide additional cavitation effect, which is beneficial for degradation in heterogeneous catalytic systems.[33] Therefore, to overcome the defects in Fe_3O_4 + H_2O_2 heterogeneous Fenton system can be promoted by combining with ultrasonic through a synergistic effect between the sonochemical and catalytic reactions. In this study,[27] unmodified palygorskite (magnesium aluminum phyllosilicate: $(Mg,Al)_2Si_4O_{10}(OH) \cdot 4(H_2O)$) and acid-leached palygorskite were used, observing that they have an adsorption of BPA in insignificant amounts. However, with the addition of hydrogen peroxide (HP) solution, it was observed that some percentage of BPA was removed. With the aid of air bubbles, there is a significant improvement of BPA degradation, resulting in up to five-fold BPA removal percentage compared to the result obtained by using H_2O_2 only.[34]

For ZVI-Com and nZVI/palygorskite composite materials, air bubbles significantly enhance the removal from 3.4% to up to 95%. The enhanced degradation with the aid of air bubbles is due to the presence of O_2 in a solution, which enhances the oxidation of organo-radicals as the radicals can react with dissolved O_2 in faster and irreversible reactions. Fe(0) also reacts with O_2 in acidic condition to form H_2O_2.[35]

More recently, heterogeneous Fenton systems, ferrate (Fe^{4+}) and ZVI (Fe(0)) applications have been developed to overcome the limitations of homogeneous treatment systems.[10,36,37] ZVI is a reactive, non toxic, abundant, relatively cheap, and easy to produce and handle metal (reduction potential = −0.44 V). In acidic medium, ZVI can degrade organic compounds in the presence of dissolved oxygen by transferring two electrons to O_2 to produce H_2O_2 and ultimately HO· by a Fenton-like treatment system.[38] In particular, Fenton, Fenton-like, and photo-Fenton treatments are known for their superiority in terms of efficient pollutant removal and detoxification.[39,40]

Girit *et al.*[41] used commercial nZVI particles to degrade the surrogate endocrine disrupting chemical BPA from aqueous solution. For this purpose, air-stable nZVI was used in combination with two common oxidants, namely HP and persulfate (PS), forming an advanced oxidative treatment system. In the first part of the study, several baseline and control experiments were conducted under varying pH, ZVI, HP, and PS concentrations at temperatures to optimize the removal of BPA and its total organic carbon (TOC) content.

In the last 10 years, sulfate radical (SO_4^-)-based advanced oxidation processes (SR-AOPs) were found to be effective for decomposition of refractory organic contaminants in contaminated soils and ground waters.[42–44] However, some intrinsic drawbacks of Fe^{2+}/PS system may restrict its practical application. For example, Fe^{2+} / PS system can only maintain high efficiency at acidic conditions.[45] In addition, SO_4^- can be consumed by reacting with excessive Fe^{2+} in water, which results in reduced degradation efficiency of target contaminants.[46]

ZVI can serve as an alternative source of Fe^{2+}. In ZVI/PS system, Fe^{2+} is produced by ZVI corrosion under oxic/anoxic condition and/or oxidation by PS.[38,47,48] Unlike homogeneous Fe^{2+} activation, Fe^{2+} is gradually released into the aqueous solution in ZVI/PS system. In this way, accumulation of excessive Fe^{2+} and the subsequent quenching of SO_4^- can be minimized.[38,47,48]

In addition, before the wide application of Fe (0) for removing various groundwater contaminants, some technological challenges, such as its poor stability, low durability, and weak mechanical strength have to be overcome. In particular, the serious agglomeration during preparation processes largely decreases the reactivity of the Fe (0) particles, and also results in poor mobility and slow transport of Fe (0) to the contaminated area for *in situ* remediation.[49] Recently, loading Fe (0) onto supporting material is demonstrated as a potential method to enhance the dispersion of Fe(0) particles.

Liu et al.[50] organo-montmorillonite-supported ZVI NPs (Fe (0)/ OMt) has been applied to effectively remove some heavy metals from aqueous systems.

Peng et al.[51] in control experiments in the absence of simple Fe (0)/OMt, tetrabromobisphenol A (TBBPA) (a typical brominated flame retardants [BFRs] with the largest production in the world), and BPA expressed almost no degradation for 60 min (Fig. 9.1). The condition (included Ar, air, and O_3) was crucial for the degradation activity of as-prepared Fe (0)/OMt catalyst. In this study, the degradation activity of simple Fe (0)/OMt was evaluated under deoxygenated (sparging with Ar) conditions and in the presence of oxidants (sparging with air and O_3). When generating Ar, no degradation of BPA occurred on Fe (0)/OMt (Fig. 9.1). In contrast, degradation of TBBPA (over 80%) proceeded gradually by spraying Ar (Fig. 9.1), suggesting that only the debromination of TBBPA occurred on Fe

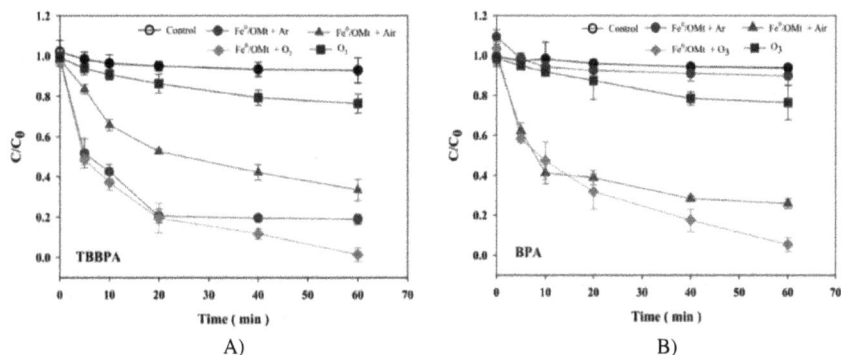

Fig. 9.1. The degradation efficiencies of TBBPA (A) and BPA (B) over sample Fe (0)/OMt under different treatment conditions.

(0)/OMt under Ar atmosphere. It was worth mentioning that adsorption of BPA and TBBPA was minor over Fe (0)/OMt in the present study, possibly because of surface deposition of ZVI on organic montmorillonite occupying the adsorption sites and affecting the hydrophobicity of the supporter.[51]

When spraying air, degradation efficiencies of TBBPA and BPA in the presence of Fe (0)/OMt were 66.6% and 74.2%, respectively (Fig. 9.1). Note that, degradation of TBBPA was slower under air bubbling condition than that under Ar bubbling condition. It is possibly that the debromination process of TBBPA was competed with the oxidation process of TBBPA under air (O_2 as oxidant) condition. In order to further enhance the degradation efficiency of TBBPA, O_3 was employed as a better oxidant (Fig. 9.1). When spraying with O_3, the degradation efficiencies of TBBPA and BPA within 60 min were both nearly 100%.[51]

Recently, Jin *et al.*[52] investigated the effect of solution pH on BPA adsorption to N-doped Fe(0)/Fe_3C@C with the initial BPA concentration of 50 mg L^{-1} and catalyst dosage of 0.2 g L^{-1} in the absence of PS (Fig. 9.2(A)). A little effect of solution pH was observed, indicating that N-doped Fe (0)/Fe_3C@C showed good adsorption performance over the entire pH range. Therefore, the initial solution pH (~7) was adopted in the following adsorption experiments.

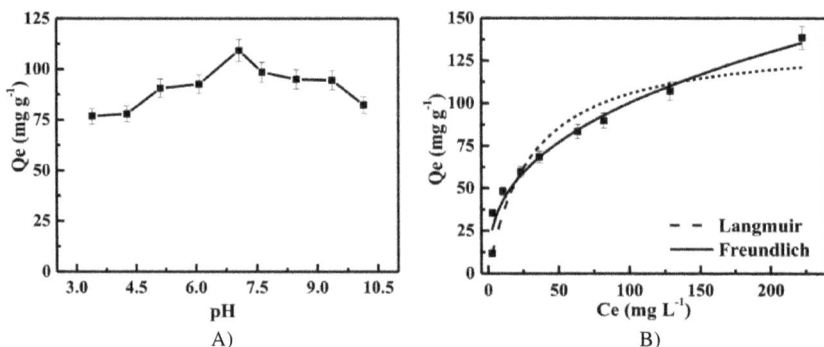

Fig. 9.2. The adsorption of BPA onto N-doped Fe (0)/Fe3C@C as a function of solution pH (A) and the adsorption isotherm of BPA (B) under the conditions of pH 7, 298 K, and 0.2 g L^{-1} of N-doped Fe (0)/Fe3C@C.

Figure 9.2 (B) shows the adsorption isotherm of BPA (5–250 mg L^{-1}, pH = 7) on N-doped Fe (0)/Fe$_3$C@C (0.2 g L^{-1}) at 298 K. With the increase of initial BPA concentration, more BPA molecules were transferred onto the surface of N-doped Fe (0)/Fe$_3$C@C micro-cubes, resulting in a higher adsorption amount.

The Langmuir and Freundlich models were used to simulate the obtained data.[53,54] The Langmuir model is an adsorption model to describe the monolayer adsorption of pollutant on homogeneous surface, while the Freundlich model is used to describe the multilayer adsorption of pollutant on heterogeneous surfaces. The correlation coefficient (R^2) value of the Freundlich model and Langmuir model were, respectively, 0.968 and 0.875, demonstrating that the adsorption of BPA on N-doped Fe (0)/Fe$_3$C@C was better fitted by the Freundlich model, indicating the heterogeneous distribution of active sites on solid surfaces. The adsorption capacity of BPA onto the N-doped Fe (0)/Fe$_3$C@C calculated by the Langmuir model was 138 mg g^{-1}, which was superior to the adsorption of BPA on magnetic non template imprinted polymers (112 mg g^{-1}, 298 K), magnetic graphene (49 mg g^{-1}, 293 K), and chi-tosan/fly-ash-cenospheres/γ-Fe$_2$O$_3$ magnetic composites (78 mg g^{-1}, 298 K),[55] indicating the superior adsorption capacity of N-doped

Fe (0)/Fe$_3$C@C. The outstanding adsorption capacity of N-doped Fe (0)/Fe$_3$C@C might be caused by the large BET surface area (Brunauer–Emmett–Teller theory), and thereby exposed more binding sites for BPA.

9.3 Degradation of Chlorinated Organic Compounds

Chlorinated organic compounds constitute (Fig. 9.3) a large group of pollutants of international concern due to their high toxicity, persistence, and various sources of distribution in the environment. In the last 10 years, nZVI metal (especially iron) particles have attracted a growing attention in groundwater remediation of chlorinated solvents.[56–58] nZVI iron particles have been used to remediate a wide range of groundwater contaminants including chlorinated compounds,[57] pesticides.[58] nZVI particles are ideal for the degradation of environmental contaminants because of their environment friendly nature, high reactivity, and low cost.[59] The mode of degradation by which the nZVI breaks down such contaminants is reductive dehalogenation and adsorption.[5]

Polychlorinated ethanes, like 1,1,1-trichloroethane (1,1,1-TCa) and 1,1,2,2-tetrachloroethane (TeCa), have been used for decades

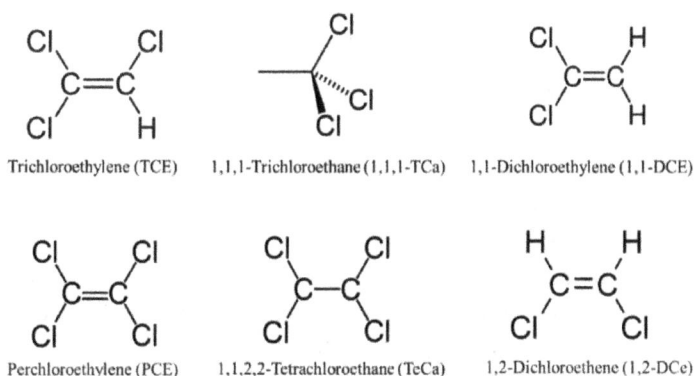

Trichloroethylene (TCE)   1,1,1-Trichloroethane (1,1,1-TCa)   1,1-Dichloroethylene (1,1-DCE)

Perchloroethylene (PCE)   1,1,2,2-Tetrachloroethane (TeCa)   1,2-Dichloroethene (1,2-DCe)

Fig. 9.3. Chemicals structure of different chlorinated organic pollutants.

as chemical intermediates, solvents, degreasing agents, and paint removers.[60] Both compounds have adverse health effects on the liver, kidneys, and the nervous and immune systems. Due to improper disposal practices and accidental releases, these contaminants are widely distributed in groundwater and soils.[61] Therefore, it is extremely important that the products generated by the degradation of these chlorinated compounds are not toxic molecules. In this regard, e.g., the development of a catalytic process for the degradation of 1,1,1-TCa, where no trace amounts of vinyl chloride or 1,1-dichloroethylene (1,1-DCE) are formed, is extremely important. In this way, the application of iron nanocatalysts has been described as excellent for the degradation of this kind of compounds.[62]

The metallic iron technology used in groundwater treatment, a striking manifestation of classic corrosion chemistry, has been extensively studied.[63] As is the case in corrosion chemistry, water and molecular oxygen typically serve as the electron acceptors in aqueous environments. Elemental iron acts as the electron donor, while relatively oxidized compounds serve as the electron acceptors.

Burgeoning research over the last 30 years effectively demonstrates that reduction of many common environmental contaminants, both organic and inorganic, can be coupled to iron oxidation.[64] Generally speaking, these contaminants are transformed into less toxic or even benign end products. For example, 1,1,1-trichloroethane (1,1,1-TCa).

Iron is oxidized more rapidly when it is attached to a less active (noble) metal (e.g., Pd, Ag, Cu, Co, Ni). Therefore, the transformation of environmental contaminants can be enhanced by coupling iron to a noble metal.[65,66] The iron-noble metal couple essentially creates numerous galvanic cells wherein iron serves as the anode and becomes preferably oxidized. Meanwhile, the noble metal (cathode) is protected and remains unchanged. Studies also suggest that noble metals such as palladium can promote dechlorination through catalytic functions such as hydrogenation.[67]

Metal iron NPs have been of interest because of their outstanding properties and potential applications in the areas of catalysis, magnetism, electronics, biomedical and environmental engineering.[68,69]

They are also used as a catalyst for chemical vapor deposition (CVD) synthesis of multi walled carbon nanotubes,[65] and in remediation of contaminants from wastewater.[66,70] As well as a new agent *in situ* subsurface remediation agent to treat various kinds of vexing environmental contaminants, including chlorinated organics such as trichloroethylene (TCE).[5,59,71]

From the materials science point of view, significant research progress has been made in understanding the important properties of bare and polymeric-modified nZVIs that affect nZVI reactivity with chlorinated organics.[56,72]

Li *et al.*[73] used the solution method and the microemulsion method to synthesize iron NPs. The microemulsion systems were characterized by measuring the conductivity and surface tension. The performance of iron NPs produced by the microemulsion method was compared with the solution method and a commercially available nanoiron product by degrading TCE.

The degradation of TCE over time by iron NPs was compared. When the metal iron to solution loading was 1.5 g L^{-1}, a complete reduction of 20 mg L^{-1} TCE was achieved in 54 h. In the case of 100 mg L^{-1} TCE, about 40% reduction was achieved after 75 h. Since complete TCE reduction was not achieved, metal loading was increased and degradation of TCE was studied. When the nano iron loading was increased to 5 g L^{-1}, 20 mg L^{-1} of TCE solution was completely degraded within an hour, and 100 mg L^{-1} of TCE was totally degraded in 105 h.[73]

The initial rate of degradation of TCE can be represented by a first-order kinetic relationship as following (1):

$$dC/dt = -k \cdot C \tag{1}$$

where C is the concentration of TCE in the aqueous phase (mg L^{-1}), k is the dichlorination rate coefficient. The above equation represents a pseudo-first-order kinetics; hence, the change in TCE concentration can be represented as (2):

$$\ln (C/C_0) = -k \cdot t \tag{2}$$

where C_0 is the initial TCE concentration. The dechlorination rate coefficient depended on the initial TCE concentration and iron loading.[73] The degradation efficiency of iron NPs by microemulsion method reduced with time partly because powder was being oxidized as was observed by the change in color.

To further investigate the performance of solution products and commercial grade iron NPs, parallel experiments were performed to quantify the degradation of TCE. The results are shown in Fig. 9.4(A) and the degradation is compared with the first-order kinetic relationship in Fig. 9.4(B). Results indicated that the initial dechlorination rate coefficient (k) for microemulsion product was about

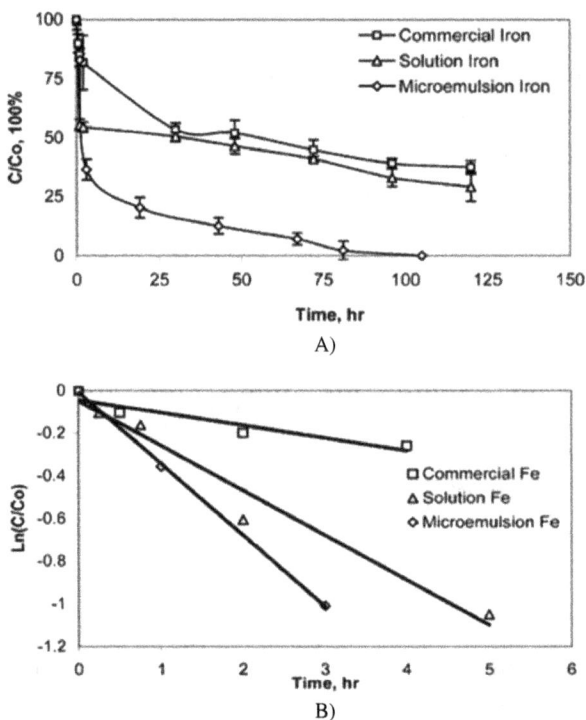

Fig. 9.4. (A) TCE reduction by three types of iron powder (iron to solution loading is 5 g L^{-1}). (B) TCE degradation by three types of iron products are compared with the first-order kinetic relationship (TCE con centration: 100 mg L^{-1}, iron to solution loading: 5 mg L^{-1}).

2.6 times higher than the solution product and 12 times that of the commercial product. The degradation rate of commercial iron was the lowest and after 100 h it degraded 60% of 100 mg L^{-1} TCE. The solution iron degraded 70% TCE in 100 h. The microemulsion product totally degraded TCE in 105 h. The microemulsion product with the smallest particles (average particle size of 10 nm) had the best results. The results showed that the smaller the iron NP, the greater its ability to degrade TCE.

The above results showed that the iron NPs synthesized by the microemulsion method were very effective in degrading TCE. Based on the degradation results, it can be concluded that the crystalline properties of the nanoiron particle were not an important factor in degrading TCE.

Sun *et al*[74] verified the reactivity of iron NPs by batch experiments with TCE. Polyvinyl alcohol-vinyl acetate-co-itaconic acid (PV3A) was evaluated for its effectiveness as a nZVI dispersant. They also evaluated the reactivity of iron NPs stabilized with PV3A for the degradation of TCE, a common contaminant in groundwater.

Figure 9.5 shows the comparison of 10 g/L bare and PV3A stabilized iron NPs for the transformation of TCE. The two batch

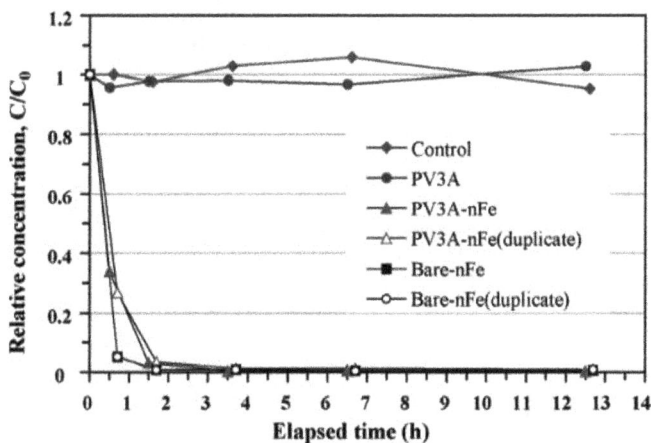

Fig. 9.5. Transformation of trichloroethene (TCE) by 10 g/L bare- and PV3A stabilized iron NPs. Initial TCE concentration (C_0) was 7 mg/L.

reactors contained the same initial concentrations of TCE ($C_0 = 7$ mg/L) and iron NPs (10 g/L). The PV3A stabilized iron and the bare iron particles exhibited similar transformation rates for TCE with over 99% removal within approximately 3 h. As no TCE decrease was observed in a control batch reactor containing ~1 mg/L PV3A (Fig. 9.5), it can be concluded that sorption or partitioning of TCE to PV3A is negligible. From the perspective of a surface-mediated reaction, reactivity is proportional to the available surface area. Assuming that TCE is not a speed-limiting factor, the near reaction rates between two nZVI suspensions can infer equal reactive surface areas. Obviously, the NPs stabilized with PV3A are much smaller and, consequently, have a much larger total surface area per unit mass. Therefore, it is possible that a large fraction of the reactive surface of the NPs stabilized with PV3A is occupied by adsorbed PV3A molecules and is not available for direct reaction with TCE, although this study demonstrated that the PV3A polyelectrolyte is capable of creating a stable dispersion of ZVI NPs.

Kim et al.[75] an efficient way to immobilize nZVI in a support material, alginate bead, using both Fe^{3+} and Ba^{2+} as the cross-linking cations. The effectiveness of the immobilized nZVI on the dechlorination of TCE was tested.

Figure 9.6(A) shows the ratio of final to initial TCE concentration by pure alginate bead and Fe/Pd-alginate with reaction time where

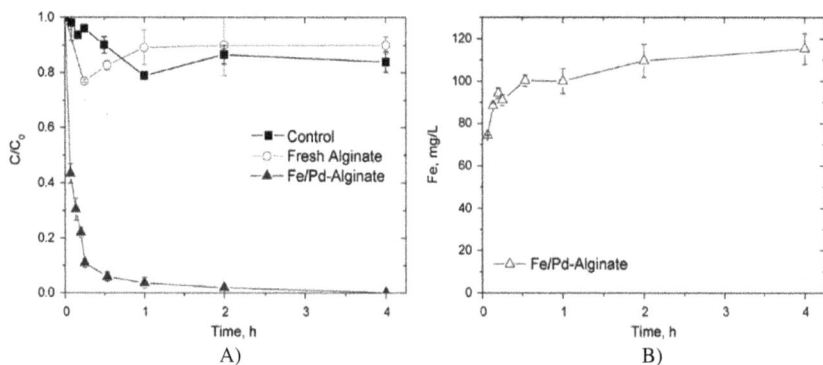

Fig. 9.6. Removal of TCE by the alginate-immobilized nZVI (A) and aqueous iron con- centration (B) with reaction time, where initial TCE concentration was 34.3 mg/L.

the initial TCE was 34.3 mg/L. The change of TCE concentration by fresh alginate was close to that in control test. So, the decreased amount of TCE (~16%) in fresh alginate was just caused by leakage from the vial. With Fe/Pd-alginate, TCE rapidly decreased during 30 min, and 96% of TCE was removed after 1 h. Removal efficiency of TCE was >99.8% by Fe/Pd-alginate within 4 h. Aqueous Fe concentration increased from 55 to 115 mg/L as reaction proceeded (Fig. 9.6(B)). The increased iron concentration may cause esthetic concern and typical treatment processes such as precipitation can be used to control the problem.[75]

When powedered activated carbon (PAC) is incorporated into Fe/Pd-alginate (Fe (0)-PAC-alginate), the final TCE concentration was lowered to 47% of initial TCE after 30 min (Fig. 9.7). At doubled Fe (0)-PAC-alginate concentration, the TCE concentration was still 38% of initial value. Compared with the results of PAC-alginates, most of TCE was removed by sorption on PAC. Although the addition of PAC resulted in the adsorption of the hydrophobic pollutant, the reactivity of the nZVI was reduced significantly in the presence of PAC. The reactivity of nZVI might be consumed by the functional groups (e.g., carboxylic, phenolic groups) of PAC. Or the strong

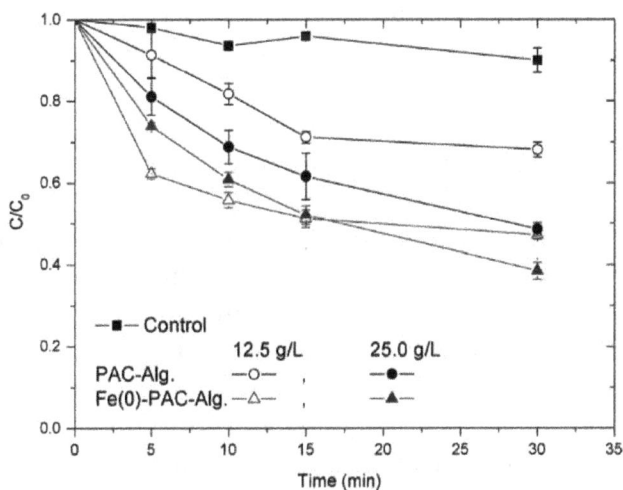

Fig. 9.7. Removal of TCE by ZVI immobilized in PAC/alginate with reaction time, where initial TCE concentration was 34.3 mg/L.

sorption of TCE by PAC might hinder the contact between TCE and nZVI.[75]

Recently, Phenrat *et al*[76] evaluated the TCE dechlorination rate and reaction by-products using poly(aspartate) (PAP)-modified and bare nZVIs in groundwater samples from actual TCE contaminated sites in Florida (FL), South Carolina (SC), and Michigan (MI).[76]

The TCE degradation kinetics and by-product formation kinetics using bare and PAP-modified nZVIs in MI groundwater as a representative example of the groundwater samples was evaluated in this study.

Acetylene was the intermediate, while ethane and ethene were the main by-products. Mass balance of TCE and dechlorination by-products was from 96% to 127% for all the cases. The presence of ionic and organic solutes in natural groundwater decreased the TCE dechlorination rates using bare nZVIs to a similar extent as that previously reported by Liu *et al.*, 2006.[57] We found that the TCE dechlorination rates in the SC and MI groundwater samples for the first TCE spike were around ~22% of the TCE dechlorination rate using bare nZVIs in dionized water (DI) water. Similarly, the TCE dechlorination rate using bare nZVIs in FL groundwater, in which the solute concentrations were in the range of those for the SC and MI groundwater, was around 13% of the TCE dechlorination rate in DI water. Presumably, the decline in the TCE dechlorination rate in groundwater is attributed to the surface complexation of the nZVIs by cationic and anionic solutes[77] and reactive site blocking via the adsorption of natural organic matter onto the surface of the nZVIs.

The TCE dechlorination rates in the second TCE spike for MI and FL groundwater were similar to the dechlorination rates of the first spike, which is in good agreement with Liu and Lowry's observation that TCE dechlorination rates remained constant over the lifetime of nZVIs under a particular solution chemistry (i.e., pH).[57] However, the TCE dechlorination rate of bare nZVIs in SC groundwater declined in the second spike. In the third TCE spike (around 30 days after the first TCE spike), bare nZVIs stopped reacting with TCE in all three groundwater Fe (0) samples, presumably due to the depletion of at the Fe (0) end of the particles (lifetime content <3% by mass for all the cases).

Groundwater solutes did not alter TCE dechlorination pathway using PAP-modified nZVI. Similar to PAP-modified nZVIs in DI water, for TCE dechlorination using PAP-modified nZVIs in the groundwater samples, acetylene was the intermediate and ethane and ethane were the main by-products.[78] For the first TCE spike, the TCE dechlorination rate using PAP-modified nZVIs was less than that of bare nZVIs in all the groundwater samples. This is because the adsorbed PAP blocks the nZVI reactive sites and decreases the aqueous TCE concentration at the nZVI surface due to the partitioning of TCE to the adsorbed PAP layer. This observation is in good agreement with our recent study.[78]

However, the adsorbed PAP layer on the nZVI surface limited the adverse effect of groundwater solutes on TCE dechlorination rates. For the first TCE spike, TCE dechlorination rates using PAP-modified nZVIs in the MI and SC groundwater samples were 70% and 85% of the TCE dechlorination rate using PAP-modified nZVIs in DI water.[78]

Overall, the TCE dechlorination rate using PAP-modified nZVIs in the ground water samples was much less affected by the presence of groundwater solutes than that using bare nZVIs.[78]

Lin *et al.*[79] performed reductive dechlorination experiments using fresh solutions of nZVI and (polyethylenimine) PEI-nZVI (surface-modified ZVI NPs; PEI) at a concentration of 2 g/L, respectively. Kinetic studies for each set of experiments were performed by adding nZVI at different concentrations (50 and 100 mg/L) of TCE solutions, perchloroethylene (PCE), and 1,2-Dichloroethene (1,2-DCe). All dechlorination reactions (reductions in TCE, PCE, and 1,2-DCe) were carried out at 25°C for 24 h.

Batch studies were performed to determine the potential for reductive elimination of PCE, TCE, and 1,2-DCe by dehalogenation at different concentrations. The results of dechlorination studies for three dense nonaqueous phase liquids (DNAPLs) using both fresh and surface modified nZVI are summarized. Careful observation of kinetic studies revealed that reaction rates varied for different

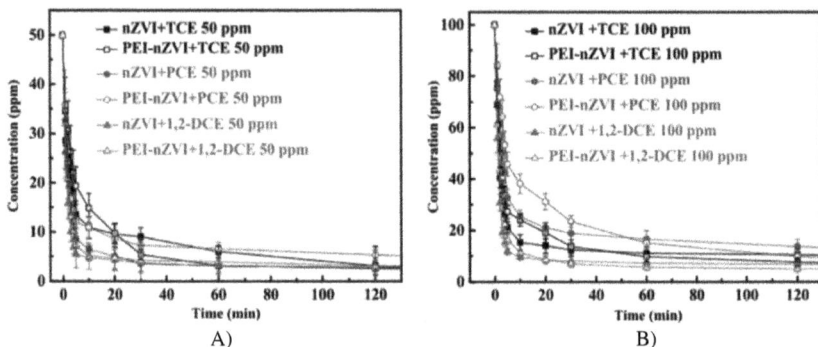

Fig. 9.8. (A) 50 and (B) 100 ppm of TCE, PCE, or 1,2-DCe over nZVI and PEI e nZVI, respectively, via reductive reaction within initial 120 min (concentration of nZVI was 2 g/L).

chemicals (PCE, TCE, and 1,2-DCe) and also for different initial concentrations. As shown in Fig. 9.8(A) over 95% of 1,2-DCe at a concentration of 50 and 100 mg/L were removed within 10 min in reaction with fresh nZVI, whereas the number for PCE and TCE was 93% and 90%, respectively. However, the degradation of a higher concentration (100 mg/L) was different, and only 90% of 1,2-DCe, 85% of TCE, and 75% of PCE were removed (Fig. 9.8(B)). Higher concentrations may elevate the passivation of nZVI due to the formation of iron oxides and contaminant immobilization on the nZVI surface.[79] The comparison of these results with the degradation rate and removal efficiency of DNAPLs by PEI-nZVI is enlightening. However, after 2 h, both reached the same values of 97%, 96%, and 90% for TCE, PCE, and 1,2-DCe, respectively.

9.4 Remediation of Heavy Metals

Heavy metals are one of the groups of contaminants that can be removed by ZVI NPs at high rates and capacities.[80]

Several authors have proposed that removal processes depend on the type of contaminant and its interaction with the NPs.[81] The mechanisms proposed include[37] surface-mediated reduction and precipitation in the NPs core (Cr^{6+}, Hg^{2+}, Cu^{2+}, etc.), adsorption plus precipitation or surface complex formation (Zn^{2+}, Cd^{2+}), sorption plus partial chemical reduction (Ni^{2+}, Pb^{2+}), and surface mineralization (H_2S). Moreover, some studies have affirmed that ZVI NPs can degrade contaminants by oxidation.

For example, among the different arsenic removal techniques, the adsorption method has received more attention due to its high efficiency and cost-effectiveness, and it is also considered the most suitable technology for use in the developing countries.[82]

Evidence that As (III) can be removed by adsorption on ZVI NPs on a minute time scale has been presented.[83] As (III) strongly absorbs on nZVI in a wide range of pHs, and various As (III) and As (V) coprecipitates on iron (III) oxide/hydroxide corrosion products are involved. The results of this study showed that ZVI NPs are an efficient material for the treatment of As (III) and also that they could be used as a new material for permeable reactive barrier walls as well as a material for *ex-situ* treatment.

A recent study demonstrated how ZVI NPs were able to perform simultaneous removal of different ions (e.g., Cu, Zn, Ni and As ions) from wastewater.[84] The removal mechanisms of ZVI NPs included reduction, adsorption, and (co)precipitation. The removal capacities of ZVI NPs reached 226 and 245 mg per g-Fe for Cu and As, with the total capacity >500 mg-metals/g-ZVI NPs for all ions in field applications. Gravitational separation and recirculation of nZVI was shown to be a feasible process of applying ZVI NPs for wastewater treatment. nZVI could be easily separated via gravitational settling and recycled via pumping. nZVI recirculation increased material efficiency and enriched the heavy metal contents in the reacted nZVI. Valuable mineral resources were harvested from the wastewater using the separation–recirculation system.

The major mechanisms of heavy metals removal by ZVI NPs is shown in Fig. 9.9. The mechanism of nZVI reaction was based on nZVI having a metallic iron core and an amorphous oxide shell.[84] Its metallic iron core possesses the well-characterized reducing or electron-donating power. The metallic iron has a standard potential

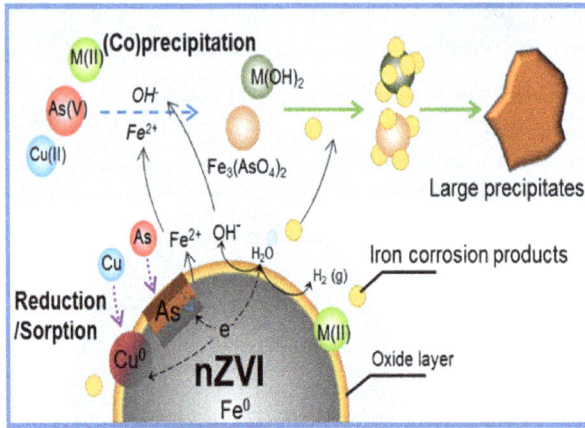

Fig. 9.9. Scheme of heavy metal removal using nZVI from Li *et al.*[84]

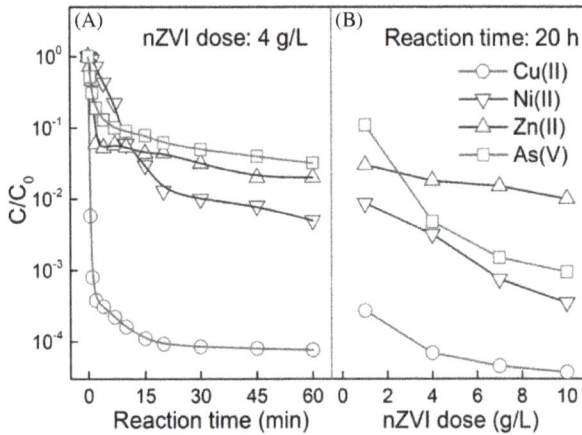

Fig. 9.10. Removal of heavy metals using nZVI from industrial wastewater as a function of time (A) an nZVI dose (B), initial pH = 5.

of −440 mV and is a fairly potent reductant in water. Its large surface area allows the quick release of its electrons; therefore, heavy metals with standard reduction potentials much more positive than that of iron, such as Cu (II), can be quickly reduced.[84]

Quick removal of the four-target ions was observed after ZVI NPs addition (Fig. 9.10(A)), with 99% of copper, 90% of arsenic, zinc, and nickel removed in less than 10 min. All the removal efficiencies were greater than 96% after 60 min. Also, increasing nZVI dosages improved the effluent quality (Fig. 9.10(B)).[84]

Compared with chemical precipitation and sorption, the chemical reduction of Cu (II) and As (V) by ZVI NPs was thermodynamically much more favorable, and less affected by pH changes and chelates.

Another important process is the removal of uranium from nuclear wastewater. In this vein, Ling and coworkers[85] used ZVI NPs with a diameter range of 20–100 nm to recover uranium (VI) from water. Around 90% of the uranium was recovered with 1 g/L of NPs in less than 2 min. Trace level (2.32–882.68 µg/L) uranium could be quickly separated from water and encapsulated at the center of the ZVI NPs. The adsorption or attachment step was reversible depending on the solution pH and very fast due to the small size and large surface area of nZVI.

Batch experiments with repeated spikes of concentrated (up to 300 mg of U/L) uranyl confirmed that the observed removal capacity was as high as 2.4 g of U/g of Fe. The effectiveness of ZVI NPs compared to conventional sorbents lies in their large driving force for fast uranium (VI) reduction by the ZVI because of the large differentiation in the standard potential (E_H^0, mV).[85] Materials based on sorption alone tend to have relatively low capacity and fast release of uranium upon even modest changes in the solution pH.

Rao *et al.*[86] used a bio nanocomposite (phyto-inspired $Fe(0)/Fe_3O_4$ nanocomposite-modified cells of *Yarrowia lipolytica*: NCIM 3589 and NCIM 3590) to evaluate its capacity to remove hexavalent chromium, which were proved to be good bioadsorbents. The sorption capacity of magnetically modified yeast cells was three times more than that of unmodified yeast cells. At the initial chromium concentration of 1 g/L and under optimum conditions, modified NCIM 3589 showed better adsorption capacity (186.32 mg/g) than modified NCIM 3590 (137.31 mg/g).

Another strategy has been described by Madhavi *et al.*,[87] which reported a single-step synthesis of nZVI at room temperature using the *Eucalyptus globulus* leaf extract. The reaction for synthesis of iron NPs was increased by adding more extract. Fourier-transform infrared (FTIR) spectroscopy provided the information about the vibrational state of adsorbed molecules and, hence, the nature of surface complexes. The phytogenic Fe (0) NPs (nZVI) were further used for the adsorption of Cr (VI) metal.

Adsorption parameters such as dose of adsorbent (nZVI), initial concentration of Cr (VI), and kinetics were also studied by batch experiments. The highest adsorption efficiency of nZVI was 98.1% at reaction time of 30 min, and dosage of nZVI was 0.8 g/L. One occurrence of particular interest was that phyto-synthesized iron NP (nZVI) were stabilized and remained in that state for up to 2 months after preparation. Likewise Savasari *et al.*[88] synthesized green nZVI by ascorbic acid, which was employed for reduction of Cd (II) from aqueous and ascorbic acid synthesized NPs proved to be stable and efficient.

In two different studies, Mystrioti *et al.*[89] produced stable colloidal suspensions of nZVI coated with polyphenol of green tea (GT) and studied their chromium removal efficiency from groundwater as well as their transport characteristics through representative porous media.[89,90] The effectiveness of the resulting GT-nZVI suspension with diameter of 5–10 nm was evaluated for the removal of hexavalent chromium Cr (VI) from polluted groundwater flowing through the permeable soil bed. GT extract is characterized as a higher antioxidant compound due to presence of polyphenols of GT. Polyphenols-enriched GT extract plays dual role in synthesis of nZVI, since they have capability to reduce ferric cations, meanwhile shield nZVI from being oxidized and agglomerated, functioning as capping agents. Column tests were performed at different flow rates in order to analyze the effect of contact time between the nZVI attached on porous media and the flow-over solution on reduction of Cr (VI). According to the results of the study, reduction and removal of Cr (VI) from the aqueous phase can be increased by increasing contact time. Leaching tests indicate that chromium in precipitated form is insoluble. In the tested soil material, the total amount of precipitated Cr was observed to be in the range between 280 and 890 mg/kg of soil, whereas the soluble Cr was less than 1.4 mg/kg of soil, which was most likely due to the presence of residual Cr (VI) solution in the porous soil. Nano zero-valent suspension is a very conductive to remediation of a contaminated aquifer, and the use of stable NPs makes this technique successful.[90]

Metals adsorbed on NPs via redox reaction, coprecipitation, or surface adsorption process.[91,92] The reactivity of iron NPs based on different factors, which ultimatly influence on removal mechanism of iron NPs with variable oxidation states, possess different chemical characteristics as well as their mechanism of reaction with contaminants might be dissimilar as described by Tang and Lo.[93]

In another recent study, Xiao *et al.*[94] employed effectively plant-mediated iron NPs for removal of chromium, synthesized by various leaf extracts. Plant were selected on the basis of their reduction potential, i.e., selected from high to low antioxidant potential such as Syzygium jambos (L.) Alston (SJA) extract with strong reducing ability, Oolong tea (OT) extract with moderate reducing ability, and *A. moluccana* (L.) Willd (AMW) extract with weak reducing ability. The study showed that removal of chromium (VI) was consistent with reducing capacity of plants extracts. One milliliter of SJA-Fe NPs colloidal were able to remove 91.9% of the Cr (VI) in 5 min and 100% in 60 min. TEM image of the SJA-Fe NPs showed that NPs were spherical with diameter about to 5 nm and amorphous in nature when studied by X-ray diffraction (XRD). However, this study lacks information on whether the removal of chromium depends on reduction potential of plants or the size of NPs produced by the extracts.

9.5 Conclusion and Perspective

Although, no doubt, there is a large amount of ground to cover, this chapter has shown the use of iron as a potential candidate for its use in the remediation of toxic pollutants such as organic, chlorinated compounds, and heavy metals. In this sense, iron emerges as a substitute for "conventional" noble metals such as gold or silver which, despite having a better performance today, have the serious drawback of a high cost. However, its overall performance still lags behind that of noble metals. It is necessary to make efforts to optimize catalysts of this type or designing catalysts on surfaces with high strength and low cost.

Acknowledgments

This work was supported by the Spanish Government the Spanish National Research Council (CSIC) (project 201980E081) and SAMSUNG electronics (GRO PROGRAM).

References

1. P. A. Frey and G. H. Reed, The ubiquity of iron, *ACS Chem. Biol.* **7**, 1477–1481 (2012).
2. G. Lemming, J. C. C. Chambon, P. J. Binning and P. L. Bjerg, Life cycle assessment combined with remedial performance modeling for assessment of the environmental impacts of remediation technologies, Abstr. 2012 World Congr. Adv. Civil, *Environ. Mater. Res.* (2012).
3. P. Xu, G. M. Zeng, D. L. Huang, C. L. Feng, S. Hu, M. H. Zhao, C. Lai, Z. Wei, C. Huang, G. X. Xie and Z. F. Liu, Use of iron oxide nanomaterials in wastewater treatment: A review, *Sci. Total Environ.* **424**, 1–10 (2012).
4. A. B. Cundy, L. Hopkinson and R. L. D. Whitby, Use of iron-based technologies in contaminated land and groundwater remediation: A review, *Sci. Total Environ.* **400**, 42–51 (2008).
5. P. G. Tratnyek and R. L. Johnson, Nanotechnologies for among the many applications of nanotechnology that have environmental, *Nano Today* **1**, 44–48 (2006).
6. X. Zhang, Y. Ding, H. Tang, X. Han, L. Zhu and N. Wang, Degradation of bisphenol A by hydrogen peroxide activated with $CuFeO_2$ microparticles as a heterogeneous Fenton-like catalyst: Efficiency, stability and mechanism, *Chem. Eng. J.* **236**, 251–262 (2014).
7. S. Aredes, B. Klein and M. Pawlik, The removal of arsenic from water using natural iron oxide minerals, *J. Clean. Prod.* **60**, 71–76 (2013).
8. C. Salazar-Camacho, M. Villalobos, M. de la L. Rivas-Sánchez, J. Arenas-Alatorre, J. Alcaraz-Cienfuegos and M. E. Gutiérrez-Ruiz, Characterization and surface reactivity of natural and synthetic magnetites, *Chem. Geol.* **347**, 233–245 (2013).
9. X. S. Wang, L. Zhu and H. J. Lu, Surface chemical properties and adsorption of Cu (II) on nanoscale magnetite in aqueous solutions, *Desalination* **276**, 154–160 (2011).

10. S. Zha, Y. Cheng, Y. Gao, Z. Chen, M. Megharaj and R. Naidu, Nanoscale zero-valent iron as a catalyst for heterogeneous Fenton oxidation of amoxicillin, *Chem. Eng. J.* **255**, 141–148 (2014).

11. X. Lv, X. Xue, G. Jiang, D. Wu, T. Sheng, H. Zhou and X. Xu, Nanoscale Zero-Valent Iron (nZVI) assembled on magnetic Fe_3O_4/ graphene for Chromium (VI) removal from aqueous solution, *J. Colloid Interface Sci.* **417**, 51–59 (2014).

12. Z. Li, H. Kirk Jones, P. Zhang and R. S. Bowman, Chromate transport through columns packed with surfactant-modified zeolite/zero valent iron pellets, *Chemosphere* **68**, 1861–1866 (2007).

13. T. Shahwan, C. Üzüm, A. E. Eroğlu and I. Lieberwirth, Synthesis and characterization of bentonite/iron nanoparticles and their application as adsorbent of cobalt ions, *Appl. Clay Sci.* **47**, 257–262 (2010).

14. X. Sun, X. Wang, J. Li and L. Wang, Degradation of nitrobenzene in groundwater by nanoscale zero-valent iron particles incorporated inside the channels of SBA-15 rods, *J. Taiwan Inst. Chem. Eng.* **45**, 996–1000 (2014).

15. C. A. Staples, P. B. Dom, G. M. Klecka, T. O. Sandra and L. R. Harris, A review of the environmental fate, effects, and exposures of bisphenol A, *Chemosphere* **36**, 2149–2173 (1998).

16. S. Flint, T. Markle, S. Thompson and E. Wallace, Bisphenol A exposure, effects, and policy: A wildlife perspective, *J. Environ. Manage.* **104**, 19–34 (2012).

17. T. Yamamoto, A. Yasuhara, H. Shiraishi and O. Nakasugi, Bisphenol A in hazardous waste landfill leachates, *Chemosphere* **42**, 415–418 (2001).

18. J. C. O'Connor and R. E. Chapin, Critical evaluation of observed adverse effects of endocrine active substances on reproduction and development, the immune system, and the nervous system, *Pure Appl. Chem.* **75**, 2099–2123 (2007).

19. L. Joseph, J. Heo, Y. G. Park, J. R. V. Flora and Y. Yoon, Adsorption of bisphenol A and 17α-ethinyl estradiol on single walled carbon nanotubes from seawater and brackish water, *Desalination* **281**, 68–74 (2011).

20. W. Han, L. Luo and S. Zhang, Adsorption of bisphenol A on lignin: Effects of solution chemistry, *Int. J. Environ. Sci. Technol.* **9**, 543–548 (2012).

21. A. S. Stasinakis, A. V. Petalas, D. Mamais and N. S. Thomaidis, Application of the OECD 301F respirometric test for the biodegradability assessment of various potential endocrine disrupting chemicals, *Bioresour. Technol.* **99**, 3458–3467 (2008).

22. P. V Nidheesh and R. Gandhimathi, Trends in electro-Fenton process for water and wastewater treatment: An overview, *Desalination* **299**, 1–15 (2012).

23. F. Gulshan, S. Yanagida, Y. Kameshima, T. Isobe, A. Nakajima and K. Okada, Various factors affecting photodecomposition of methylene blue by iron-oxides in an oxalate solution, *Water Res.* **44**, 2876–2884 (2010).

24. N. Das, Recovery of precious metals through biosorption - A review, *Hydrometallurgy* **103**, 180–189 (2010).

25. R. Cheng, C. Cheng, G. H. Liu, X. Zheng, G. Li and J. Li, Removing pentachlorophenol from water using a nanoscale zero-valent iron / H_2O_2 system, *Chemosphere* **141**, 138–143 (2015).

26. X. Li, Y. Zhao, B. Xi, X. Mao, B. Gong, R. Li, X. Peng and H. Liu, Removal of nitrobenzene by immobilized nanoscale zero-valent iron: Effect of clay support and efficiency optimization, *Appl. Surf. Sci.* **370**, 260–269 (2016).

27. R. Huang, Z. Fang, X. Yan and W. Cheng, Heterogeneous sono-Fenton catalytic degradation of bisphenol A by Fe_3O_4 magnetic nanoparticles under neutral condition, *Chem. Eng. J.* **197**, 242–249 (2012).

28. N. Wang, L. Zhu, D. Wang, M. Wang, Z. Lin and H. Tang, Sono-assisted preparation of highly-efficient peroxidase-like Fe_3O_4 magnetic nanoparticles for catalytic removal of organic pollutants with H_2O_2, *Ultrason. Sonochem.* **17**, 526–533 (2010).

29. S. Shin, H. Yoon and J. Jang, Polymer-encapsulated iron oxide nanoparticles as highly efficient Fenton catalysts, *Catal. Commun.* **10**, 178–182 (2008).

30. H. Wei and E. Wang, Fe_3O_4 magnetic nanoparticles as peroxidase mimetics and their applications in H_2O_2 and glucose detection, *Anal. Chem.* **80**, 2250–2254 (2008).

31. D. A. Links, Carbon nanodots as peroxidase mimetics and their applications to glucose detection, *Chem. Commun.* **47**, 6695–6697 (2011).

32. S. Zhang, X. Zhao, H. Niu, Y. Shi, Y. Cai and G. Jiang, Superparamagnetic Fe_3O_4 nanoparticles as catalysts for the catalytic oxidation of phenolic and aniline compounds, *J. Hazard. Mater.* **167**, 560–566 (2009).

33. Y. Segura, R. Molina, F. Martínez and J. A. Melero, Ultrasonics sono-chemistry integrated heterogeneous sono—Photo fenton processes for the degradation of phenolic aqueous solutions, *Ultrason. Sonochem.* **16**, 417–424 (2009).

34. Y. Xi, Z. Sun, T. Hreid, G. A. Ayoko and R. L. Frost, Bisphenol A degradation enhanced by air bubbles via advanced oxidation using in situ generated ferrous ions from nano zero-valent iron/palygorskite composite materials, *Chem. Eng. J.* **247**, 66–74 (2014).

35. J. A. Elías-Maxil, F. Rigas, M. T. O. de Velásquez and R. M. Ramírez-Zamora, Optimization of Fenton's reagent coupled to dissolved air flotation to remove cyanobacterial odorous metabolites and suspended solids from raw surface water, *Water Sci. Technol.* **64**, 1668–1674 (2011).

36. L. F. Greenlee, J. D. Torrey, R. L. Amaro and J. M. Shaw, Kinetics of zero valent iron nanoparticle oxidation in oxygenated water, *Environ. Sci. Technol.* **46**, 12913–12920 (2012).

37. F. Fu, D. D. Dionysiou and H. Liu, The use of zero-valent iron for groundwater remediation and wastewater treatment: A review, *J. Hazard. Mater.* **267**, 194–205 (2014).

38. I. Hussain, Y. Zhang, S. Huang and X. Du, Degradation of p-chloro-aniline by persulfate activated with zero-valent iron, *Chem. Eng. J.* **203**, 269–276 (2012).

39. A. R. Fernández-Alba, D. Hernando, A. Agüera, J. Cáceres and S. Malato, Toxicity assays: A way for evaluating AOPs efficiency, *Water Res.* **36**, 4255–4262 (2002).

40. I. A. Alaton and S. Teksoy, Acid dyebath effluent pretreatment using Fenton's reagent: Process optimization, reaction kinetics and effects on acute toxicity, *Dye. Pigment.* **73**, 31–39 (2007).

41. B. Girit, D. Dursun, T. Olmez-Hanci and I. Arslan-Alaton, Treatment of aqueous bisphenol A using nano-sized zero-valent iron in the presence of hydrogen peroxide and persulfate oxidants, *Water Sci. Technol.* **71**, 1859–1868 (2015).

42. M. Usman, P. Faure, C. Ruby and K. Hanna, Application of magnetite-activated persulfate oxidation for the degradation of PAHs in contaminated soils, *Chemosphere* **87**, 234–240 (2012).

43. Y. C. Lee, S. L. Lo, P. T. Chiueh and D. G. Chang, Efficient decomposition of perfluorocarboxylic acids in aqueous solution using microwave-induced persulfate, *Water Res.* **43**, 2811–2816 (2009).

44. S. H. Liang, C. M. Kao, Y. C. Kuo and K. F. Chen, Application of persulfate-releasing barrier to remediate MTBE and benzene contaminated groundwater, *J. Hazard. Mater.* **185**, 1162–1168 (2011).

45. A. Tsitonaki, B. Petri, M. Crimi, H. Mosbaek, R. L. Siegrist and P. L. Bjerg, In situ chemical oxidation of contaminated soil and groundwater using persulfate: A review, *Environ. Sci. Technol.* **40**, 55–91 (2010).

46. X. R. Xu and X. Z. Li, Degradation of azo dye Orange G in aqueous solutions by persulfate with ferrous ion, *Sep. Purif. Technol.* **72**, 105–111 (2010).

47. C. Liang and Y. Y. Guo, Mass transfer and chemical oxidation of naphthalene particles with zerovalent iron activated persulfate. *Environ. Sci. Technol.* **44**, 8203–8208 (2010).

48. S. Y. Oh, S. G. Kang and P. C. Chiu, Degradation of 2,4-dinitrotoluene by persulfate activated with zero-valent iron, *Sci. Total Environ.* **408**, 3464–3468 (2010).

49. B. Schrick, B. W. Hydutsky, J. L. Blough and T. E. Mallouk, Delivery vehicles for zerovalent metal nanoparticles in soil and groundwater, *Chem. Mater.* **16**, 2187–2193 (2004).

50. C. Liu, P. Wu, Y. Zhu and L. Tran, Simultaneous adsorption of Cd2+ and BPA on amphoteric surfactant activated montmorillonite, *Chemosphere* **144**, 1026–1032 (2016).

51. X. Peng, Y. Tian, S. Liu and X. Jia, Degradation of TBBPA and BPA from aqueous solution using organo-montmorillonite supported nanoscale zero-valent iron, *Chem. Eng. J.* **309**, 717–724 (2017).

52. Q. Jin, S. Zhang, T. Wen, J. Wang, P. Gu, G. Zhao, X. Wang, Z. Wang, T. Hayat and X. Wang, Simultaneous adsorption and oxidative degradation of Bisphenol A by zero-valent iron/iron carbide nanoparticles encapsulated in N-doped carbon matrix, *Environ. Pollut.* **243**, 218–227 (2018).

53. S. Veli and B. Alyüz, Adsorption of copper and zinc from aqueous solutions by using natural clay, *J. Hazard. Mater.* **149**, 226–233 (2007).

54. X. L. Wu, L. Wang, C. L. Chen, A. W. Xu and X. K. Wang, Water-dispersible magnetite-graphene-LDH composites for efficient arsenate removal, *J. Mater. Chem.* **21**, 17353–17359 (2011).

55. J. Pan, H. Yao, X. Li, B. Wang, P. Huo, W. Xu, H. Ou and Y. Yan, Synthesis of chitosan/γ-Fe$_2$O$_3$/fly-ash-cenospheres composites for the fast removal of bisphenol A and 2, 4, 6-trichlorophenol from aqueous solutions, *J. Hazard. Mater.* **190**, 276–284 (2011).

56. Y. Liu, S. A. Majetich, R. D. Tilton, D. S. Sholl and G. V. Lowry, TCE dechlorination rates, pathways, and efficiency of nanoscale iron particles with different properties, *Environ. Sci. Technol.* **39**, 1338–1345 (2005).

57. Y. Liu and G. V. Lowry, Effect of particle age (Fe (0) content) and solution pH on NZVI reactivity: H$_2$ evolution and TCE dichlorination, *Environ. Sci. Technol.* **40**, 6085–6090 (2006).

58. J. M. Thompson, B. J. Chisholm and A. N. Bezbaruah, Reductive dechlorination of chloroacetanilide herbicide (alachlor) using zero-valent iron nanoparticles, *Environ. Eng. Sci.* **27**, 227–232 (2010).

59. W. X. Zhang, Nanoscale iron particles for environmental remediation: An overview, *J. Nanoparticle Res.* **5**, 323–332 (2003).

60. A. Fallis, Toxicological profile for 1,1,2,2-tetrachloroethane, *J. Chem. Inf. Model.* **53**, 1689–1699 (2013).

61. P. Bhatt, M. S. Kumar, S. Mudliar and T. Chakrabarti, Biodegradation of chlorinated compounds: A review, *Crit. Rev. Environ. Sci. Technol.* **37**, 165–198 (2007).

62. B. K. Salunke and B. S. Kim, Facile synthesis of graphene using a biological method, *RSC Adv.* **6**, 17158–17162 (2016).

63. R. W. Gillham and S. F. O'Hannesin, Enhanced degradation of halogenated aliphatics by zero-valent iron, *Groundwater* **32**, 958–967 (1994).

64. T. L. Johnson, W. Fish, Y. A. Gorby and P. G. Tratnyek, Degradation of carbon tetrachloride by iron metal: Complexation effects on the oxide surface, *J. Contam. Hydrol.* **29**, 379–398 (1998).

65. Y. Li, J. Liu, Y. Wang and Zhong Lin Wang, Preparation of monodispersed Fe-Mo nanoparticles as the catalyst for CVD synthesis of carbon nanotubes, *Chem. Mater.* **13**, 1008–1014 (2001).

66. H. L. Lien and W. Zhang, Nanoscale iron particles for complete reduction of chlorinated ethenes, *Colloids Surfaces A Physicochem. Eng. Asp.* **191**, 97–105 (2001).

67. K. T. Park, K. Klier, C. B. Wang and W. X. Zhang, Interaction of tetra-chloroethylene with Pd (100) studied by high-resolution X-ray photoemission spectroscopy, *J. Phys. Chem. B.* **27**, 5420–5428 (1997).

68. E. E. Carpenter, J. A. Sims, J. A. Wienmann, W. L. Zhou and C. J. O'Connor, Magnetic properties of iron and iron platinum alloys synthesized via microemulsion techniques, *J. Appl. Phys.* **87**, 5615–5617 (2002).

69. M. Mikhaylova, D. K. Kim, N. Bobrysheva, M. Osmolowsky, V. Semenov, T. Tsakalakos and M. Muhammed, Superparamagnetism of magnetite nanoparticles: dependence on surface modification, *Langmuir.* **6**, 2472–2477 (2004).

70. S. M. Ponder and J. G. Darab, Remediation of Cr (VI) and Pb (II) Aqueous solutions using supported, nanoscale zero-valent iron, *Environ. Sci. Technol.* **34**, 2564–2569 (2000).

71. M. Wiesner and J. Y. Bottero, *Environmental Nanotechnology*, McGraw-Hill Professional Publishing, US (2007).

72. T. Phenrat, H. J. Kim, F. Fagerlund, T. Illangasekare, R. D. Tilton and G. V. Lowry, Particle size distribution, concentration, and magnetic attraction affect transport of polymer-modified Fe (0) nanoparticles in sand columns, *Environ. Sci. Technol.* **43**, 5079–5085 (2009).

73. F. Li, C. Vipulanandan and K. K. Mohanty, Microemulsion and solution approaches to nanoparticle iron production for degradation of trichloroethylene, *Colloids Surfaces A Physicochem. Eng. Asp.* **223**, 103–112 (2003).

74. Y. P. Sun, X. Q. Li, W. X. Zhang and H. P. Wang, A method for the preparation of stable dispersion of zero-valent iron nanoparticles, *Colloids Surfaces A Physicochem. Eng. Asp.* **308**, 60–66 (2007).

75. H. Kim, H. J. Hong, J. Jung, S. H. Kim and J. W. Yang, Degradation of trichloroethylene (TCE) by nanoscale zero-valent iron (nZVI) immobilized in alginate bead, *J. Hazard. Mater.* **176**, 1038–1043 (2010).

76. T. Phenrat, D. Schoenfelder, T. L. Kirschling, R. D. Tilton and G. V. Lowry, Adsorbed poly(aspartate) coating limits the adverse effects of dissolved groundwater solutes on Fe (0) nanoparticle reactivity with trichloroethylene, *Environ. Sci. Pollut. Res.* **25**, 7157–7169 (2018).

77. G. V. Lowry, Effect of TCE concentration and dissolved groundwater solutes on NZVI-promoted TCE dechlorination, *Environ. Sci. Technol.* **41**, 7881–7887 (2007).

78. T. Phenrat, Y. Liu, R. D. Tilton and G. V. Lowry, Adsorbed polyelectrolyte coatings decrease Fe (0) nanoparticle reactivity with TCE in water: Conceptual model and mechanisms, *Environ. Sci. Technol.* **43**, 1507–1514 (2009).

79. K. S. Lin, N. V. Mdlovu, C. Y. Chen, C. L. Chiang and K. Dehvari, Degradation of TCE, PCE, and 1,2–DCE DNAPLs in contaminated groundwater using polyethylenimine-modified zero-valent iron nanoparticles, *J. Clean. Prod.* **175**, 456–466 (2018).

80. X. Q. Li and W. X. Zhang, Sequestration of metal cations with zerovalent iron nanoparticles - A study with high resolution x-ray photoelectron spectroscopy (HR-XPS), *J. Phys. Chem. C* **111**, 6939–6946 (2007).

81. Y. Su, A. S. Adeleye, A. A. Keller, Y. Huang, C. Dai, X. Zhou and Y. Zhang, Magnetic sulfide-modified nanoscale zerovalent iron (S-nZVI) for dissolved metal ion removal, *Water Res.* **74**, 47–57 (2015).

82. Wang, X. et al. Effect of the nature of the spacer on the aggregation properties of gemini surfactants in an aqueous solution, *Langmuir.* **20**, 53–56 (2004).

83. S. R. Kanel, B. Manning, L. Charlet and H. Choi, Removal of arsenic (III) from groundwater by nanoscale zero-valent iron, *Environ. Sci. Technol.* **39**, 1291–1298 (2005).

84. S. Li, W. Wang, F. Liang and W. X. Zhang, Heavy metal removal using nanoscale zero-valent iron (nZVI): Theory and application, *J. Hazard. Mater.* **322**, 163–171 (2017).

85. L. Ling and W. X. Zhang, Enrichment and encapsulation of uranium with iron nanoparticle, *J. Am. Chem. Soc.* **137**, 2788–2791 (2015).

86. A. Rao, A. Bankar, A. Ravi, S. Gosavi and S. Zinjarde, Removal of hexavalent chromium ions by Yarrowia lipolytica cells modi fi ed with phyto-inspired Fe (0)/ Fe$_3$O$_4$ nanoparticles, *J. Contam. Hydrol.* **146**, 63–73 (2013).

87. V. Madhavi, T. N. V. K. V. Prasad, A. V. B. Reddy, B. R. Reddy and G. Madhavi, Application of phytogenic zerovalent iron nanoparticles in the adsorption of hexavalent chromium, *Spectrochim. Acta Part A Mol. Biomol. Spectrosc.* **116**, 17–25 (2013).

88. M. Savasari, M. Emadi and M. Ali, Optimization of Cd (II) removal from aqueous solution by ascorbic acid- stabilized zero valent iron nanoparticles using response surface methodology, *J. Ind. Eng. Chem.* **21**, 1403–1409 (2015).

89. C. Mystrioti, N. Papassiopi, A. Xenidis, D. Dermatas and M. Chrysochoou, Column study for the evaluation of the transport properties of polyphenol-coated nanoiron, *J. Hazard. Mater.* **281**, 64–69 (2015).

90. P. Taylor, C. Mystrioti, A. Xenidis and N. Papassiopi, Reduction of hexavalent chromium with polyphenol-coated nano zero-valent iron: Column studies, *Desal. Water Treat.* **56**, 37–41 (2014).

91. S. M. Ponder, J. G. Darab and T. E. Mallouk, Remediation of Cr (VI) and Pb (II) aqueous solutions using supported, nanoscale zero-valent iron, *Environ. Sci. Technol.* **34**, 2564–2569 (2000).

92. Y. S. Keum and Q. X. Li, Reduction of nitroaromatic pesticides with zero-valent iron, *Chemosphere* **54**, 255–263 (2004).

93. S. C. N. Tang and I. M. C. Lo, Magnetic nanoparticles: Essential factors for sustainable environmental applications, *Water Res.* **47**, 2613–2632 (2013).

94. A. T. Yeung, Remediation technologies for contaminated sites, *Adv. Environ. Geotech.* 328–369 (2009).

Index

(-)-acetoxy-*p*-menthane, 175
α-azidobiaryls, 234
Aconitase, 116
acrolein, 39, 43, 44
acrylic acid, 43
active site, 83, 84, 90, 97, 113, 116, 118, 120, 121
additives, 8, 12, 17, 21, 37, 206, 256, 301
adducts, 137, 140, 149, 155, 158, 165, 168, 171, 177, 180
adsorbate, 62, 63
adsorption, 45, 50, 60, 61, 69, 263, 283, 284, 301, 305, 307, 317, 321
α-Fe, 23
α-Fe$_2$O$_3$, 2, 3, 6, 7, 12, 16
    supported on different materials, 6
    nanoplates, 7
    nanorings, 6
α-FeOOH, 22
agglomeration, 6, 10, 17, 304
agrochemicals, 204
alkene, 35, 52, 115, 135, 178
    oxides, 35, 52
allene migratory insertion, 142

allosteric, 115
alpha, 104, 152, 154, 158
amidation, 81, 166, 167, 230, 242, 243, 245
aminoacids, 78
ammonia solution, 13
ammonium hydroxide, 11
amorphous, 259, 260, 267, 274, 275, 321
anilines, 164
animals, 57, 59, 64, 99, 113
annulation reaction, 161
anodes, 253, 258, 259, 284
antiferromagnetic, 3
aquatic mammals, 101
arenes, 22, 133, 134, 137, 151, 154, 164, 178
arsenic, 318
artificial, 77–82, 84–86, 89
artificial hydrogenase, 85
aspartic acid, 114, 116
atmospheric pressure, 98
ATP, 106, 110, 115, 117
austenite, 275
azetidine, 237
azides, 167, 203, 208, 230, 235, 237, 245

bacteria, 5, 57, 59, 106, 112–114
BASF, 38
benzene, 144, 182, 183, 210, 243
    benzenediamines, 165
    diphenylphosphino, 140
    tricarboxylate, 181
benzoate, 172
benzoxazoles, 139, 171
Berenguer-Murcia Á., 35
BET surface, 307
β-FeOOH, 19, 20
    TiO$_2$ heterostructure metals, 21
β-hydride elimination, 136, 145
binding sites, 104, 105, 307
bioadsorbents, 319
bioavailability, 62
biocompatible, 57
bioinformatics, 84
biological processes, 77, 87, 302
biomass, 263, 266–268, 270, 281
bionanohybrids, 77
bioremediation, 83
biosensor, 10, 18, 24
biotin, 85
blood, 99, 101, 103, 105
Boc$_2$O, 240
Bohr effect, 100
boronic acid, 134, 183
borylation, 177–179
bottom-up, 16
BPA, 301–303, 305, 306
Buchner reaction, 144
bulk metals, 87

C–H, 40, 79–83, 120, 127–129, 139,
    148, 155, 157, 180, 203, 229, 236,
    242
    Activation, 127, 129
    bond, 120, 150

C–H bonds, 130, 152, 184,
    204, 210
    functionalization, 83, 127, 128,
    133, 135, 148, 179, 184, 203,
    208, 214, 229
    synthases, 82
capping, 5, 320
    agents, 5, 320
carbazole heterocycles, 234
carbon dioxide, 100
carbon materials, 253, 259, 260,
    264, 269, 278, 284
carbon nanotubes, 259, 309
carbon oxides (COx), 39
carboxylate, 119
cascade processes, 90
catalase, 99, 109, 110
catalysis, 98, 100, 117, 133, 136,
    137, 157, 179, 259, 308
catalyst loading, 175, 272, 278, 284
catalysts, 1, 5, 20, 35, 37, 40, 41, 58,
    70, 77, 97, 127, 203, 300
catechol, 118
Cazorla-Amorós, D., 35
C–C bond reactions, 22
cell, 98, 106, 111, 255, 258, 283,
    301
cellulose, 265–268, 270, 276
cementite carbide, 276
chelate, 52, 97, 319
chemisorption, 62
chlorobenzene, 131
chlorohydrin, 38
chromium, 319–321
C–N adduct, 172
Co, 41, 271, 273, 275, 278, 308
coatings, 301
coenzyme Q, 115
coenzyme Q reductase succinate, 114

cofactors, 97, 117, 120
composites, 7, 69, 70, 258, 280, 306
conformational changes, 100
conjugation, 77, 84
conversion, 41, 43, 44, 46, 47, 50,
   83, 110, 183, 217, 258, 274
coordination, 47, 50, 60, 78, 99,
   102, 173, 212
copper, 106, 120, 205, 256, 318
coprecipitation, 8, 9, 13, 272, 321
core-shell structure, 15, 18, 258
corn, 268
    cob, 268
    straw, 268
correlation coefficient, 306
corrosion, 304, 308, 317
cost-effective, 1, 70, 208, 300
cost-effectiveness, 44, 317
cross-coupling reaction, 165, 209
crystallinity, 261, 274
crystal structure, 3, 61
C–S bond, 176
Csp^2–H bonds, 136
Csp^2–S, 165
Csp^2–H, 133, 145, 163, 167, 176
Csp^3–H, 147, 148, 152, 154, 163, 174
CTAB, 12
Cu (II), 318, 319
cyanobacteria, 100
cyclization, 117, 140, 171, 227, 237,
   265
cyclopentadienone, 85
cysteine, 107, 114, 116, 119
cytochrome P450 enzyme, 80, 229

dative, 85
DDQ, 182
debromination, 304, 305
dechlorination, 310, 312, 314, 315

by-products, 314
   experiments, 315
   pathway, 315
   rate, 310, 314, 315
   reactions, 315
   studies, 315
degreasing agents, 308
dehalogenation, 307
dehydration, 109
detoxification, 205, 303
diameter, 7, 9, 12, 16, 18–24,
   319–321
diastereoselectivity, 209, 237, 238
diazoalkanes, 205, 206
dicumyl peroxide, 135
diffuse layer, 63
dimerization, 109
dioxygen, 118, 119
dioxygenases, 118, 245
directed evolution, 78, 79, 81, 218,
   228, 245
discharge capacity, 282
disintegration, 17
dissociation, 60, 61, 128, 220
dissolution–precipitation, 274
δ-lactams, 81
DMF, 139
d-orbitals, 271
DOW, 38
drugs, 204
DTBP, 159

Earth, 36, 58, 205, 299
EDL, 61, 63
efficiency, 1, 24, 39, 64, 65, 67,
   83, 105, 181, 262, 270, 283, 304,
   317
effluents, 300
electrical charge density, 61

electrochemical properties, 262, 279–281, 283

electrolyte, 61, 255, 256, 283

electron-rich, 141, 151, 154, 155, 159, 174, 181, 213, 232

electron-withdrawing, 211

electrophiles, 137

electrospinning, 23

element, 36, 58, 299

Empel, C., 203

enantioselectivity, 205

energy, 254, 258, 259, 264, 269, 278
  abundance, 258
  consumption, 253, 254, 264
  efficiencies, 258
  for crystallization, 269
  from food, 105
  levels, 278
  nongraphitizing, 268
  release, 106
  savings, 269
  storage devices, 254
  storage systems, 254, 259, 265
  of sunlight, 105
  upon cooling, 276

environmental remediation, 15, 299

enzymatic transformations, 203

epoxidation, 37, 39, 41, 46, 47, 120

epoxy resins, 301

erythrocytes, 104

Escherichia coli, 83, 216

ethyl α-diazoacetate, 83, 206, 210, 215–217, 220, 228

ethylene-forming enzyme, 245

ethylene glycol, 10

ethylene oxide, 40

ethyl phenyldiazoacetate, 207

EtOAc, 172

European Environment Agency (EEA), 300

exothermic process, 62

E/Z-stereoselectivity, 136

FAD, 110, 115, 120

Fe (0), 2, 15, 23

Fe_2O_3, 18

Fe_3O_4, 2, 8, 9, 11, 12, 14, 16
  NPs, 11
  PB, 10

$Fe(acac)_3$, 44, 134–137, 139

$FeBr_2$, 242

$FeCO_3$, 21

Fe(IV), 162, 183

Fe(IV)-E, 109

Fenton, 20, 58, 64, 65, 70, 173, 302

$Fe(OTf)_2$, 153, 157, 177

Fernández-Catalá, J, 35

ferredoxin, 112–114

ferric chloride, 4–6, 14, 17, 20, 23, 24

ferromagnetic, 11, 12

Fe–S clusters, 111

fluoracetate, 117

footprint, 205

formaldehyde, 270

formaldehyde-derived, 280

Fossil fuels, 254

FTIR, 10, 320

functionalization, 135

galvanostatic experiments, 262

García-Aguilar, J., 35

García Mancheño, O., 127

gas phase, 35, 37, 39, 40, 46, 266

γ-Fe_2O_3, 2, 4, 16

glycoporphyrins, 235

goethite, 58, 59, 65–67
goethite (FeOOH), 16
gold, 35, 39, 40, 52, 321
Gomez-Martin, A., 253
Gómez-Martínez, M., 127
graphene, 6, 13–15, 69, 87, 89,
    259–261, 306
graphene oxide, 6, 13, 17, 69, 87
    carbon nanotubes, 87
graphite, 253, 257–261, 263, 267,
    284
graphitization, 253, 264, 269, 270,
    272, 273, 284
Grignard, 130–133, 135, 140, 146
groundwater, 300, 302, 304

Haber–Bosch ammonia synthesis,
    37
Heckase, 89
hematite, 2–4, 7, 23, 49, 58, 65
hemerythrins, 119
hemicellulose, 265, 266
hemoglobin, 36, 99–101, 103, 105
hemoproteins, 82, 99, 100, 107
heterogeneous, 22, 38, 65, 87, 88,
    129, 179–181, 269, 302, 303
heterogeneous nanomaterial, 88
heterogonous nanobiohybrid, 88
histidine, 116
histidine residue, 102–104, 118
holoenzymes, 98
homogenous catalysts, 85
horseradish peroxidase, 65, 87
human, 36, 70, 105
hydrocarbons, 37, 86, 107, 120,
    169, 173, 265
hydrogen, 119, 263, 265, 266
hydrogenation, 85, 308

hydrogen peroxide, 109, 110, 174,
    176, 183, 303
hydrolysis, 107
hydrophobicity, 305
hydroxyl, 11, 50, 60, 61, 64–66,
    121, 302
    (Fe–OH) groups, 11
    groups, 50, 60, 61
    radicals, 64–66, 121, 302
    surface groups, 60
hysteresis, 263

ICP-AES, 181
imidazolium salt, 138
impregnation, 46, 47, 50, 272, 274,
    276
indoles, 83, 138, 154, 179, 214,
    217, 231, 245
indoline, 167, 235
interfacial, 61
intermediate, 45, 142, 143, 150,
    153, 161, 207, 210, 277
intermolecular, 79–81, 156, 166,
    167
interplanar distance, 268–271
Ir, 127
Ir(III), 205
iron, 1
    atom, 102
    boryl species, 178
    carbide, 275
    catalysis in environmental
        remediation, 299
    catalysis in metal-ion
        batteries, 253
    catalyst, 1, 46
    catalytic C–H activation, 129
    catalyzed C–H arylation, 137

catalyzed carbene and nitrene
transfer reactions, 203
catalyzed C–H
functionalization reaction of
indoles, 215
complex Bu4N[Fe(CO)$_3$–
(NO)]—TBA[Fe], 234
complexes, 129
core, 86
dicarbonyl silyl complex, 179
dipyrrinato, 236
enzymes, 77
hydroxide, 12
hydroxides (FeOOH), 2
imido complex, 232
iron nanostructured, 1
iron nanostructures, 1
metalloenzymes, 77
nanoflowers, 24
nanomaterial, 7
nanoparticles, 2, 3
nanoplates, 8
nanostructures, 2, 3, 7
naphthyl-azo complex catalyst,
183
nitrate, 229, 272
nitrenes, 229
nitrene complex, 166
nitrenoid, 80
oxide nanoenzymes, 57
oxides, 2, 275
oxide species, 3
phthalocyanine complex, 227
porphyrin complex, 206
precursor, 18
protoporphyrin IX, 106
salt, 3
species, 2, 4

sulfur proteins, 111
supported on the surface of
mesoporous silica, 45
triflate, 230
iron biphenyl-4,4'-dicarboxylate–
organic framework, 181
iron nanostructured, 25, 299
iron nanostructured catalysts, 1
irreversible, 62, 132, 259, 303
isoenzymes, 116
isomerization, 109, 116, 131
isoquinolinones, 142

Jana, S., 203
jellyfish mesoglea (JF), 9

kinetics, 62, 261, 283, 309, 314,
320
Koenigs, R. M., 203
Krebs cycle, 115, 116

Langmuir model, 306
L-Arg, 111
large-scale, 59, 254
lattice fringes, 11
L-citrulline, 111
Lewis acids, 60
Lewis base, 60
LFe(OTf)$_2$, 144
ligand, 138, 139, 146, 164, 167,
206, 207, 210, 227, 238
light, 67, 70, 302
lignin, 265, 266, 270, 273
lipid bilayer, 115
liquid phase, 38
lithium cobalt oxide, 254
lithium-ion, 57, 253, 254, 270, 284
batteries, 57, 253, 254, 270

living organisms, 36, 57, 59, 97, 98, 101, 109, 122
loops, 102
Losada-García, N, 299
LyondellBasell, 38

magnetic resonance imaging, 3
magnetization, 4
malonate, 115
mangosteen shell, 268
Mateo, C., 97
m-chloroperoxybenzoic acid (CPBA), 175
mechanism, 108, 111, 128, 132, 141, 151, 211, 261, 317
mesoporous aluminosilicate, 180
mesoporous silica, 41
metabolite, 108
metallic, 90, 106, 129, 254, 255, 257, 258, 275
   activity, 90
   atom, 106
   cation, 258
   Fe, 275
   iron, 308
   lithium, 254, 255, 257
   oxidant, 129
metalloids, 300
metalloproteins, 36, 99
metals, 11, 14, 25, 35, 40, 52, 78, 88, 129, 205, 270, 284, 299
Michaelis constant, 87
microorganisms, 59, 64, 109, 113
mitochondria, 101
Mondaca, F, 57
monolayer, 62
mononuclear, 47, 117
monooxygenase, 40, 79, 107, 120

morpholine, 164
morphologies, 3, 21, 25
Mtz-Enriquez, A. I., 57
multilayers, 62
muscle, 101
mutagenesis, 78, 218, 243, 245
m-xylene, 242
myoglobin, 82, 99, 101, 104, 105, 216

$Na_2S_2O_8$, 177
NADPH, 79, 107, 109–111
nanobiohybrids, 21, 79, 89
nanocubes, 4
nanoflowers, 2, 23, 25
nanohybrid, 9, 10, 15, 18
nanomaterials, 4, 6, 8, 19, 59, 87
nanoparticles, 12
nanopores, 263
nanorods, 2
nanoscale, 16, 57, 302
Nanostructured, 1
nanowires, 2
NaOH, 23, 172
N-bromosuccinimide (NBS), 166
N–H cyclization, 171
N-heterocyclic carbene, 240
Ni, 14, 41, 272, 273, 275–278, 308, 317
nitric oxide, 110, 111
nitrogenases, 85, 86
NO_2, 35, 37
Novolac resin, 274
N-synthase, 118
nucleophilic attack, 119, 220

1,1,1-trichloroethane (1,1,1-TCa), 307, 308

1,1,2,2-tetrachloroethane (TeCa), 307

1,1-dichloroethylene (1,1-DCE), 308

O_2, 17, 35, 37

olefin, 139, 157, 162

organic peroxides, 38

organic synthesis, 1, 128, 177, 184, 204

organoboron, 130, 133

organoiron, 142, 146, 173, 208

Organometallics, 77

Organometallics-Protein, 84

organozinc, 130, 133, 137, 147

ortho-arylation, 131

ortho-metalation, 130

oxaloacetate, 117

oxidase-like activity, 15

oxidation states, 15, 113, 321

oxide shell, 16, 17, 19, 317

oxygenases, 117

Palomo, Jose M., 1, 77, 299

Pariona, N, 57

PCE, 316

p-Cl-toluene, 213

Pd, 14, 127, 205, 308, 312
  FePd nanoflowers, 25

PDB, 102, 104

PEG, 9, 10

peptide, 102, 103, 104, 114

peroxidase-like, 15, 58, 64, 65, 67, 302

persulfate, 158, 303

phosphorylation, 117

photocatalyst, 67, 70

photoelectrochemical systems, 57

photosynthesis, 105, 106

pinacolborane, 178

plants, 5, 57–59, 64, 105, 113, 114, 265, 321

pollutants, 66, 67, 69, 70, 84, 299, 300, 301, 307, 321

polyacrylates, 301

porphyrin, 99, 100, 104, 105, 152, 167, 206, 207, 235

powedered activated carbon (PAC), 313

precursor, 4, 20, 37, 144, 167, 203, 217, 253

propene, 39, 42, 52

propylene, 9, 10, 35, 37–40, 42, 44, 46, 49

prosthetic group, 98, 99, 102, 110

protein, 36, 59, 77, 79, 84, 86, 98, 100, 102, 105, 110, 115, 216

protein engineering, 89

Pt, 7, 8, 37

PVP, 9, 17, 21, 23

quantum chemical studies, 207

quaternary structure, 104

Ramirez-Rico, J., 253

reactivity, 43, 60, 78, 129, 174, 175, 184, 204, 246, 304, 311

redox, 36, 65, 100, 108, 111, 113, 150, 164, 167, 171, 238, 258

redox potential, 15, 108

reductases, 89

reduction, 22, 23, 25, 47, 51, 64, 79, 86, 108, 110, 120, 133, 245, 258, 261, 303, 309, 317, 321

regioisomer, 232

regioselectivity, 169

rhodamine B, 67

ROS, 70
rubredoxins, 112, 113

saccharide moiety, 235
scaffold, 78, 86, 232, 265, 279
scanning electron microscopy, 4
screening, 79, 218, 243
selective monoarylation, 134
selectivity, 78, 79, 82, 85, 89, 128, 184
serine, 116, 227
silver, 35, 40, 52, 321
silylperoxides, 177
size, 2, 5, 8, 20, 21, 23, 59, 63, 66, 86, 88, 102, 180, 237, 272, 319
S-nitrosylation, 111
society, 36
sodium borohydride, 16, 17
sodium dodecyl sulfate (SDS), 69
solid phase, 13, 17, 266
spirocyclic systems, 226, 240
stacking direction, 268
stem, 280
stereoselectivity, 135, 246
Stern layer, 63
streptavidin, 85
subunit, 101, 104
succinate dehydrogenase, 114
sulfite reductase, 84
supercapacitor, 254, 270, 278
superoxide, 108, 109, 113, 119
superparamagnetic, 3, 12, 13, 18, 22
surface mineralization, 317
surfaces, 3, 13, 60, 260, 261, 306
surfactant, 11, 69
synthase, 82, 99

[2 + 2 + 2]-annulation reaction, 141
2,6-butylated hydroxytoluene (BHT), 136
2-azido-2-methyl-1-(pyrrolidin-1-yl) propan-1-one, 240
2D, 2, 13, 25
3,3',5,5'tetramethylbenzidine, 87
3D, 13, 14, 15
TBHP, 154, 168, 170, 173, 176, 181
*t*BuOH, 172
TEA, 12, 13
template, 82, 84, 86, 306
tertiary structure, 101
tetrabromobisphenol A (TBBPA), 304, 305
tetrahydrobiopterin, 110, 118
thioaldehyde, 119
thiophene, 159
TiO_2, 39
$Ti\text{-}SiO_2$, 42
transmetalation, 133, 141
transmission electron microscopy, 5, 275
transporter, 101
trichloroethylene (TCE), 309–316
turnover, 81, 209, 227
    numbers (TON), 80, 209
    number (TTN), 81, 227

ultrasonic process, 14
unnatural, 89
uranium, 319

van der Waals, 62, 259
variant, 80–84, 217, 227–229, 242, 245
versatility, 80, 81, 107

vitamin D, 114
voltammetry, 261, 262

wastewater, 62, 64, 65, 67, 301, 309,
   317, 319
water, 4, 7, 8, 11, 58, 70, 80, 107,
   265, 301, 315
Wheland-type intermediate, 212
wüstite, 16, 58

xerogels, 281
X-ray, 18, 120, 267

X-ray photoelectron spectroscopy,
   9
X-ray powder diffraction, 7

Yarrowia lipolytica, 319

zero-valent, 299, 320
zero-valent iron (ZVI), 15, 299, 300
zinc, 146, 147, 318
ZnBr$_2$•TMEDA, 146
ZnCl$_2$•TMEDA, 134, 135, 139
Zn-free transformations, 130

CATALYTIC SCIENCE SERIES

(Continued from page ii)

Vol. 9   *Deactivation and Regeneration of Zeolite Catalysts*
         edited by M. Guisnet and F. R. Ribeiro

Vol. 8   *Petrochemical Economics: Technology Selection in a*
         Carbon Constrained World
         by D. Seddon

Vol. 7   *Combinatorial Development of Solid Catalytic Materials:*
         Design of High-Throughput Experiments, Data Analysis, Data Mining
         edited by M. Baerns and M. Holeňa

Vol. 6   *Catalysis by Gold*
         edited by G. C. Bond, C. Louis and D. T. Thompson

Vol. 5   *Supported Metals in Catalysis*
         edited by J. A. Anderson and M. F. García

Vol. 4   *Isotopes in Heterogeneous Catalysis*
         edited by Justin S. J. Hargreaves, S. David Jackson and Geoff Webb

Vol. 3   *Zeolites for Cleaner Technologies*
         edited by Michel Guisnet and Jean-Pierre Gilson

Vol. 2   *Catalysis by Ceria and Related Materials*
         edited by Alessandro Trovarelli

Vol. 1   *Environmental Catalysis*
         edited by F. J. J. G. Janssen and R. A. van Santen